Lecture Notes in Artificial Intell

Edited by R. Goebel, J. Siekmann, and W. Wahlster

Subseries of Lecture Notes in Computer Science

Petra Perner (Ed.)

Advances in Data Mining

Applications and Theoretical Aspects

9th Industrial Conference, ICDM 2009
Leipzig, Germany, July 20-22, 2009
Proceedings

 Springer

Series Editors

Randy Goebel, University of Alberta, Edmonton, Canada
Jörg Siekmann, University of Saarland, Saarbrücken, Germany
Wolfgang Wahlster, DFKI and University of Saarland, Saarbrücken, Germany

Volume Editor

Petra Perner
Institute of Computer Vision
and Applied Computer Sciences, IBaI
Kohlenstr. 2
04107 Leipzig, Germany
E-mail: pperner@ibai-institut.de

Library of Congress Control Number: Applied for

CR Subject Classification (1998): I.2.6, I.2, H.2.8, K.4.4, J.3, I.4, J.1

LNCS Sublibrary: SL 7 – Artificial Intelligence

ISSN   0302-9743
ISBN-10   3-642-03066-1 Springer Berlin Heidelberg New York
ISBN-13   978-3-642-03066-6 Springer Berlin Heidelberg New York

springer.com

© Springer-Verlag Berlin Heidelberg 2009
Printed in Germany

Typesetting: Camera-ready by author, data conversion by Scientific Publishing Services, Chennai, India
Printed on acid-free paper   SPIN: 12718375   06/3180   5 4 3 2 1 0

Preface

This volume comprises the proceedings of the Industrial Conference on Data Mining (ICDM 2009) held in Leipzig (www.data-mining-forum.de).

For this edition the Program Committee received 130 submissions. After the peer-review process, we accepted 32 high-quality papers for oral presentation that are included in this book. The topics range from theoretical aspects of data mining to applications of data mining, such as on multimedia data, in marketing, finance and telecommunication, in medicine and agriculture, and in process control, industry and society.

Ten papers were selected for poster presentations that are published in the ICDM Poster Proceedings Volume by *ibai-publishing* (www.ibai-publishing.org).

In conjunction with ICDM two workshops were run focusing on special hot application-oriented topics in data mining. The workshop Data Mining in Marketing DMM 2009 was run for the second time. The papers are published in a separate workshop book "Advances in Data Mining on Markting" by *ibai-publishing* (www.ibai-publishing.org). The Workshop on Case-Based Reasoning for Multimedia Data CBR-MD ran for the second year. The papers are published in a special issue of the *International Journal of Transactios on Case-Based Reasoning* (www.ibai-publishing.org/journal/cbr).

We are pleased to announce that we gave out the best paper award for ICDM fourth time. More details are mentioned at www.data-mining-forum.de. The final decision was made by the Best Paper Award Committee based on the presentation by the authors and the discussion with the auditorium. The ceremony took place at the end of the conference. This prize is sponsored by ibai solutions (www.ibai-solutions.de) one of the leading data mining companies in data mining for marketing, Web mining and E-commerce.

The conference was rounded up by a session on new challenging topics in data mining before the Best Paper Award Ceremony.

We also thank the members of the Institute of Applied Computer Sciences, Leipzig, Germany (www.ibai-institut.de) who handled the conference as secretariat. We appreciate the help and understanding of the editorial staff at Springer, and in particular Alfred Hofmann, who supported the publication of these proceedings in the LNAI series.

Last, but not least, we wish to thank all the speakers and participants who contributed to the success of the conference. The next ICDM will take place in Berlin in 2010.

July 2009

Petra Perner

Industrial Conference on Data Mining, ICDM 2009

Chair

Petra Perner   IBaI Leipzig, Germany

Committee

Klaus-Peter Adlassnig	Medical University of Vienna, Austria
Andrea Ahlemeyer-Stubbe	ENBIS, Amsterdam
Klaus-Dieter Althoff	University of Hildesheim, Germany
Chid Apte	IBM Yorktown Heights, USA
Eva Armengol	IIA CSIC, Spain
Bart Baesens	KU Leuven, Belgium
Isabelle Bichindaritz	University of Washington, USA
Leon Bobrowski	Bialystok Technical University, Poland
Marc Boullé	France Télécom, France
Henning Christiansen	Roskilde University, Denmark
Shirley Coleman	University of Newcastle, UK
Juan M. Corchado	Universidad de Salamanca, Spain
Da Deng	University of Otago, New Zealand
Antonio Dourado	University of Coimbra, Portugal
Peter Funk	Mälardalen University, Sweden
Brent Gordon	NASA Goddard Space Flight Center, USA
Gary F. Holness	Quantum Leap Innovations Inc., USA
Eyke Hüllermeier	University of Marburg, Germany
Piotr Jedrzejowicz	Gdynia Maritime University, Poland
Janusz Kacprzyk	Polish Academy of Sciences, Poland
Mehmed Kantardzic	University of Louisville, USA
Ron Kenett	KPA Ltd., Israel
Mineichi Kudo	Hokkaido University, Japan
Eduardo F. Morales	INAOE, Ciencias Computacionales, Mexico
Stefania Montani	Università del Piemonte Orientale, Italy
Jerry Oglesby	SAS Institute Inc., USA
Eric Pauwels	CWI Utrecht, The Netherlands
Mykola Pechenizkiy	Eindhoven University of Technology, The Netherlands
Ashwin Ram	Georgia Institute of Technology, USA
Tim Rey	Dow Chemical Company, USA
Rainer Schmidt	University of Rostock, Germany
Yuval Shahar	Ben Gurion University, Israel
David Taniar	Monash University, Australia

Stijn Viaene   KU Leuven, Belgium
Rob A. Vingerhoeds   Ecole Nationale d'Ingénieurs de Tarbes, France
Claus Weihs   University of Dortmund, Germany
Terry Windeatt   University of Surrey, UK

Table of Contents

Data Mining in Process Control, Industry and Society

Data Mining on Multimedia Data

Theoretical Aspects of Data Mining

Distances in Classification

Claus Weihs and Gero Szepannek

Department of Statistics
University of Dortmund
44227 Dortmund

Abstract. The notion of distance is the most important basis for classification. This is especially true for unsupervised learning, i.e. clustering, since there is no validation mechanism by means of objects of known groups. But also for supervised learning standard distances often do not lead to appropriate results. For every individual problem the adequate distance is to be decided upon. This is demonstrated by means of three practical examples from very different application areas, namely social science, music science, and production economics. In social science, clustering is applied to spatial regions with very irregular borders. Then adequate spatial distances may have to be taken into account for clustering. In statistical musicology the main problem is often to find an adequate transformation of the input time series as an adequate basis for distance definition. Also, local modelling is proposed in order to account for different subpopulations, e.g. instruments. In production economics often many quality criteria have to be taken into account with very different scaling. In order to find a compromise optimum classification, this leads to a pre-transformation onto the same scale, called desirability.

1 Introduction

The notion of distance is the most important basis for classification. This is especially true for unsupervised learning, i.e. clustering, since there is no validation mechanism by means of objects of known groups. But also for supervised learning standard distances often do not lead to appropriate results. For every individual problem the adequate distance is to be decided upon. Obviously, the choice of the distance measure determines whether two objects naturally go together (Anderberg, 1973). Therefore, the right choice of the distance measure is one of the most decisive steps for the determination of cluster properties. The distance measure should not only adequately represent the relevant scaling of the data, but also the study target to obtain interpretable results.

Some classical distance measures in classification are discussed in the following. In supervised statistical classification distances are often determined by distributions. A possible distance measure treats each centroid and covariance matrix as the characteristics of a normal distribution for that class. For each new data point we calculate the probability that point came from each class; the

P. Perner (Ed.): ICDM 2009, LNAI 5633, pp. 1–12, 2009.

data point is then assigned to the class with the highest probability. A simplified distance measure assumes that the covariance matrices of each class are the same. This is obviously valid if the data for each class is similarly distributed, however, nothing prevents from using it if they are not. Examples for the application of such measures are **Quadratic and Linear Discriminant Analysis** (QDA and LDA) (Hastie et al., 2001, pp. 84). For a more general discussion of distance measures in supervised classification see Gnanadesikan (1977).

With so-called kernels, e.g., like in **Support Vector Machines** (SVM) (Hastie et al., 2001, p. 378) standard transformations were explicitly introduced in classification methods, in order to transform the data so that the images can be separated linearly as with LDA.

In unsupervised classification Euclidean distance is by far the most chosen distance for metric variables. One should notice, however, that the Euclidean distance is well-known for being outlier sensitive. This might lead to switching to another distance measure like, e.g., the **Manhattan-distance** (Tan et al., 2005). Moreover, one might want to discard correlations between the variables and to restrict the influence of single variables. This might lead to transformations by means of the covariance or correlation matrices, i.e. to **Mahalanobis-distances** (Tan et al., 2005). Any of these distances can then be used for defining the distance between groups of data. Examples are minimum distance between the elements of the groups (**single linkage**), maximum distance (**complete linkage**), and average distance (**average linkage**) (Hastie et al., 2001, p. 476).

For non-metric variables often methods are in use, which, e.g., count the number of variables with matching values in the compared objects, examples are the **Hamming-, the Jaccard- and the simple matching distances** (Tan et al., 2005).

Thus, data type is an important indicator for distance selection. E.g., in Perner (2002), distance measures for image data are discussed. However, distance measures can also be related to other aspects like, e.g., application. E.g. time-series representing music pieces need special distances (Weihs et al. 2007). Other important aspects of distance are translation, size, scale and rotation invariance, e.g. when technical systems are analysed (Perner, 2008).

Last but not least, **variable selection** is a good candidate to identify the adequate space for distance determination for both supervised and unsupervised classification.

In practice, most of the time there are different plausible distance measures for an application. Then, quality criteria are needed for distance measure selection. In supervised classification the misclassification error rate estimated, e.g., on learning set independent test sets, is the most accepted choice. In unsupervised learning, one might want to use background information about reasonable groupings to judge the partitions, or one might want to use indices like the ratio between within and between cluster variances which would also be optimized in discriminant analysis in the supervised case.

In what follows examples are given for problem specific distances. The main ideas are as follows. Clusters should often have specific properties which are not related to the variables that are clustered, but to the space where the clusters are represented. As an example city districts are clustered by means of social variables, but represented on a city map. Then, e.g., the connection of the individual clusters may play an important role for interpretation. This may lead to an additional objective function for clustering which could be represented by a distance measure for unconnected cluster parts. These two objective functions or distance measures could be combined to a new measure. Another, much simpler, possibility would be, however, just to include new variables in the analysis representing the district centres. By differently weighting the influence of these variables the effect of these variables can be demonstrated. This will be further discussed in section 2.1.

Often, the observed variables are not ideal as a basis for classification. Instead, transformations may be much more sensible which directly relate to a redefinition of the distance measure. Also, in supervised classification the observed classes may not have the right granularity for assuming one simple distance measure per class. Instead, such distances may be more adequate for subclasses, which may be, e.g., defined by known subpopulations across the classes or by unknown subclasses of the classes. Distances then relate to, e.g., distributions in subclasses, i.e. to mixtures of distributions in classes. This will be further discussed in section 2.2.

Another example for more than one objective function is given for production economics. Typically, for more than one objective function there is the problem of weighting the different targets. In contrast to section 2.1 this can also be achieved by transformation to a common scale by means of different so-called desirability functions. The overall distance is then typically related to some combination of the different desirabilities in a so-called desirability index. This will be further discussed in section 2.3.

2 Case-Based Distance Measures

2.1 Additional Variables

In social science clustering is often applied to spatial regions with very irregular borders. Then adequate spatial distances may have to be taken into account for clustering. Clusters of spatial regions should most of the time represent similar properties of the predefined regions. However, for better interpretation the question arises as well whether the resulting clusters are connected in space. Then, two different kinds of distances have to be compared, namely the distance of regions in clusters related to given properties and the spatial dispersion of the clusters.

Assume that spatial regions are predefined, e.g. as city districts. Consider the case where some clusters are already defined, e.g. by means of social properties in the regions. In more detail, social milieus were clustered by means of six social variables (after variable selection), namely "fraction of population of 60-65", "moves to district per inhabitant", "apartments per house", "people per apartment", "fraction of welfare recipients" and "foreigners share of employed people". Then, the question arises whether clusters represent connected regions in space. Among others, there are the following possibilities to measure the distance of two unconnected cluster parts:

- One could rely on the minimum Euclidean distance between two regions in different parts of the cluster defined by the minimum distance of points $||.||_2$ in the regions (single-linkage distance), or
- one could use the minimum distance measured by the number of borders $||.||_b$ between such regions (cp. Sturtz, 2007).

The former distance $||.||_2$ reflects the idea that region borders could be mainly ignored in the distance definition, or that regions mainly have the form of balls in space. The latter distance reflects the assumption that the regions are thoughtfully fixed and can have all forms not necessarily approximately similar to a ball.

In Figure 1 you can find a typical partition of a city into districts (please ignore the colours and the numbering for the moment). Obviously, the districts have all kinds of forms, not all similar to balls.

In order to define the **dispersion of one cluster** (say d_2 or d_b relying on $||.||_2$ and $||.||_b$, respectively) first define sequences of most neighboured connected parts of the cluster, and then sum up the distances between all sequential pairs. The dispersion may be defined as the minimum such sum over all possible sequences.

Consider the partition in Figure 1 obtained by a clustering algorithm based on social properties of the city districts of Dortmund (cp. Roever and Szepannek, 2005). Is this clustering ready for interpretation? How well are the clusters connected? Ignoring the white regions which were not clustered, Table 1 gives the dispersions d_2 and d_b of the four clusters. As an example, please consider the ordering of the connected parts of the ▩ - cluster as indicated in Figure 1. Obviously the ▩ - cluster is very much connected, whereas the other clusters are much more dispersed.

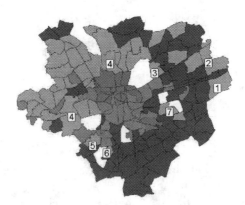

Fig. 1. Clusters of districts of the City of Dortmund (Germany)

Another possible, but simpler, dispersion measure would be the percentage pc of districts in the maximum connected part of a cluster. With this measure, the ▨▨▨▨ - cluster is the least dispersed (see Table 1).

Table 1. Dispersion of clusters

cluster	d_2	d_b	p_c
▨▨▨▨	1.1	4	0.88
▨▨▨▨	5.9	14	0.83
▨▨▨▨	6.5	18	0.79
▨▨▨▨	7.8	13	0.90

In Roever and Szepannek, 2005, dispersion was not utilized for clustering. However, there would be the option to use dispersion as a penalty (or complexity) term for clustering. Roever and Szepannek minimize the Classification Entropy

$$CE = -\frac{1}{N} \sum_{i=1}^{N} \sum_{j=1}^{k} (u_{ij} \log_2 u_{ij}),$$

where N = number of observations, k = number of clusters, u_{ij} = probability that observation i belongs to cluster j. Using this fitness function and some variables' subgrouping, $k = 4$ clusters were produced similar to Figure 1 by means of genetic programming. However, this fitness-function could have been supplemented with a dispersion measure to force non-dispersed clusters. For this, the dispersions for the individual clusters should be combined to one measure D. For ease, one could use the simple measure

D_c = percentage of districts in the maximum connected parts of all clusters.

A possible combination of fitness functions is then $CE - c \cdot D_c$, where $c > 0$ is a constant to be fixed, e.g., so that the two parts of the fitness function are well-balanced.

Another option would be to base optimization on two different fitness functions, CE and D_c, where CE is to be minimized, and D_c to be maximized, and combine them, e.g., by means of a desirability index (cp. section 2.3).

However, for this paper we have tried to take into account the cluster dispersion in a different way. We introduced new variables representing the x- und y-coordinates of the district centres. By this, distance of district centres are also taken into account with clustering. When these centre variables were weighted only 20% of the other variables the result was hardly influenced (Figure 2, left). After they were weighted twice as much as the other variables, however, the result was totally different and the clusters were much more connected (Figure 2, right).

Fig. 2. Clusters with 20%- (left) and 200%- (right) weighting of district centres

2.2   Transformations and Local Modelling

In statistical musicology the main problem is often to find the right transformation of the input time series adequate for analysis. Also, local modelling is proposed in order to account for different subpopulations, e.g. instruments.

This example of distance definition concerns supervised classification. In music classification the raw input time series are seldom the right basis for analysis. Instead, various transformations are in use (see, e.g., Weihs et al., 2007). Since with music frequencies play a dominant role, periodograms are a natural representation for observations. From the periodogram corresponding to each tone, voice characteristics are derived (cp. Weihs and Ligges, 2003). For our purpose we only use the mass and the shape corresponding to the first 13 partials, i.e. to the fundamental frequency (FF) and the first 12 overtones (OTs), in a pitch independent periodogram (cp. Figure 3). Mass is measured as the sum of the percentage share (%) of the peak, shape as the width of the peak in parts of half tones (pht) between the smallest and the biggest involved frequency.

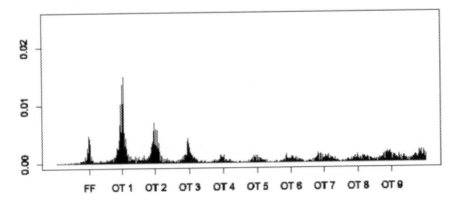

Fig. 3. Pitch independent periodogram (professional bass singer)

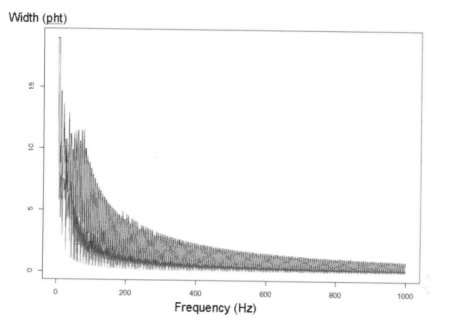

Fig. 4. Width measured in parts of half-tone (pht) dependent on frequency (upper line = fundamental frequency, lower line = first overtone)

These 26 characteristics were determined for each individual tone, as well as averaged characteristics over all involved tones leading to only one value for each characteristic per singer or instrument. LDA based on these characteristics results in an astonishingly good prediction of register (classes low / high) (Weihs et al., 2005). The register of individual tones are predicted correctly in more than 90% of the cases for sung tones, and classification is only somewhat worse if instruments are included in the analysis. Even better, if the characteristics are averaged over all involved tones, then voice type (high or low) can be predicted without any error.

However, this classification appeared, in a way, to be too good so that it was suspected that mass and/or width might somewhat reflect frequency and thus register though the pitch independent periodogram was used. And indeed, simulations showed that width is frequency dependent because it is measured in number of half tones (s. Figure 4). However, if the absolute width in number of involved Fourier-Frequencies is used instead, then this dependency is dropped leading, though, to poorer classification quality. This example distinctly demonstrates an effect of choosing a wrong transformation, and thus a wrong distance measure.

In subsequent analyses (Weihs et al., 2006, Szepannek et al., 2008) this redefined width characteristics was applied to a data set consisting of 432 tones (= observations) played / sung by 9 different instruments / voices. In order to admit different behaviour for different instruments, so-called **local modelling** was applied building local classification rules for each instrument separately. For this, we consider the population to be the union of subpopulations across the classes high / low. Then, a mixture distribution is assumed for each class. The problem to be solved consists in register prediction for a new observation if the instrument (and thus the choice of the local model) is not known. This task can be formulated as some globalization of local classification rules. A possible solution is to identify first the local model, and further work only with the parts of the mixtures in the classes corresponding to this model.

Imagine all local (subpopulation-) classifiers return local class posterior probabilities $P(k|l,x)$, where $k = 1,\ldots,K$ denotes the class, x is the actual observation and $l = 1,\ldots,L$ is the index of the local model, i.e. the instrument in our case. The following **Bayes Rule**

$$\hat{k} = \arg\max_k \sum_l P(k|l,x)P(l|x)$$

showed best performance for the musical register classification problem. To implement this, an additional classifier has to be built to predict the presence of each local model l for a given new observation x. Using LDA for both classification models, the local models and the global decision between the local models, leads to the best error rate of 0.263 on the data set. Note that - since only posterior probabilities are used to build the classification rule - all models can be built on different subsets of variables, i.e. subpopulation individual variable selection can be performed. This may lead to individual distance measures for the different localities (voices, instruments) and for the global decision.

2.3 Common Scale

In production economics often many quality criteria have to be taken into account with very different scaling. In order to find a compromise optimum, a pre-transformation, called desirability, onto the same scale may be used.

In a specific clustering problem in production economic product variants should be clustered to so-called product families so that production interruptions caused by switching between variants (so-called machine set-up times) are minimal (Neumann, 2007). Three different distance measures (Jaccard, simple-matching, and Euclidean) and many different clustering methods partly based on these distance measures are compared by means of four competitive criteria characterizing the goodness of cluster partitions, namely the similarity of the product variants in the product families, the number of product families, the uniformity of the dispersion of the product variants over the product families, and the number of product families with very few product variants. Therefore, partition quality is measured by $d = 4$ criteria. Overall, the problem is therefore to identify the cluster method and the corresponding distance measure, as well as the number of clusters, i.e. the number of product families, optimal to all four criteria. In order to rely on only one compromise criterion a so-called desirability index is derived.

In order to transform all these criteria to a common scale, the four criteria are first transformed to so-called **desirabilities**, w_i a value in the interval $[0, 1]$, where 1 stands for best and 0 for worst, unacceptable quality. In order to join the criteria to one objective function, a so-called **desirability index** W (Harrington, 1965) is defined

$$W : \{w_1, w_2, \ldots, w_d\} \rightarrow [0, 1].$$

Harrington 1965 suggests the geometric mean for W:

$$W(w_1, \ldots, w_d) = \sqrt[d]{\prod_{i=1}^{d} w_i}.$$

This choice has the advantage that $W = 0$ already if one desirability $w_i = 0$, and $W = 1$ only if all $w_i = 1$. Another reasonable index choice would be $\min(w_1, \ldots, w_d)$ with the same properties. The geometric mean will be used here.

In order to minimize the average machine set-up time the following desirability is defined:

$$w_1(C^{(k)}) = 1 - \sum_{i=1}^{k} \sum_{X_j, X_l \in C_k} d(X_j, X_l),$$

where $C^{(k)}$ is a partition with k clusters, and $d(X_j, X_l)$ is the machine set-up time between product variants X_j and X_l measured by one of the above distance measures (Jaccard, simple-matching, Euclidean).

In this application, for the number of product families a certain range is assumed to be optimal. This lead to the desirability function w_2 indicated in Figure 5, where the number of product families with desirability $= 1$ are considered optimal.

Desirability

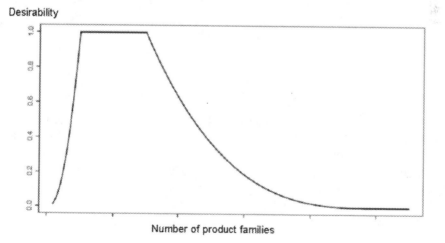

Number of product families

Fig. 5. Desirability function w_2

Desirability index

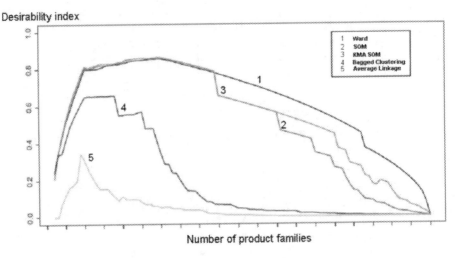

Fig. 6. Desirability index for different cluster methods

For application roughly equal sized clusters are of advantage. This leads to a criterion based on the number n_w of within cluster distances of a partition, i.e. the number of distances between objects in the same cluster. When $\min C^{(k)}(n_w)$ is the minimal number of distances over all possible partitions of size k with n objects, and $\max C^{(k)}(n_w)$ the corresponding maximum, this leads, e.g., to the following criterion to measure how imbalanced the cluster sizes are:

$$w_3(C^{(k)}) = 1 - \frac{n_w - \min C^{(k)}(n_w)}{\max C^{(k)}(n_w) - \min C^{(k)}(n_w)}.$$

Product families with less than five product variants are not desirable. This leads, e.g., to the criterion:

$$w_4(C^{(k)}) = 2^{-a}$$

with

$$a = \text{number of product families with less or equal five variants.}$$

Some results of different cluster methods (for each method based on the most appropriate distance measure) evaluated with the desirability index of the four desirability criteria are shown in Figure 6. Obviously, Ward clustering (Ward, 1963) appears to be best, and for about the intended number of product families the index is maximal.

3   Conclusion

In section 2 it is demonstrated by means of examples from very different application areas that various transformations might be necessary to be able to use an adequate distance measure for unsupervised and supervised classification. In section 2.1 additional variables were added with tentative weights, in section 2.2 the original variables were transformed before application of standard methods and local measures appeared adequate, in section 2.3 original criteria were transformed to a common scale and combined to one criterion used for optimal clustering. All these examples showed that application of standard methods to originally observed variables might not be adequate for problem solution.

Acknowledgments

The authors thank cand. Stat. O. Mersmann for conducting the cluster analyses including district centres. Also, financial support of the Deutsche Forschungsgemeinschaft (SFB 475, "Reduction of complexity in multivariate data structures") is gratefully acknowledged.

References

Anderberg, M.R.: Cluster Analysis for Applications. Acadamic Press, New York (1973)

Gnanadesikan, R.: Methods for Statistical Data Analysis of Multivariate Observations. Wiley, New York (1977)

Hastie, T., Tibshirani, R., Friedman, J.: The Elements of Statistical Learning - Data Mining, Inference and Prediction. Springer, New York (2001)

Harrington, J.: The desirability function. Industrial Quality Control 21(10), 494–498 (1965)

Neumann, C.: Einsatz von Clusterverfahren zur Produktfamilienbildung. Diploma Thesis, Department of Statistics, TU Dortmund (2007)

Perner, P.: Case-based reasoning and the statistical challenges. Journal Quality and Reliability Engineering International 24(6), 705–720 (2008)

Perner, P. (ed.): Data Mining on Multimedia Data, vol. 2558. Springer, Heidelberg (2002)

Roever, C., Szepannek, G.: Application of a Genetic Algorithm to Variable Selection in Fuzzy Clustering. In: Weihs, C., Gaul, W. (eds.) Classification - the Ubiquitous Challenge, pp. 674–681. Springer, Heidelberg (2005)

Sturtz, S.: Comparing models for variables given on disparate spatial scales: An epidemiological example. PhD Thesis, Department of Statistics, TU Dortmund, p. 38 (2007)

Szepannek, G., Schiffner, J., Wilson, J., Weihs, C.: Local Modelling in Classification. In: Perner, P. (ed.) ICDM 2008. LNCS, vol. 5077, pp. 153–164. Springer, Heidelberg (2008)

Tan, P.-N., Steinbach, M., Kumar, V.: Introduction to Data Mining. Addison-Wesley, Reading (2005)

Ward, J.H.: Hierarchical grouping to optimize an objective function. Journal of the American Statistical Association 58, 236–244 (1963)

Weihs, C., Ligges, U., Mörchen, F., Müllensiefen, D.: Classification in Music Research. Advances in Data Analysis and Classification (ADAC) 1(3), 255–291 (2007)

Weihs, C., Szepannek, G., Ligges, U., Lübke, K., Raabe, N.: Local models in register classification by timbre. In: Batagelj, V., Bock, H.-H., Ferligoj, A., Ziberna, A. (eds.) Data Science and Classification, pp. 315–332. Springer, Heidelberg (2006)

Weihs, C., Reuter, C., Ligges, U.: Register Classification by Timbre. In: Weihs, C., Gaul, W. (eds.) Classification - The Ubiquitous Challenge, pp. 624–631. Springer, Berlin (2005)

Weihs, C., Ligges, U.: Voice Prints as a Tool for Automatic Classification of Vocal Performance. In: Kopiez, R., Lehmann, A.C., Wolther, I., Wolf, C. (eds.) Proceedings of the 5th Triennial ESCOM Conference, Hanover University of Music and Drama, Germany, September 8-13, pp. 332–335 (2003)

Electronic Nose Ovarian Carcinoma Diagnosis Based on Machine Learning Algorithms

José Chilo[1], György Horvath[2], Thomas Lindblad[3], and Roland Olsson[4]

[1] Center for RF Measurement Technology, University of Gävle, S-801 76 Gävle, Sweden
jco@hig.se
[2] Department of Oncology, Sahlgrenska University Hosp. Gothenburg, Sweden
gyorgy.horvath@oncology.gu.se
[3] Department of Physics, Royal Institute of Technology, S-106 91 Stockholm, Sweden
lindblad@particle.kth.se
[4] Department of Computer Science, Ostfold University College, N-1757 Halden, Norway
Roland.Olsson@hiof.no

Abstract. Ovarian carcinoma is one of the most deadly diseases, especially in the case of late diagnosis. This paper describes the result of a pilot study on an early detection method that could be inexpensive and simple based on data processing and machine learning algorithms in an electronic nose system. Experimental analysis using real ovarian carcinoma samples is presented in this study. The electronic nose used in this pilot test is very much the same as a nose used to detect and identify explosives. However, even if the apparatus used is the same, it is shown that the use of proper algorithms for analysis of the multi-sensor data from the electronic nose yielded surprisingly good results with more than 77% classification rate. These results are suggestive for further extensive experiments and development of the hardware as well as the software.

Keywords: Machine learning algorithms, odor classification, ovarian carcinoma, medicine.

1 Introduction

Although most people would agree on the fact that there is no "artificial nose" [1] with the same sensitivity as that of a dog. It is claimed that a dog can detect less than 100 molecules per cubic meter. A sophisticated gas chromatograph with the proper injection system can maybe detect a 1000 molecules. However, it is obvious that there are advantages with a small and simple electronic device, even if its performance is not as good. A few years ago the authors presented an electronic nose for detecting explosives. This nose could in no way compete with trained "bomb dogs" to sense the presence of explosives, but it could distinguish between various types of chemicals (alcohols) and explosives [2-4]. The nose is relatively small and is shown in Fig. 1. The square tube holds four types of sensors, each operating at four different temperatures (ranging from 20 – 750 deg C). The sensors are mounted on the sides of the tube with pertinent electronics and support system directly on a printed circuit board. The

P. Perner (Ed.): ICDM 2009, LNAI 5633, pp. 13–23, 2009.

Fig. 1. The electronic nose with its holder (cirkle to the left), the square tube with four sensors and electronis on each side and the exhaust fan to the right

fan and exhaust tube are shown to the right in Fig. 1. A small modification has been made to carry out this study: it is a combined holder and inlet (made of teflon an easily replaceable) shown to the left with a circle.

Many cancers are detected in late stages with consequential high mortality rates. For example, the majority of human ovarian carcinomas are diagnosed in stage III or IV, and 70% of these patients will die within five years. Thus, it is essential to develop inexpensive and simple methods for early diagnosis.

It has been proposed, and also to some extent demonstrated, that electronic noses could be used in medical diagnostics [1]. Most recently, it has also been shown that a dog can be trained to distinguish between histopathological types and grades of ovarian carcinomas, including borderline tumors, from healthy control samples [5]. This study clearly demonstrated that human ovarian carcinomas may be characterised by a specific odor. To detect this odor, a trained dog was used. However, dogs may be influenced by several factors before and during their work, leading to changes in the accuracy rates. Use of an artificial nose with adequate sensitivity should limit the possible diagnostic errors and might be developed for early diagnosis. We will investigate here whether our electronic nose could carry out such a task.

2   Experimental Details

The electronic nose consists of four different types of gas-sensors from Figaro [3]. The four types of sensors are: TGS2600 different "air pollutions", TGS2610 combustible gases, TGS2611 combustible gases and TGS2620 alcohols of different kinds. The gas-sensors work by the changing resistance induced by gas molecules that bind on two electrodes that are separated a small distance. The response of each individual sensor is strongly influenced by temperature, so the Nose is actually an assembly of four individuals of the four above sensor types, each held at a different internal temperature, producing in total 16 independent analog signals, each varying with time and smell [4]. The gas-sensors, which change conductivity upon exposure to the odor under test, are used in a voltage divider circuit to provide and output voltage which increases with gas concentration. The voltage is sampled at 10 Hz and the raw data file is compacted to a data matrix consisting of 16 columns by 500 rows. Hence, what we measure is simply the slowly varying DC voltage. The measurement starts a few seconds before the sample is being analyzed, since (as shown in the case of the bomb nose) the rise time of the signals from the sensors holds much information.

In the first study pieces of human seropapillary adenocarcinomas were used, these were clearly smaller (a few millimeter) than in the case of explosives (a few centimeter). Tumors were taken from three different individuals and sized to 3x3x3 mm. Two similar sizes of human tuba obtained from two different healthy individuals were also used as control samples.

In the second study, tumors from 6 different individuals were divided into 21 pieces and analyzed by 92 runs. Myometrium control samples from 3 different individuals were divided into 23 pieces and analyzed by 70 runs.

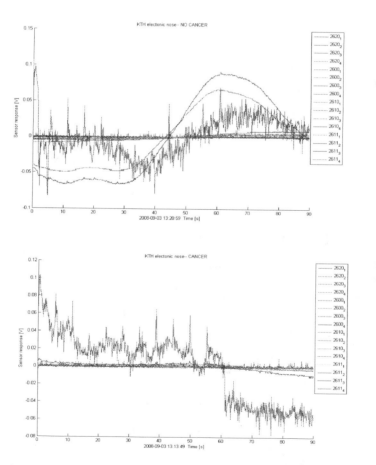

Fig. 2. Signals from the 16 sensors of the electronic nose. The horizontal axis is the time scale and the vertical scale is the DC voltage. Cases are shown for healthy tissue (up) and for tissue diagnosed as cancer (bottom).

Both ovarian cancer samples and controls were kept at -80^0C in our tumor bank (Ethical Committee license number: S-154-02). Samples were thawed to room temperature for 15-30 minutes before being used.

As mentioned, the primary data are the 16 signals from the various sensors (4 different sensors operating at 4 different temperatures each) as digitized and stored for each sample. An example of the 16 sensor outputs when the array is exposed to healthy tissue and to tissue diagnosed as cancer are shown in Fig. 2.

3 Evaluation of the Primary Signals with WEKA

In this work we use the Waikato Environment for Knowledge Analysis (WEKA) [6]. This is an open source data mining toolbox (written in Java) developed by Ian Witten's group at the University of Waikato. It provides tools for all the tasks usually performed in data mining, including numerous algorithms for pre-processing, classification, regression and clustering.

Here we utilized 19 different classification algorithms in WEKA. We used their default parameters unless otherwise stated. These algorithms are grouped into five groups in WEKA according to the models they create. Below we give a brief summary of these and some pertinent references.

Bayes includes algorithms where learning results in Bayesian models. In our study we use BayesNet and NaiveBayes algorithms. NaiveBayes is an implementation of the standard naïve Bayes algorithm, where normal distribution is for numerical features. BayesNet creates a Bayesian Network with the ability to represent the same model as NaiveBayes or other more complex models where the independence between features is not assumed.

Lazy is comprised of algorithms that delay construction of classifiers until classification time. The IB1, IBK, Kstar and LWL algorithms were used in this work. IB1 is a nearest-neighbor algorithm that classifies an instance according to the nearest neighbor identified by the Euclidean distance as defined in [7]. IBK is similar to IB1 except that the K- nearest neighbors is used instead of only one. We determined the appropriate number of neighbors using leave-one-out cross-validation. The Kstar algorithm uses entropic distance measure, based on the probability of transforming one instance into another by randomly choosing between all possible transformations [8] and turns out to be much better than Euclidean distance for classification. The LWL (Locally weighted learning) algorithm differs from the other three algorithms in using only a nearest-neighbor algorithm to weight the instances in the training set before applying another classification algorithm to them.

Rules contains methods which creates classification rules. We use NNge, JRip, Ridor and PART algorithms. NNge is a nearest-neighbor algorithm which learns rules based on the hyper rectangles that it divides the instance space into [9]. JRip is an implementation of Cohen's RIPPER. Ridor creates first a default rule and then recursively develops exceptions to it and PART constructs rules based on partial decision trees.

Functions are algorithms that can be represented mathematically. In this work we use MultilayerPerceptron, RBFNetwork, SimpleLogistic and SMO algorithms. RBFNetwork is an implementation of radial basis functions, and SimpleLogistic constructs linear logistic regression models. SMO is a sequential minimum optimization algorithm for building Support Vector Machine (SVM) [10]. We used a polynomial kernel, which is default in WEKA.

Trees includes algorithms that creates trees as models. The ADTree and J48 algorithms were used in this study. The ADTree is similar to options trees and the J48 is an implementation of the popular C4.5 [11].

Miscellaneous contains simply the rest of algorithms that do not fit into any of the other groups. VFI, which was used in our work, finds intervals for each feature, and attributes each class according to number of instances with the class in the training set for the specific interval. Voting is used to select the final class for an instance.

In this paper, the following features were used to form a feature vector that in total has 48 components as inputs to the classifiers: *transient slope* (TS), *saturation slope* (SS) and *maximum slope* (MS) when the sample is closed of each sensor.

In Table 1 we give the results for 24 runs (15 cancer tissues and 9 healthy tissues) from the first experiment. In Table 2 we give the results for 162 runs (92 cancer tissues and 70 healthy tissues) from the second experiment. We used ten-fold cross validation in our experiments, which means that each dataset was divided into ten equal sized folds and ten independent runs of each algorithm were conducted for each dataset. For the ith run, the ith fold was designated as the test set and the patterns in the remaining nine folds were used for training. At the end of training the classifier's generalization was measured on the test set.

Table 1. Classification results

	Cancer tissues	Healthy tissues	Total (%) Correctly Classified
Bayes Network	11/15	2/9	54
Naive Bayes	15/15	6/9	88
Multilayer Perceptron	9/15	4/9	54
RBF Network	15/15	5/9	83
SimpleLogistic	11/15	2/9	54
SMO	9/15	5/9	58
IB1	11/15	8/9	79
KNN	11/15	8/9	79
KStar	13/15	7/9	83
LWL	14/15	6/9	83
ClassificationVia Regression	13/15	7/9	83
ThresholdSelector	12/15	7/9	89
VFI	11/15	8/9	79
ADTree	13/15	7/9	83
J48	15/15	5/9	83
JRip	15/15	8/9	95
NNge	14/15	3/9	71
PART	15/15	5/9	83
Ridor	12/15	5/9	71

Table 2. Classification results

	Cancer tissues	Healthy tissues	Total (%) Correctly Classified
Bayes Network	83/92	66/70	92
Naive Bayes	84/92	63/70	91
Multilayer Perceptron	82/92	61/70	88
RBF Network	81/92	57/70	85
SimpleLogistic	86/92	64/70	93
SMO	84/92	62/70	91
IB1	75/92	50/70	77
KNN	75/92	50/70	77
KStar	81/92	61/70	88
LWL	82/92	58/70	86
ClassificationVia Regression	83/92	59/70	88
ThresholdSelector	77/92	57/70	83
VFI	72/92	61/70	82
ADTree	84/92	66/70	93
J48	79/92	53/70	81
JRip	78/92	62/70	86
NNge	83/92	61/70	89
PART	79/92	65/70	89
Ridor	79/92	59/70	85

4   Evaluation of the Data with the ADATE Code

4.1   A Brief Introduction to ADATE

Automatic Design of Algorithms through Evolution (ADATE) [12] is a system for general automatic programming in a first order, purely functional subset of Standard ML. ADATE can synthesize recursive programs for standard algorithm design problems such as sorting, searching, string processing and many others. It has also successfully been used to generate programs for more advanced tasks such as segmentation of noisy images [13] and driving a robot car.

However, ADATE is also well suited to a more traditional machine learning problem such as analyzing data from an electronic nose to diagnose cancer and offers several advantages in comparison with the standard methods in the WEKA toolbox, such as a better ability to find compact and yet descriptive models.

The models generated by ADATE are formulated in a general programming language which is more expressive than any of the various formalisms used in WEKA discussed above. This means that programs generated by ADATE may be more compact than any of the WEKA models. Compact and still accurate models are important both to avoid overfitting, to enhance readability and above all to give clues for further optimization and redesign of the electronic nose so that it becomes better at cancer detection.

ADATE maintains a hierarchically structured so-called "kingdom of programs". The most basic principle used to organize the kingdom is that each program must be better than all smaller ones found so far. Thus, ADATE generates a progression of gradually bigger and more accurate programs, where each program is optimized over and over again to be the best for its size on the training data.

Program transformations in varying combinations are employed to produce new programs that become candidates for insertion into the kingdom. The search for program transformations is mostly systematic and does not rely on randomization for purposes other than introducing new floating point constants.

ADATE has the ability to define new auxiliary functions "on-the-fly". However, the effectiveness of its program synthesis may strongly depend on the set of predefined functions that it is allowed to use. For the experiments reported in this paper, we included addition, multiplication, subtraction and division of floating point numbers in this set and also the hyberbolic tangent function *tanh* that is commonly used in neural networks.

Since ADATE is able to effectively introduce and optimize floating-point constants on its own, there was no need to include any special, predefined constants.

The above set of predefined functions is a superset of what is needed to implement standard feed-forward neural networks with any number of hidden layers, which can express quite good approximations to any non-linear function [14]. Therefore, the same result holds for our evolved programs.

In practice, however, the limiting factor for most neural and evolutionary computation techniques is not the theoretical expressiveness of the languages that they employ but their ability to avoid entrapment in local optima in the search space. Another key limiting factor is overfitting. We believe that ADATE excels at both reducing overfitting and avoiding local optima, but we do not have space here for a discussion of the many mechanisms employed to do so [12].

4.2 ADATE Experiments for Analyzing Data from the Electronic Nose

Given that the e-nose employs four types of sensors and that each sensor is measured at four different temperatures as discussed above, we have a total of 16 time series for each sample. Each time series was converted to three parameters as described above, giving a total of 48 floating point inputs to a program to be generated by ADATE. These inputs are called TS_0, TS_1, ..., TS_{15}, SS_{16}, SS_{17}, ..., SS_{31}, MS_{32}, MS_{33}, ..., MS_{47}.

We first conducted a simple ADATE run where half of the patients, randomly chosen, were used for training and the other half for testing. Thus, we had 81 patients in the training set and 81 in the test set and obtained overfitting characteristics as shown in Fig. 3. The horizontal axis in that figure shows the size of the programs, measured in bits, whereas the vertical axis shows the number of correctly classified instances for training and test data respectively.

When just saying that all patients in the training set have cancer, we obtain 46 correct classifications, but when instead classifying according to the following simple rule, we suddenly obtain 71 correctly classified patients.

if $SS_{23} < SS_{16}$ then healthy else cancer

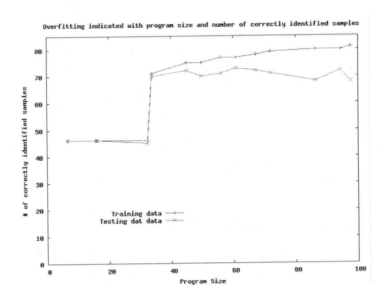

Fig. 3. Overfitting shown with program size and number of correctly identified samples in the training and testing data set

If a small increase in model complexity gives a big increase in classification accuracy, we typically have a change in the model without any overfitting. In other words, if a small amount of extra theory can explain many more observations, that extra theory is believed to be generally valid. As can be seen in Fig. 3, there is an equally big jump in accuracy for both training and testing data when moving to the simple rule above, which has a size of about 34 bits according to ADATE's built in syntactic complexity measure.

The rule above correctly classifies 71 out of 81 training cases and 70 out of 81 test cases, giving an accuracy of 86.4% on the test data and a 95% confidence interval between 77% and 94%. Note that WEKA was run using ten-fold cross validation, which means that 90% of the data were used for training instead of only 50% as in the ADATE experiments. But even if ADATE was given much less training data, it still created results comparable with those of WEKA given in Table 2 and additionally a very simple model that is easy to understand and use for optimization of the Enose.

5   A Pilot Gas Chromatography Experiment

To show that there really is a difference between the healthy and the cancer sample, an extended gas chromatography plus mass spectroscopy has been initiated [15]. The preliminary results of this study will be published soon and here we only present two spectra to show that there is a significant difference between the samples (Fig. 4).

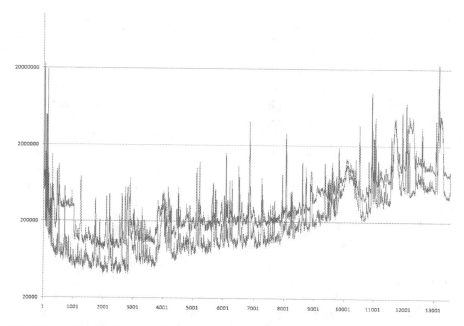

Fig. 4. It is clearly seen from the chromatogram obtained from the healthy sample tissue (upper chromatogram) and one from the cancer tissue that there are differences in the occurrence of certain peaks as well as in their intensities. This indicates that there is a reason for the electronic nose to react differently to the different tissues.

6   Summary, Conclusions and Future Work

The hardware in the present investigation is the same as in the case of a bomb nose. However, the feature extraction and analysis is different. In the first case we simply used rise times and saturation points and the PCA approach to define the regions of interest. In the present case we have tested several algorithms, e.g. the WEKA ones. Hence, from Table 1 we may possibly conclude that the "best" algorithm is JRip. This class implements a propositional rule learner, Repeated Incremental Pruning to Produce Error Reduction (RIPPER). As we can see 68 % of the used machine learning algorithms classifies correctly at least 79 %. When we extend the study to include 162 tests, the "best" algorithms are SimpleLogistic and ADTree with 93% correctly classified.

The results from the ADATE test are interesting and suggestive. They tell us which sensors operated at which temperature are important. Hence some sensors of the original bomb nose may be changed and a more redundant and efficient system could be designed.

The results show that the proposed method, although simple and inexpensive, is probably a rather efficient ovarian carcinoma identification system. It should be stressed again that the sensors are probably not optimal for the present samples. This means that we need to study the results further. We need to test on several more cases with tissues of the same character to see if there is a difference between healthy tissue-samples. Improved multivariate analysis combined a sophisticated gas chromatography and mass spectroscopy test is a logical next step. Although such spectra will most likely show hundreds of lines, it may give hints on which sensors to use. We further need to see in detail if there are any systematic effects. Again, it would be desirable to study the effects on the confusion matrices and to reduce the errors on the "healthy tissues" even if the "cancer tissues" will yield a larger uncertainty. We tried reducing the dimensionality of the inputs using PCA, with varying numbers of principal components, but this did not yield any increase in the classification accuracy obtained with WEKA. However, it would be quite interesting to try some non-linear dimensionality reduction algorithms such as Isomap or auto-encoders optimized with ADATE. Even if one needs to elaborate on modifications of the present e-nose system, one should, indeed, recall that it was originally designed to detect various explosives. From this point of view the "nose is doing quite well".

References

1. Wang, P., Chen, X., Xu, F.: Development of Electronic Nose for Diagnosis of Lung Cancer at Early Stage. In: 5th Conference on Information Technology and Application in Biomedicine, Shenzhen, China, pp. 588–591 (2008)
2. Roppel, T., Dunman, K., Padgett, M., Wilson, D., Lindblad, Th.: Feature-level signal processing for odor sensor array. In: Proc. Ind. Electronics Conf. IECON 1997, IEEE catalog No 97CH36066, pp. 212–221 (1997)
3. Waldemark, J., Roppel, T., Padgett, M., Wilson, D., Lindblad, Th.: Neural Network and PCA for Determining Region of Interest in Sensory Data Pre-processing. In: Virtual Intelligence/Dynamic Neural Networks Workshop 1998 SPIE, vol. 3728, pp. 396–405 (1998)
4. Kermit, M., Eide, Å.J., Lindblad, Th., Agehed, K.: Intelligent Machine Olfaction, IASTED. In: Int. Conf. on Artificial Intelligent Machine Olfaction and Computational Intelligence (ACI 2002), Tokio, Japan, pp. 25–27 (2002)
5. Horvath, G., Järverud, G.K.: Human ovarian carcinomas detected by specific odor. Integr. Cancer Ther. 7(2), 76–80 (2008)
6. Witten, I.H., Frank, E.: Data mining: Practical Machine Learning Tools and Techniques, 2nd edn. Morgan Kaufmann Publishers, San Mateo (2005)
7. Aha, D.W.: Tolerating noisy, irrelevant, and novel attributes in instance-based learning algorithms. International Journal of Man-Machine Studies 36(2), 267–287 (1992)
8. Cleary, J.G., Trigg, L.E.: K*: and instance-based learner using an entropic distance measure. In: Proceedings of the 12th International Conference on Machine Learning, pp. 108–114 (1995)
9. Cohen, W.W.: Fast Effective Rule Induction. In: Proceedings of the 12th International Conference on Machine Learning, pp. 115–123 (1995)
10. Platt, J.C.: Using Analytic QP and Sparseness to Speed Training of Support Vector Machines. In: NIPS conference, pp. 557–563 (1999)
11. Quinlan, R.: C4.5: Programs for Machine Learning. Morgan Kaufmann Publishers, San Mateo (1993)

12. Olsson, R.: Inductive functional programming using incremental program transformation. Artificial Intelligence 1, 55–83 (1995)
13. Berg, H., Olsson, R., Lindblad, Th.: Automatic Design of Pulse Coupled Neurons for Image Segmentation. Neurocomputing - Special Issue for Vision Research (2008)
14. Mitchell, T.: Machine Learning. McGraw-Hill Companies, Inc., New York (1997)
15. Chilo, J., Horvath, G., Lindblad, Th., Olsson, R., Redeby, J., Roeraade, J.: A Flexible Electronic Nose for Ovarian Carcinoma Diagnosis in Real Time. Accepted to IEEE NPSS Real Time Conference, Beijing, China (2009)

Data Mining of Agricultural Yield Data: A Comparison of Regression Models

Georg Ruß

Otto-von-Guericke-Universität Magdeburg

Abstract. Nowadays, *precision agriculture* refers to the application of state-of-the-art GPS technology in connection with small-scale, sensor-based treatment of the crop. This introduces large amounts of data which are collected and stored for later usage. Making appropriate use of these data often leads to considerable gains in efficiency and therefore economic advantages. However, the amount of data poses a data mining problem – which should be solved using data mining techniques. One of the tasks that remains to be solved is *yield prediction* based on available data. From a data mining perspective, this can be formulated and treated as a multi-dimensional regression task. This paper deals with appropriate regression techniques and evaluates four different techniques on selected agriculture data. A recommendation for a certain technique is provided.

Keywords: Precision Agriculture, Data Mining, Regression, Modeling.

1 Introduction

In the past decades, information technology (IT) has become more and more part of our everyday lives. With IT, improvements in efficiency can be made in almost any part of industry and services. Nowadays, this is especially true for agriculture. Due to the modernization and better affordability of state-of-the-art GPS technology, a farmer nowadays harvests not only crops but also growing amounts of data. These data are precise and small-scale – which is essentially why the combination of GPS, agriculture and data has been termed *precision agriculture*.

However, collecting large amounts of data often is both a blessing and a curse. There is a lot of data available containing information about a certain asset – here: soil and yield properties – which should be used to the farmer's advantage. This is a common problem for which the term *data mining* has been coined. Data mining techniques aim at finding those patterns or information in the data that are both valuable and interesting to the farmer.

A common specific problem that occurs is yield prediction. As early into the growing season as possible, a farmer is interested in knowing how much yield he is about to expect. In the past, this yield prediction has usually relied on farmers' long-term experience for specific fields, crops and climate conditions. However, this knowledge might also be available, but hidden, in the small-scale, precise data which can nowadays be collected in-season using a multitude of sensors. These sensors essentially aim to measure a field's heterogeneity.

P. Perner (Ed.): ICDM 2009, LNAI 5633, pp. 24–37, 2009.

Therefore, the problem of yield prediction encountered here is one of data mining and, specifically, multi-dimensional regression. This article should serve as an overview on the capabilities of different regression techniques used on agricultural yield data. Furthermore, this article can be seen as a continuation of [26]: in the previous article artificial neural networks have been evaluated and established as a well-suited reference model, which further models would have to compete against. The current work compares this particular neural network model with suitable further techniques (such as regression trees or support vector machines) to find the best prediction model. To accomplish this, the model output on site-year data from different years and sites is compared. Results on the parameterization of the different models are presented.

1.1   Research Target

The overall research target is to find those indicators of a field's heterogeneity which are suited best to be used for a yield prediction task. The sub-task here is one of multi-dimensional regression – predicting yield from past and in-season attributes. Furthermore, from the agricultural perspective, it is interesting to see how much the factor "fertilization" influences the yield in the current site-year. For this purpose, modeling techniques can be used, but have to be evaluated first. Therefore, this work aims at finding suitable data models that achieve a high accuracy and a high generality in terms of yield prediction capabilities. For this purpose, different types of regression techniques will be evaluated on different data sets.

Since models usually are strongly parameterized, an additional question is whether the model parameters can be carried over from one field to other fields which are comparable in (data set) size. This issue will also be addressed in this work. This is especially useful when new data have to evaluated using one of the presented models.

1.2   Article Structure

Section 2 lays out the data sets that this work builds upon. The attributes and their properties will be presented shortly. Section 3 presents four selected regression techniques from the data mining area which will be used for yield prediction. The free model parameters will be described. Section 4 shows the results from the modeling steps and provides answers to the aforementioned research questions. At the end of this article, future work is pointed out and implementation details are provided.

2   Data Description

The data available in this work have been obtained in the years 2003–2006 on three fields near Köthen, north of Halle, Germany[1]. All information available for these 65-, 72- and 32-hectare fields[2] was interpolated using kriging [30] to a grid with 10 by 10 meters grid cell sizes. Each grid cell represents a record with all available information. During the growing season of 2006, the latter field was subdivided into different

---

[1] GPS: Latitude N 51 40.430, Longitude E 11 58.110.
[2] called *F04*, *F330* and *F131*, respectively.

strips, where various fertilization strategies were carried out. For an example of various managing strategies, see e.g. [27], which also shows the economic potential of PA technologies quite clearly. The field grew winter wheat, where nitrogen fertilizer was distributed over three application times during the growing season.

Overall, for each field there are seven input attributes – accompanied by the respective current year's yield (2004 or 2006) as the target attribute. Those attributes will be described in the following. In total, for the F04 field there are 5241 records, for F131 there are 2278 records, for F330 there are 4578 records, thereof none with missing values and none with outliers. In addition, a subset for F131 was available: in this subset, a special fertilization strategy was carried out which used a neural network for prediction and optimization – this data set is called F131net and has 1144 records of the following attributes.

2.1   Nitrogen Fertilizer – N1, N2, N3

The amount of fertilizer applied to each subfield can be easily measured. It is applied at three points in time into the vegetation period, which is the standard strategy for most of Northwest Europe [20].

2.2   Vegetation – REIP32, REIP49

The *red edge inflection point* (REIP) is a second derivative value calculated along the red edge region of the spectrum, which is situated from 680 to 750nm. Dedicated REIP sensors are used in-season to measure the plants' reflection in this spectral band. Since the plants' chlorophyll content is assumed to highly correlate with the nitrogen availability (see, e.g. [18]), the REIP value allows for deducing the plants' state of nutrition and thus, the previous crop growth. For further information on certain types of sensors and a more detailed introduction, see [13] or [32]. Plants that have less chlorophyll will show a lower REIP value as the red edge moves toward the blue part of the spectrum. On the other hand, plants with more chlorophyll will have higher REIP values as the red edge moves toward the higher wavelengths. For the range of REIP values encountered in the available data, see Tables 1(b) and 1(c). The numbers in the REIP32 and REIP49 names refer to the growing stage of winter wheat, as defined in [16].

2.3   Electric Conductivity – EM38

A non-invasive method to discover and map a field's heterogeneity is to measure the soil's conductivity. Commercial sensors such as the EM-38[3] are designed for agricultural use and can measure small-scale conductivity to a depth of about 1.5 metres. There is no possibility of interpreting these sensor data directly in terms of its meaningfulness as yield-influencing factor. But in connection with other site-specific data, as explained in the rest of this section, there could be coherences. For a more detailed analysis of this particular sensor, see, e.g. [4]. For the range of EM values encountered in the available data, see Tables 1(a) to 1(c).

---

[3] trademark of Geonics Ltd, Ontario, Canada.

2.4   YIELD

Here, yield is measured in metric tons per hectare ($\frac{t}{ha}$) For the yield ranges for the respective years and sites, see Tables 1(b) and 1(c). It should be noted that for the F131 and F330 data sets the yield was reduced significantly due to bad weather conditions (lack of rain) during the growing season 2006.

2.5   Data Overview

In this work, data sets from three different fields are being evaluated. A brief summary of the available data attributes for both data sets is given in Tables 1(a) to 1(c). On each field, different fertilization strategies have been used. One of those strategies is based on a technique that uses a multi-layer perceptron (MLP) for prediction and optimization. This technique has been presented and evaluated in, e.g., [25,26] or [32]. For each field, one data set will contain all records, thus containing all the different fertilization strategies. In addition, a subset of F131 has been chosen to serve as a fourth data set to be evaluated.

Table 1. Overview of the F04, F131 and F330 data sets. The additional data set *F131net*, which is a subset of *F131*, is not shown as its statistical properties are very similar to those of *F131*.

(a) Data overview, F04

F04	min	max	mean	std
YIELD03	1.19	12.38	6.27	1.48
EM38	17.97	86.45	33.82	5.27
N1	0	100	57.7	13.5
N2	0	100	39.9	16.4
N3	0	100	38.5	15.3
REIP32	721.1	727.2	725.7	0.64
REIP49	722.4	729.6	728.1	0.65
YIELD04	6.42	11.37	9.14	0.73

(b) Data overview, F131

F131	min	max	mean	std
YIELD05	1.69	10.68	5.69	0.93
EM38	51.58	84.08	62.21	8.60
N1	47.70	70	64.32	6.02
N2	14.80	100	51.71	15.67
N3	0	70	39.65	13.73
REIP32	719.6	724.4	722.6	0.69
REIP49	722.3	727.9	725.8	0.95
YIELD06	1.54	8.83	5.21	0.88

(c) Data overview, F330

F330	min	max	mean	std
YIELD05	4.64	14.12	10.62	0.97
EM38	25.08	49.48	33.69	2.94
N1	24.0	70	59.48	14.42
N2	3.0	100	56.38	13.35
N3	0.3	91.6	50.05	12.12
REIP32	719.2	724.4	721.5	1.03
REIP49	723.0	728.5	726.9	0.82
YIELD06	1.84	8.27	5.90	0.54

2.6 Fertilization Strategies

There were three different strategies that have been used to guide the nitrogen fertilization of the fields. The three strategies are described in the following. A subset of the F131 data set was chosen where the strategy was "N". This is used as the smallest data set in this work that models will be built upon.

F – uniform distribution of fertilizer according to long-term experience of the farmer
N – fertilizer distribution was guided by an economic optimization with a multi-layer perceptron model; the model was trained using the above data with the current year's yield as target variable that is to be predicted (see, e.g., [26]).
S – based on a special nitrogen sensor – the sensor's measurements are used to determine the amount of nitrogen fertilizer that is to be applied.

3 Advanced Regression Techniques

As mentioned in the preceding section, the task of yield prediction is essentially a task of multi-dimensional regression. Therefore, this section will serve as an overview about different regression techniques that are applicable to the yield data sets. It is aimed to evaluate these techniques on the data sets presented in the preceding section.

The regression task can be formalized as follows: the training set

$$T = \{\{x_1, \ldots, x_n\}, y_i\}_{i=1}^{N} \qquad (1)$$

is considered for the training process, where $x_i, i = 1, \ldots, n$ are continuous input values and $y_i, i = 1 \ldots, N$ are continuous output values. Given this training set, the task of the regression techniques is to approximate the underlying function sufficiently well. The quality of the approximated function can be measured by error values, some of which are specified in section 3.6.

3.1 Introduction to Regression Techniques

Since one particular technique, namely MLPs, has been used successfully in previous work [24,26], it is used as a reference model here. Three additional modeling techniques will be presented that are suitable for the task of yield prediction.

In the past, numerous regression techniques have been used successfully on data from agriculture. Neural networks have shown to be quite effective in modeling yield of different crops ([7,28]). In [31] and [32], artificial neural networks, namely multi-layer perceptrons (MLPs) have been trained to predict wheat yield from fertilizer and additional sensor input. The basic framework for MLPs will be given in section 3.2.

Radial basis function (RBF) networks are similar to multi-layer perceptrons in that they can also be used to model non-linear relationships between input data. Nevertheless, there has been almost no research into RBF networks when applying them to agriculture data. Some of the theoretical properties and differences between MLPs and RBFs will be pointed out in section 3.2.

Regression trees have seen some usage in agriculture [6,12,14]. Essentially, they are a special case of decision trees where the outcome (in the tree leaves) is a continuous function instead of a discrete classification. Further details can be found in section 3.3.

A fourth technique that has, to the best of the author's knowledge, not been used on similar yield data, but for similar regression tasks, is a derivative of support vector machines (SVMs). Similar to decision trees, if the target attribute is discrete, SVMs would solve a classification task, whereas in the case of a continuous attribute, a regression task would be solved. Hence, support vector regression (SVR) will be explained in section 3.4.

The aforementioned techniques have not been compared to each other when used with different data sets in the agriculture context. This section presents the necessary background for each of the techniques before they will be evaluated in section 4.

3.2 Neural Networks

In previous work multi-layer perceptrons (MLPs), a type of neural networks, have been used for a modeling task [24,26] similar to the one laid out in the preceding section. The MLP model has been established as a reference model against which further regression techniques would have to compete. Hence, the MLP will be explained shortly in the following section. Furthermore, a different type of neural network, a radial basis function (RBF) network, will be presented since it is well-suited to the regression task.

MLP. According to [19], "neural networks provide a general, practical method for learning [...] vector-valued functions from examples." In the previous work multi-layer perceptrons (MLPs) with backpropagation learning have been used to learn from agricultural data and predict yield. Generally, MLPs can be seen as a practical vehicle for performing a non-linear input-output mapping [10]. The results from [24,26] lead us to assume that the extension to more than one hidden layer only marginally increases the generalization performance of MLPs, but rather drastically increases the computation time for the backpropagation algorithm. Hence, here it is assumed that one hidden layer is sufficient to approximate the underlying function sufficiently well. For a more detailed and formal description of MLP neural networks, it is referred to [9] or [10].

Once the number of hidden layers of the network is fixed, there remain a few parameters to be determined. Similar to the remaining modeling techniques, those are usually determined experimentally. This also means that often a large parameter space has to be searched. For standard MLP networks, the size of the hidden layer, the learning rate, the activation function and the minimum gradient are the most important parameters that have to be set. In this case, the matlab implementation for the MLP network was used: newff.[4]

RBF. While the MLP networks in the preceding section had a variable number of layers, the number of layers in an RBF network is fixed. There are three layers of neurons that constitute an RBF network and perform different roles. While the input layer is the same as in an MLP network, the only hidden layer applies a nonlinear transformation

---

[4] For details on matlab implementations and scripts see the link contained in the *acknowledgements* section.

from the input space to the hidden space. The output layer is again linear. The idea behind this approach is that a regression (or classification) problem is much more likely to be solvable in a high-dimensional space than in a low-dimensional space [5]. The main difference to MLPs is in the hidden layer. The activation function of each hidden unit in the RBF network computes the Euclidean norm, i.e. the distance, between the input vector and the center of that unit. In MLP networks, the activation function computes the inner product of the input vector and the synaptic weight vector of that unit.

One of the simpler learning algorithms that can be employed for RBF networks is described in the following.

1. The network is simulated: for all training examples, the output of the network is compared to the actual target value of the respective example.
2. The input vector with the greatest error is determined.
3. An RBF neuron is added to the hidden layer with weights equal to that vector.
4. The connection weights from the hidden layer to the output layer are adapted to minimize the error.

According to the above algorithm, the RBF network training algorithm has at least the following parameters: a) an error goal that must be met, b) a radius (or spread) of the radial basis function and c) a maximum number of neurons that should be added before stopping. These parameters are usually determined experimentally, although some strategies for computing them are presented in [10]. Since the current approach aims to compare four basic techniques for non-linear regression, it was chosen to employ the above training algorithm without further tweaking. It has been implemented in matlab's newrb function and the parameters have been determined experimentally.

3.3 Regression Tree

Learning decision trees is a paradigm of *inductive learning*: a model is built from data or observations according to some criteria. The model aims to learn a general rule from the observed instances. Decision trees can therefore accomplish two different tasks, depending on whether the target attribute is discrete or continuous. In the first case, a classification tree would result, whereas in the second case a regression tree would be constructed. Since the focus is on solving a regression task, the regression tree will be explained shortly in the following.

Regression trees approximate learning instances by sorting them down the tree from the root to some leaf node, which provides the value of the target attribute. Each node in the tree represents a split of some attribute of the instance and each branch descending from that node corresponds to one part left or right of the split. The value of the target attribute for an instance is determined by starting at the root node of the tree and testing the attribute specified by this node. This determines whether to proceed left or right of the split. Then the algorithm moves down the tree and repeats the procedure with the respective subtree. In principle, there could be more than one split in a tree node, which would result in more than two subtrees per node. However, in this application scenario, regression trees with more than two subtrees per split node are not taken into consideration.

Regression as well as decision trees are usually constructed in a top-down, greedy search approach through the space of possible trees [19]. The basic algorithms for constructing such trees are CART [2], ID3 [22] and its successor C4.5 [23]. The idea here is to ask the question "which attribute should be tested at the top of the tree?" To answer this question, each attribute is evaluated to determine how well it is suited to split the data. The best attribute is selected and used as the test node. This procedure is repeated for the subtrees. An attribute selection criterion that is employed by ID3 and C4.5 is the entropy and, resulting from it, the information gain. Entropy is a measure from information theory that describes the variety in a collection of data points: the higher the entropy, the higher the variety. An attribute split aims to lower the entropy of the two resulting split data sets. This reduction in entropy is called the information gain. For further information it is referred to [19].

However, if the addition of nodes is continued without a specific stopping criterion, the depth of the tree continues to grow until each tree leaf covers one instance of the training data set. This is certainly a perfect tree for the training data but is likely to be too specific – the problem of overlearning occurs. For new, unseen data, such a specific tree will probably have a high prediction error. Therefore, regression trees are usually pruned to a specific depth which is a trade-off between high accuracy and high generality. This can easily be achieved by setting a lower bound for the number of instances covered by a single node below which no split should occur. For this work the standard matlab implementation of `classregtree` was used.

3.4 Support Vector Regression

Support Vector Machines (SVMs) are a supervised learning method discovered by [1]. However, the task here is regression, so the focus is on support vector regression (SVR) in the following. A more in-depth discussion can be found in [8]. Given the training set, the goal of SVR is to approximate a linear function $f(x) = \langle w, x \rangle + b$ with $w \in \mathbb{R}^N$ and $b \in \mathbb{R}$. This function minimizes an empirical risk function defined as

$$R_{emp} = \frac{1}{N} \sum_{i=1}^{N} L_\varepsilon(\hat{y} - f(x)), \tag{2}$$

where $L_\varepsilon(\hat{y} - f(x)) = \max((|\xi| - \varepsilon), 0)$. $|\xi|$ is the so-called slack variable, which has mainly been introduced to deal with otherwise infeasible constraints of the optimization problem, as has been mentioned in [29]. By using this variable, errors are basically ignored as long as they are smaller than a properly selected ε. The function here is called ε-insensitive loss function. Other kinds of functions can be used, some of which are presented in chapter 5 of [8].

To estimate $f(x)$, a quadratic problem must be solved, of which the dual form, according to [17] is as follows:

$$max_{\alpha, \alpha^*} - \frac{1}{2} \sum_{i=1}^{N} \sum_{j=1}^{N} (\alpha_i - \alpha_i^*)(\alpha_j - \alpha_j^*) K(x_i, x_j) - \varepsilon \sum_{i=j}^{N} (\alpha_i + \alpha_i^*) + \sum_{i=1}^{N} y_i(\alpha_i - \alpha_i^*) \tag{3}$$

with the constraint that $\sum_{j=1}^{N} (\alpha_i - \alpha_i^*) = 0, \alpha_i, \alpha_i^* \in [0, C]$. The regularization parameter $C > 0$ determines the tradeoff between the flatness of $f(x)$ and the allowed number of points with deviations larger than ε. As mentioned in [8], the value of ε is inversely proportional to the number of support vectors. An adequate setting of C and ε is necessary for a suitable solution to the regression problem.

Furthermore, $K(x_i, x_j)$ is known as a kernel function which allows to project the original data into a higher-dimensional feature space where it is much more likely to be linearly separable. Some of the most popular kernels are radial basis functions (equation 4) and a polynomial kernel (equation 5):

$$K(x, x_i) = e^{-\frac{||x - x_i||^2}{2\sigma^2}} \tag{4}$$

$$K(x, x_i) = (\langle x, x_i \rangle + 1)^p \tag{5}$$

The parameters σ and p have to be determined appropriately for the SVM to generalize well. This is usually done experimentally. Once the solution for the above optimization problem in equation 3 is obtained, the support vectors can be used to construct the regression function:

$$f(x) = \sum_{i=1}^{N} (\alpha_i - \alpha_i^*) K(x, x_i) + b \tag{6}$$

In the current experiments, the SVMtorch implementation from [3] has been used, which also points out further details of the SVR process.

3.5 Performance Measurement

The performance of the models will be determined using the root mean squared error (RMSE) and the mean absolute error (MAE). For the RMSE, first the difference between an actual target value y_a and the model output value y is computed. This difference is squared and averaged over all training examples before the root of the mean value is taken, see equation 7. The MAE is computed similarly, see equation 8.

$$RMSE = \sqrt{\frac{1}{n} \sum_{i=j}^{n} (y_i - y_{a,i})} \tag{7}$$

$$MAE = \frac{1}{n} \sum_{i=1}^{n} |y_i - y_{a,i}| \tag{8}$$

3.6 Model Parameter Estimation

Each of the aforementioned four different models will be evaluated on the same data sets. One of the research goals is to establish whether a model that has been used on one data set can be used on a different data set without changing its parameters. This would mean that comparable fields could use the same prediction model. Hence, the *F04* data set is used to determine the model parameters experimentally. Afterwards, the models are re-trained on the remaining data sets using the settings determined for *F04*. The parameter settings are given in section 4.

For training the models, a cross-validation approach is taken. As mentioned in e.g. [11], the data will be split randomly into a training set and a test set. The model is trained using the training data and after each training iteration, the error on the test data is computed. During training, this error usually declines towards a minimum. Beyond this minimum, the error rises – overlearning (or overfitting) occurs: the model fits the training data perfectly but does not generalize well. Hence, the model training is stopped when the error on the test set starts rising. A size ratio of 9:1 for training and test set is used. The data sets are partitioned randomly 20 times, the models are trained and the error values are collected.

4 Results

The models are run with the following parameter settings, which were determined experimentally on *F04* and carried over to the remaining data sets.

MLP. For the multi-layer perceptron model, a relatively small number of 10 hidden neurons is used and the network is trained until a minimum gradient of 0.001 is reached, using a learning rate of 0.25 and the *tangens hyperbolicus* sigmoid activation function.
RBF. For the radial basis function network, a radius of 1 is used for the radial basis neurons in the hidden layer. The algorithm, which incrementally adds neurons until the error goal of 0.001 is met, uses a maximum number of 70 neurons, which results in a relatively long training time.
RegTree. For the regression tree, the default settings of `classregtree` perform optimal; the full tree is pruned automatically and the minimum number of training examples below which no split of a tree node should be done is 10.
SVR. For the support vector regression model, the radial basis function kernel yields the best results, using the parameter $C = 60$ (tradeoff between training error and margin), and the standard deviation $\sigma = 4.0$. The slack variable $\xi = 0.2$ is also determined as yielding the best results on *F04*.

Table 2. Results of running different models on different data sets. The best predictive model for each data set is marked in **bold** font.

Error Measure / Model	F04	F131	F131net	F330
MAE MLP:	0.3706	0.2468	0.2300	0.3576
RMSE MLP:	0.4784	0.3278	0.3073	0.5020
MAE RBF:	0.3838	0.2466	0.2404	0.3356
RMSE RBF:	0.5031	0.3318	0.3205	**0.4657**
MAE REGTREE:	0.4380	0.2823	0.2530	0.4151
RMSE REGTREE:	0.5724	0.3886	0.3530	0.6014
MAE SVR:	**0.3446**	**0.2237**	**0.2082**	**0.3260**
RMSE SVR:	**0.4508**	**0.3009**	**0.2743**	0.4746

4.1 Detailed Results

Considering the results in Table 2, support vector regression obviously performs best on all but one of the data sets, regarding both error measures. Furthermore, SVR also is the model taking the least amount of computation time (not shown in table). Hence, the slight difference between the RMSE of SVR and RBF on the *F330* data set may be considered insignificant in practice when computational cost is also taken into account when deciding for a model.

Regarding the understandability of the generated models, it would certainly be desirable to have the regression tree as the best model since simple decision rules can easily be generated from the tree. However, the regression tree performs worst in all of the cases. On the other hand, when comparing the hitherto reference model MLP with the current best model SVR, the understandability of both models is equally limited. Further research into understanding these models has been and should be undertaken.

Figure 1 shows eight graphs which depict the MAE and RMSE for the different data sets vs. different models, when run on a total of 20 random cross-validation splits. Except for the regression tree, which often produced much higher errors than the other models, the three remaining models usually agree in their error values – with slight, but constant differences in favor of the SVR.

4.2 Conclusion

The results clearly show that support vector regression can serve as a better reference model for yield prediction. It is computationally less demanding, at least as understandable as the hitherto multi-layer perceptron and, most importantly, produces better yield predictions.

The results also show that model parameters which have been established on one data set can be carried over to different (but similar with respect to the attributes) data sets.

4.3 Future Work

One aspect that should be considered in future work is the understandability of the model. While regression trees would be the easiest to understand, they bear the burden of providing worse results compared to SVR. There has already been quite a lot of work towards understanding the inner workings of an MLP, but it remains one of the more intransparent models. The same holds for SVR, but there has been some work using the support vectors for visualization, such as [33,15].

Rather than selecting one of the four presented models, there might be a certain combination of models that performs better than a single one. Some prerequisites would have to be fulfilled – such as the error for a certain data record would have to be alternatively low in one model and high in another. A similar idea is presented in [21].

The most time-consuming part of this article is the manual determination of parameters for one or more models since a large parameter space has to be considered. Even though it has been established that the model parameters which have been determined on one data set can be carried over to different data sets, heuristics for model parameters might reduce the size of the parameter space which has to be searched.

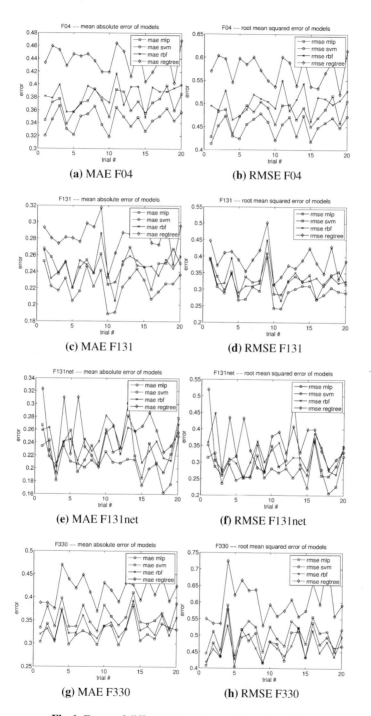

Fig. 1. Errors of different data sets vs. different models

Acknowledgements

Experiments have been conducted using Matlab 2008a and the corresponding Neural Network Toolbox. The field trial data came from the experimental farm Görzig of Martin-Luther-University Halle-Wittenberg, Germany. The trial data have kindly been provided by Martin Schneider and Prof. Dr. Peter Wagner[5] of the aforementioned institution. Supplementary material, such as Matlab scripts and plots can be found at the author's research site http://research.georgruss.de/?cat=20.

References

1. Boser, B.E., Guyon, I.M., Vapnik, V.N.: A training algorithm for optimal margin classifiers. In: Proceedings of the 5th Annual ACM Workshop on Computational Learning Theory, pp. 144–152. ACM Press, New York (1992)
2. Breiman, L., Friedman, J., Olshen, R., Stone, C.: Classification and Regression Trees. Wadsworth and Brooks, Monterey (1984)
3. Collobert, R., Bengio, S., Williamson, C.: Svmtorch: Support vector machines for large-scale regression problems. Journal of Machine Learning Research 1, 143–160 (2001)
4. Corwin, D.L., Lesch, S.M.: Application of soil electrical conductivity to precision agriculture: Theory, principles, and guidelines. Agron. J. 95(3), 455–471 (2003)
5. Cover, T.M.: Geometrical and statistical properties of systems of linear inequalities with applications in pattern recognition. IEEE Transactions on Electronic Computers EC-14, 326–334 (1965)
6. Crone, S.F., Lessmann, S., Pietsch, S.: Forecasting with computational intelligence - an evaluation of support vector regression and artificial neural networks for time series prediction. In: International Joint Conference on Neural Networks, 2006. IJCNN 2006, pp. 3159–3166 (2006)
7. Drummond, S., Joshi, A., Sudduth, K.A.: Application of neural networks: precision farming. In: International Joint Conference on Neural Networks, IEEE World Congress on Computational Intelligence, vol. 1, pp. 211–215 (1998)
8. Gunn, S.R.: Support vector machines for classification and regression. Technical Report, School of Electronics and Computer Science, University of Southampton, Southampton, U.K. (1998)
9. Hagan, M.T.: Neural Network Design (Electrical Engineering). Thomson Learning (December 1995)
10. Haykin, S.: Neural Networks: A Comprehensive Foundation, 2nd edn. Prentice Hall, Englewood Cliffs (1998)
11. Hecht-Nielsen, R.: Neurocomputing. Addison-Wesley, Reading (1990)
12. Huang, C., Yang, L., Wylie, B., Homer, C.: A strategy for estimating tree canopy density using landsat 7 etm+ and high resolution images over large areas. In: Proceedings of the Third International Conference on Geospatial Information in Agriculture and Forestry (2001)
13. Liu, J., Miller, J.R., Haboudane, D., Pattey, E.: Exploring the relationship between red edge parameters and crop variables for precision agriculture. In: 2004 IEEE International Geoscience and Remote Sensing Symposium, vol. 2, pp. 1276–1279 (2004)
14. Lobell, D.B., Ivan Ortiz-Monasterio, J., Asner, G.P., Naylor, R.L., Falcon, W.P.: Combining field surveys, remote sensing, and regression trees to understand yield variations in an irrigated wheat landscape. Agronomy Journal 97, 241–249 (2005)

---

[5] {martin.schneider,peter.wagner}@landw.uni-halle.de

15. Maszczyk, T., Duch, W.: Support Vector Machines for Visualization and Dimensionality Reduction. In: Kůrková, V., Neruda, R., Koutník, J. (eds.) ICANN 2008, Part I. LNCS, vol. 5163, pp. 346–356. Springer, Heidelberg (2008)
16. Meier, U.: Entwicklungsstadien mono- und dikotyler Pflanzen. Biologische Bundesanstalt für Land- und Forstwirtschaft, Braunschweig, Germany (2001)
17. Mejía-Guevara, I., Kuri-Morales, Á.F.: Evolutionary feature and parameter selection in support vector regression. In: Gelbukh, A., Kuri Morales, Á.F. (eds.) MICAI 2007. LNCS, vol. 4827, pp. 399–408. Springer, Heidelberg (2007)
18. Middleton, E.M., Campbell, P.K.E., Mcmurtrey, J.E., Corp, L.A., Butcher, L.M., Chappelle, E.W.: "Red edge" optical properties of corn leaves from different nitrogen regimes. In: 2002 IEEE International Geoscience and Remote Sensing Symposium, vol. 4, pp. 2208–2210 (2002)
19. Mitchell, T.M.: Machine Learning. McGraw-Hill Science/Engineering/Math (March 1997)
20. Neeteson, J.J.: Nitrogen Management for Intensively Grown Arable Crops and Field Vegetables, ch. 7, pp. 295–326. CRC Press, Haren (1995)
21. Orr, M., Hallam, J., Murray, A., Ninomiya, S., Oide, M., Leonard, T.: Combining regression trees and radial basis function networks. International Journal of Neural Systems 10 (1999)
22. Quinlan, J.R.: Induction of decision trees. Machine Learning 1(1), 81–106 (1986)
23. Quinlan, R.J.: C4.5: Programs for Machine Learning. Morgan Kaufmann Series in Machine Learning. Morgan Kaufmann, San Francisco (1993)
24. Ruß, G., Kruse, R., Schneider, M., Wagner, P.: Estimation of neural network parameters for wheat yield prediction. In: Bramer, M. (ed.) Artificial Intelligence in Theory and Practice II. IFIP International Federation for Information Processing, vol. 276, pp. 109–118. Springer, Heidelberg (2008)
25. Ruß, G., Kruse, R., Schneider, M., Wagner, P.: Optimizing wheat yield prediction using different topologies of neural networks. In: Verdegay, J.L., Ojeda-Aciego, M., Magdalena, L. (eds.) Proceedings of IPMU 2008, pp. 576–582. University of Málaga (June 2008)
26. Ruß, G., Kruse, R., Schneider, M., Wagner, P.: Data mining with neural networks for wheat yield prediction. In: Perner, P. (ed.) ICDM 2008. LNCS, vol. 5077, pp. 47–56. Springer, Heidelberg (2008)
27. Schneider, M., Wagner, P.: Prerequisites for the adoption of new technologies - the example of precision agriculture. In: Agricultural Engineering for a Better World, Düsseldorf. VDI Verlag GmbH (2006)
28. Serele, C.Z., Gwyn, Q.H.J., Boisvert, J.B., Pattey, E., Mclaughlin, N., Daoust, G.: Corn yield prediction with artificial neural network trained using airborne remote sensing and topographic data. In: 2000 IEEE International Geoscience and Remote Sensing Symposium, vol. 1, pp. 384–386 (2000)
29. Smola, A.J., Olkopf, B.S.: A tutorial on support vector regression. Technical report, Statistics and Computing (1998)
30. Stein, M.L.: Interpolation of Spatial Data: Some Theory for Kriging. Springer Series in Statistics. Springer, Heidelberg (1999)
31. Wagner, P., Schneider, M.: Economic benefits of neural network-generated site-specific decision rules for nitrogen fertilization. In: Stafford, J.V. (ed.) Proceedings of the 6th European Conference on Precision Agriculture, pp. 775–782 (2007)
32. Weigert, G.: Data Mining und Wissensentdeckung im Precision Farming - Entwicklung von ökonomisch optimierten Entscheidungsregeln zur kleinräumigen Stickstoff-Ausbringung. PhD thesis, TU München (2006)
33. Wu, S., Chow, T.W.S.: Support vector visualization and clustering using self-organizing map and vector one-class classification. In: Proceedings of the International Joint Conference on Neural Networks, vol. 1, pp. 803–808 (2003)

Study of Principal Components on Classification of Problematic Wine Fermentations

Alejandra Urtubia U.[1,*] and J. Ricardo Pérez-Correa[2]

[1] Escuela de Ingeniería Industrial. Facultad de Ciencias Económicas y Administrativas.
Universidad de Valparaíso,
Las Heras N°6. Valparaíso, Chile.
Tel.: +56-32-2507788; Fax: +56-32-2507958
[1] Departamento de Ingeniería Química y Ambiental,
Universidad Técnica Federico Santa María
Av. España 1680, Casilla 110-V, Valparaíso, Chile.
Tel.: +56-32-2654258; Fax: +56-32-2654478
alejandra.urtubia@usm.cl
[2] Departamento de Ingeniería Química y Bioprocesos, Escuela de Ingeniería,
Pontificia Universidad Católica de Chile.
Vicuña Mackenna 4860, Casilla 306, Santiago 22, Chile
Tel.: +56-2-3544258; Fax: +56-2-3545803
perez@ing.puc.cl

Abstract. Data mining techniques have already shown useful to classify wine fermentations as problematic. Then, these techniques are a good option for winemakers who currently lack the tools to identify early signs of undesirable fermentation behavior and, therefore, are unable to take possible mitigating actions. In this study we assessed how much the performance of a clustering K-means fermentation classification procedure is affected by the number of principal components (PCs), when principal component analysis (PCA) is previously applied to reduce the dimensionality of the available data. It was observed that three PCs were enough to preserve the overall information of a dataset containing reliable measurements only. In this case, a 40% detection ability of problematic fermentations was achieved. In turn, using a more complete dataset, but containing unreliable measurements, the number of PCs yielded different classifications. Here, 33%f the problematic fermentations were detected.

Keywords: clustering, data mining, k-means, PCA, stuck, sluggish.

1 Introduction

Early detection and diagnosis of critical operating conditions in an industrial process is essential to ensure process safety and productivity, and to keep the product within specifications. Nowadays instrumentation and data acquisition systems are able to

---

* Corresponding author.

P. Perner (Ed.): ICDM 2009, LNAI 5633, pp. 38–43, 2009.
© Springer-Verlag Berlin Heidelberg 2009

measure hundreds of variables per second, generating massive amounts of data. Nevertheless, it is not evident how to assess the real status of the process directly from these measurements. Therefore, specific statistical techniques for the particular process should be applied to unveil useful information from this data.

Wine production is a significant and growing business in Chile, although it is facing a highly competitive international market. Hence, the wine industry is focused on reducing costs and improving wine quality. Since operating problems that are detrimental to these objectives arise at various stages of the winemaking process, enologists require tailor-made instrumentation, statistical and management tools. Specifically, the fermentation stage is often problematic, causing major product losses and product downgrading, which can be minimized or even averted if they are corrected on time. Thus, here we explore ways to optimize the application of statistical tools to detect problematic fermentations early.

Cluster analysis is a robust classification method and it has been employed in numerous fields such as gene expression analysis, [1], [5], [6], [11]. In addition, several authors have employed principal component analysis (PCA), to retrieve useful information from fermentation databases [2], [3], [4] [7], [10]. For some applications of PCA, most of the variability in the data can be captured in two or three dimensions, and the process variability can be visualized with a single plot. But, there are other cases when most of the data variations cannot be captured in two or three dimensions and process monitoring is more difficult. For example, in [7] a signal to noise ratio (SNR) method was defined and proposed as an index of fault detection ability. The method applied PCA to reduce the data dimensionality. It was found that the fault detection ability was sensitive to the number of PCs. Thus, if a priori faulty operation data is available, optimization of the number of PCs to maximize fault detection ability is straightforward. In case not such data exists, running in parallel multiple models with different numbers of PCs is the next best strategy. Other authors [4] have applied PCA to remove outliers from a product quality historical data set in a chemical industry. They successfully used T^2 statistic and contribution charts for fault detection and identification. Some benefits obtained were: immediate alarm when a newly manufactured lot of material is out of norm; fast identification and root cause follow-up of production and raw material failures; and establishment of a data base of normal and abnormal product lots to compare with future customer complaints.

In this work we study the impact that the number of principal components (PCs) has on the classification of problematic wine fermentations using clustering k-means.

2 Methodology

2.1 Datasets

Between 30 and 35 samples were collected from each of 24 normal and abnormal operation industrial wine fermentations of Cabernet-Sauvignon. Each sample was analyzed for sugars, alcohols, and organic acids [8]. In addition, nitrogen-rich compounds were analyzed, though these measurements were less reliable. In all, 28 compounds were analyzed per sample resulting in 22,000 measurements. Dataset A includes only reliable measurements, i.e., sugars, alcohols, and organic acids, while dataset E includes these plus the nitrogen-rich compounds.

2.2 Data Processing and Analysis

For classification, the common K-means clustering technique was used. The method assumes that dataset X contains k clusters X_i, which can be represented by their mid value μ_i. The Euclidian distance between the observed data and its cluster's center, is used here as a measure of similarity [9]. The procedure can be summarized as follows:

Step 1: Random placement of the initial centers.
Step 2: Assign each data point to its closest cluster. After all the assignments are completed, redefine the center of the cluster so as to minimize function J_i.

$$J_i = \sum_{j=1}^{N_i} d(x_{ij}, \mu_i) \qquad x_{ij} \in X_i, N_i \neq \# X_i \qquad (1)$$

And

$$d(x, \mu_i)^2 = (x - \mu_i)^T (x - \mu_i) \qquad (2)$$

Where subscript $_i$ represents the cluster or group, μ_i is the center of the cluster, and $d(x, \mu_i)$ the distance from observation x to the center.

Step 3: the new center positions are taken as the initial points for a new iteration starting at step 2. The procedure is repeated until the positions of the centers have minor variations or no longer change between successive iterations. Using this procedure, convergence to a local minimum of function J is guaranteed.

Identified clusters were analyzed to determine which normal fermentations behaved similarly and to explore any associations between these and problematic fermentations.

For data preprocessing, PCA was applied to reduce the dimensionality of the 22,000 measurements, while retaining relevant patterns. Eigenvalue limit criterion was used to select the number of PCs. Over 70% of the total variance of the raw data was explained with the PCA model. The impact that the number of PCs has on the classification is analyzed here.

Previous works observed that classifications with the procedure described above using measurements from the first three days of fermentation (77 samples) were similar to those obtained with the entire data set (570 samples) [9]. Hence, in this study we included only samples of the first three days of fermentation.

3 Results

Here, fermentations that finished with more than 2 g/L of sugar or lasted more than 13 days were arbitrarily defined as problematic. Using this criterion, of the 24 fermentations studied, 9 were normal (1 to 9) and 15 problematic. The classification procedure was applied to all the measurements taken during the first three days of fermentation and no distinction was a priori made if the samples were from a normal or a problematic fermentation. First, PCA was applied to reduce the dimensionality of the dataset A (sugars, alcohols, organic acids) and E (dataset A plus nitrogen-rich compounds). The effect of 3, 5 and 8 PCs over clustering K-means was studied. Three clusters were used in the classification procedure: blue (B), red (R) and pink (P). The results of classification are presented in tables 1 and 2.

Table 1. Classification results of clustering K-means using 3 clusters and 3, 5 or 8 PC, for dataset A

# PC	Cluster					
	Blue		**Red**		**Pink**	
	Normal	Problematic	Normal	Problematic	Normal	Problematic
3	2, 5, 6, 7, 8	10, 11, 12, 13, 14, 15, 16, 17, 18, 20, 22	1, 3, 4, 6, 7, 8, 9	10, 12, 13, 15, 17, 18, 19, 20, 21, 23, 24	1, 2, 3, 4, 7, 8, 9	10, 14, 15, 20, 23, 24
5	2, 5, 6, 7, 8	10, 11, 12, 13, 14, 16, 17, 18, 20, 22	1, 3, 4, 6, 7, 8, 9	10, 12, 13, 14, 15, 17, 19, 21, 23, 24	1, 2, 3, 4, 6, 7, 8, 9	10, 15, 18, 19, 20, 23, 24
8	2, 5, 6, 8	10, 11, 12, 13, 14, 16, 17, 18, 20, 22	1, 3, 4, 6, 7, 8, 9	10, 12, 13, 14, 15, 17, 18, 19, 20, 21, 23, 24	1, 2, 3, 4, 7, 8, 9	10, 15, 20, 21, 23, 24

Table 2. Classification results of clustering k-means using 3 clusters, and 3, 5 or 8 PC, for dataset E

# PC	Cluster					
	Blue		**Red**		**Pink**	
	Normal	Problematic	Normal	Problematic	Normal	Problematic
3	1, 2, 5, 6, 8	10, 11, 13, 14, 15, 16, 17, 21, 22	1, 2, 3, 4, 5, 6, 7, 8, 9	10, 12, 13, 15, 16, 17, 18, 19, 20, 21, 22, 23, 24	2, 3, 6, 8, 9	13, 14, 19, 20, 22
5	1, 3, 4, 6, 8	10, 12, 13, 14, 15, 17, 21	2, 4, 5, 6, 7	11, 12, 13, 15, 16, 17, 18, 22, 23, 24	1, 2, 3, 6, 7, 8, 9	10, 13, 14, 15, 16, 17, 18, 19, 20, 21, 22
8	1, 6, 8	10, 13, 14, 15, 16, 17, 21	2, 4, 5, 6, 7	11, 12, 13, 15, 16, 18, 20, 22, 23, 24	1, 3, 6, 7, 8, 9	10, 12, 13, 14, 15, 16, 17, 18, 19, 20, 21

Regarding classification quality, results can be analyzed in two steps. First, we analyze only those fermentations that classified all their samples in one cluster. Then, we analyze the complete set of fermentations, independently of the number of clusters required to classify the respective samples. However, the more the samples of a given fermentation are spread in different clusters, the more difficult the interpretation of the classification is. Hence, we tried first to maximize the number of fermentations with all their samples in one cluster. Then, we look for classifications that minimize the spread of samples. Using these criteria, we expect to simplify the detection of problematic fermentations.

Three PCs explained 74% of the total variance for dataset A and only 47% for dataset E. To explain 75% of the total variance in dataset E, 8 PCs were needed. Clustering K-means was then applied with the defined number of PCs. Fermentations were classified in the 3 clusters (B, R, P), however only few of them classified all samples in one cluster.

For dataset A, 7 fermentations - including normal and problematic- classified all their samples in the same clusters when 3 PCs were used (5, 11, 14, 16, 19, 21, 22) and when 5 or 8 PCs were used, six fermentations classified all their samples in one cluster. Hence, in this case, adding more PCs obscure the detection of problematic fermentations.

For dataset E, results were different. Since the number of fermentations –including normal and problematic- that classified all their samples in the same cluster was 7 using 3 (4, 7, 11, 12, 18, 23, 24) or 5 PCs (5, 9, 11, 19, 20, 23, 24), and 10 using 8 PCs (2, 3, 4, 5, 9, 11, 19, 22, 23, 24). Contrary to dataset A, more PCs improve the total classification quality of dataset E according to the first criteria. Furthermore, even considering that dataset E contains biased measurements, a better classification than that with dataset A is achieved. However, it must be analyzed the ability to detect problematic fermentation with representative information. Considering only one-cluster fermentations, a 40% detection ability of problematic fermentations is achieved using 3 PCs (11, 14, 16, 19, 21, 22) for dataset A. In turn, for dataset E, it is possible to detect the 33% of total problematic fermentations independently of the number of PCs used. In this case, a better ability of problem detection with dataset A is achieved.

Studying the differences in the classification between 3, 5 and 8 PCs, and including all fermentations in the analysis, 67% of the fermentations (1, 2, 3, 4, 5, 8, 9, 10, 11, 12, 13, 16, 17, 22, 23, 24) presented the same classification independently of the number of PCs used for dataset A. These included 60% of the total problematic fermentations (10, 11, 12, 13, 16, 17, 22, 23, 24). Besides, seven of the nine normal fermentations classified all their samples in the same clusters (1, 2, 3, 4, 5, 8, 9). For dataset E, that includes less reliable measurements, results were different. Only 21% of fermentations presented the same classification using 3, 5 or 8 PCs (6, 13, 14, 23, 24), although almost all of them were problematic (13, 14, 23, 24).

Dataset A with reliable information gives better results with dataset E, which includes nitrogen rich compounds, independently of the used criterion. That it means that the quality of data is relevant in the results. However, with more information it is possible classify 42% of fermentations (10/24 fermentations), 56% of normal fermentations and 33% of problematic fermentations.

4   Conclusions

In the case studied here, three PCs were enough to preserve the relevant information contained in the raw data and achieve optimum classification using a dataset with reliable information. In addition, when unreliable measurements were usedmore PCs were needed to obtain better results. However these measurements degraded the classification which was also dependent on the number of PCs. Consequently, the classification algorithm is highly sensitive to the number of PCs when the dataset is contaminated with noisy data. New methodologies must be explored considering the complex structure of database. Finally, discriminate analysis, MPCA and MPLS are being explored and their results are promising.

Acknowledgements

The author A. Urtubia thanks the financial support from FONDECYT Project (Number 11070251).

References

1. Al-Harbi, S., Rayward-Smith, V.: The use of a supervised k-means algorithm on real-valued data with applications in health. In: Chung, P.W.H., Hinde, C.J., Ali, M. (eds.) IEA/AIE 2003. LNCS (LNAI), vol. 2718, Springer, Heidelberg (2003)
2. Chiang, L.H., Russell, E.L., Braatz, R.D.: Fault Detection and Diagnosis in Industrial Systems. Springer, Heidelberg (2001)
3. Chiang, L.H., Leardi, R., Pell, R., Seasholtz, M.B.: Industrial experiences with multivariate statistical analysis of batch process data. Chemometrics and Intelligent Laboratory Systems 81, 109–119 (2006)
4. Chiang, L.H., Colegrove, L.I.: Industrial implementation of on-line multivariate quality control. Chemometrics and Intelligent Laboratory Systems 88(2), 143–153 (2007)
5. Fx, W., Zhang, W.J., Kusalik, A.J.: A genetic K-means clustering algorithm applied to gene expression data. In: Xiang, Y., Chaib-draa, B. (eds.) Canadian AI 2003. LNCS (LNAI), vol. 2671, pp. 520–526. Springer, Heidelberg (2003)
6. Kamimura, R., Bicciato, S., Shimizu, H., Alford, J., Stephanopoulos, G.: Mining of Biological Data II: Assessing Data Structure and Class Homogeneity by Cluster Analysis. Metab. Eng. 2, 228–238 (2000)
7. Tamura, M., Tsujita, S.: A study on the number of principal components and sensitivity of fault detection using PCA. Computers and Chemical Engineering 31, 1035–1046 (2006)
8. Urtubia, A., Pérez-Correa, R., Meurens, M., Agosin, E.: Monitoring large scale wine fermentations with infrared spectroscopy. Talanta 64, 778–784 (2004)
9. Urtubia, A., Pérez-C, J., Soto, A., Pszczólkowski, P.: Using data mining techniques to predict industrial wine problem fermentations. Food Control 18(12), 1512–1517 (2007)
10. Vlasides, S., Ferrier, J., Block, D.: Using Historical Data for Bioprocess Optimization: Modeling Wine Characteristics Using Artificial Neural Networks and Archives Process Information. Biotechnol. and Bioeng. 73(1), 55–68 (2001)
11. Yoshioka, T., Morioka, R., Kobayashi, K., Oba, S., Ogawsawara, N.: Clustering of gene expression data by mixture of PCA models. In: Dorronsoro, J.R. (ed.) ICANN 2002. LNCS, vol. 2415, pp. 522–527. Springer, Heidelberg (2002)

A Data Mining Method for Finding Hidden Relationship in Blood and Urine Examination Items for Health Check

Kazuhiko Shinozawa[1,2], Norihiro Hagita[1,2], Michiko Furutani[1],
and Rumiko Matsuoka[1]

[1] International Research and Educational Institute for Integrated Medical Science
(IREIIMS), Tokyo Women's Medical University, 8-1 Kawada-cho, Shinjuku-ku, Tokyo
162-8666 Japan
[2] ATR Intelligent Robotics and Communication Laboratories, 2-2-2, Hikaridai,
Seika-cho, Soraku-gun, Kyoto 619-0288 Japan

Abstract. Our periodic health examination often describes whether each examination item in blood and urine takes in the reference range of each examination item and a simple summary report on checks in everyday life and the possibility of suspicious diseases. However, it uses n variable items such as AST(GOT), ALT(GPT) which are less correlated, and often includes expensive tumor markers. Therefore, this paper proposes a data mining method for finding hidden relationships between these items in order to reduce the examination fee and giving a report depending on individuals. Since low correlation coefficients are shown in most pairs of items over all clients, a set of item's values in consecutive health examinations of each client is investigated for data mining. Four groups are formed according to the frequency taking outside the reference range in an item for three consecutive examinations, and average values of the other items included in each group are calculated in all pairs of items. The experiment results for three consecutive health examinations show that a lot of item pairs have positive or negative correlations between different frequencies with an item and the averages with the other item despite the fact that their correlation coefficients are small. The result shows both possible reducing of reducing the examination fee as inexpensive as possible and the possibility of a health-care report reflecting individuals.

1 Introduction

— We have a health checkup periodically for health care. It examines blood and urine examination items, such as AST(GOT), ALT(GPT), blood pressure, X-ray, etc. The check report shows each examination item's value and the reference range for each item in blood and urine, and marks the items that are outside the reference range. As a result, the reference range is available for an index of our health state. However, in terms of data mining, this reference range in current health examination corresponds to one-dimensional analysis despite the fact

P. Perner (Ed.): ICDM 2009, LNAI 5633, pp. 44–50, 2009.
© Springer-Verlag Berlin Heidelberg 2009

that n variable analysis could be made for more detailed care information. On the other hand, in the field of clinical research, n variable analysis in the investigate items is made by using conventional statistical methods, such as correlation coefficient, principal component analysis, and machine learning one[2,3,4]. Our previous work on n variable analysis based on machine learning algorithm shows that predominant relationships between items could hardly be founded when we used the Cyber Integrated Medical Infrastructure(CIMI) database that has 117 examination items including 30 tumor markers in blood and urine from 300 clients[1]. Therefore, this paper proposes a data mining method for finding hidden relationships between these items for thorough health examination. We use the CIMI database for data mining that has 117 items including 24 tumor markers periodically examined with 687 clients. If we can obtain hidden relationship using them, it makes our normal health examination effective. When we apply statistical analysis to the data, it shows low correlation coefficients in most pairs of items over all clients. Next, we focus on a set of item's values in consecutive health checks of each client for data mining. Two kinds of features are proposed. The first feature represents a frequency taking outside the reference range of an item in the set is selected. An average of the other item is obtained as the second feature from all clients with the frequency of the item. Two features help finding hidden relationships between the examination items.

2   Relation Extraction

All client data are divided into M groups according to a ratio of outside-state of each examination item. Each examination item's value has a reference range that includes 95% of total. Most health checks investigate whether each item value is inside or outside of the reference range . If the value is outside, there is suspicious of some diseases. An examination item's value has fluctuation that is caused with factors that do not related to diseases. So, we think that several examinations are needed and outside-state ratios are important for finding possibility of diseases. So, we focus on a set of item's values in consecutive examinations of each client and try to find relationship between outside-state ratios of an item and the other values. Examination item data consists of $x_{j,i}(t)$ that means the j-th client's value of i examination item at t examination time. The vector of data can be shown as the following equation;

$$\mathbf{X}_j(t) = \{x_{j,i}(t) | 1 \leq i \leq N, 1 \leq t \leq T_j, 1 \leq j \leq H\}. \tag{1}$$

Here, T_j means total numbers of examination times of j-th client, H means total numbers of clients. Each examination item has independent reference range that is corresponding to normal ranges or normal values and we can convert item's values into outside-states with reference ranges . The vector $\mathbf{B}_j$ that show each outside-state of examination item is

$$\mathbf{B}_j(t) = \{b_{j,i}(t) | 1 \leq i \leq N, 1 \leq t \leq T_j, 1 \leq j \leq H\}$$

$$b_{j,i}(t) = \begin{cases} 0 \ if \ x_{j,i}(t) \ within \ a \ reference \ range \\ 1 \ if \ x_{j,i}(t) \ without \ a \ reference \ range \end{cases}$$

We can obtain an outside-state ratio $(r_{j,i})$ of each examination item according to

$$\mathbf{R}_j = \{r_{j,i}|1 \leq i \leq N\} \tag{2}$$

$$r_{j,i}^\tau = \frac{\sum_{t=\tau}^{\tau+M-2} b_{j,i}(t)}{M-1}. \tag{3}$$

We use G_i^k for presenting a group with an outside-state of i examination item in $k(1 \leq k \leq M)$ levels. Then, each item set can be assigned to G_i^k according to the following conditions;

$$G_i^k = \{j, \tau | r_{j,i}^\tau = \frac{k-1}{M-1}\} \tag{4}$$

An average of examination item value in each group is calculated.

$$\bar{x_{i,k}} = \frac{\sum_{j,\tau \in G_i^k} \sum_{t=\tau}^{\tau+M-2} x_{j,i}(t)}{(M-1)\sum_{j,\tau \in G_i^k}}$$

Then, we choose examination item's combination that satisfies the following two conditions;

1. The average monotonically increases or decreases according to k variable that means reference range level.
2. Each difference in averages is statistically significant.

When we investigate all combinations, we can obtain a relation matrix that consists of predominant relationships between examination items.

3   Analysis Results with Actual Data

We estimated the relation matrix with actual data from CIMI data base. The CIMI data base had 3633 records obtained from 687 persons until Oct. 2008. Male and Female ratio is 333/354. One record has 117 examination items. Figure 1 shows age and male/female distribution of CIMI data base acquired until Oct. 2008.

3.1   Comparison with Correlation

Figure 2 shows male's raw relationship between β2-microglobulin urine (BMG urine) and blood glucose. Its correlation coefficient is 0.17. The coefficient is too small to support its relationship. Proposed method clarified positive relationship between them. Figure 3 shows a relationship between BMG urine's outside-state ratio and averages of blood glucose's examination values with our method ($M = 4$). Reference range consists of upper and lower bound value. Here, we focus on upper bound of reference range. When BMG urine's value is over upper bound, the state becomes outside-state. Male's upper bound of

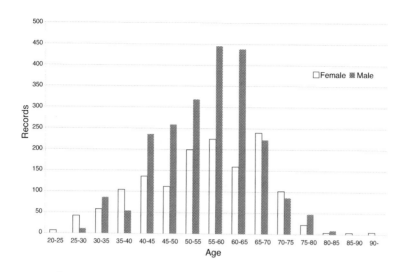

Fig. 1. Distribution of CIMI data base acquired until Oct. 2008

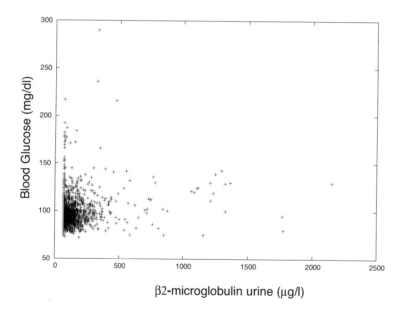

Fig. 2. Correlation between BMG and Blood glucose

BMG urine is 200 ($\mu g/l$). Differences between all average values are statistically significant (ANOVA $F = 66.2967, p < 0.01$, multiple comparison with scheffé's method, $p < 0.01$). The figure shows that an average of blood glucose's examination value in each client increase as outside-state ratios of BMG urine increase. Estimating regression line's parameters, its coefficients are 5.10741 and

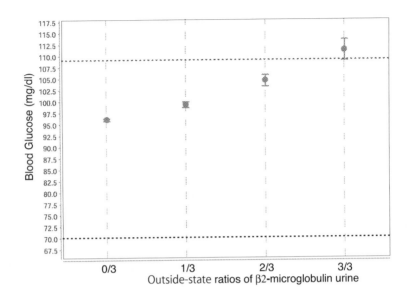

Fig. 3. Relationship BMG urine's outside-state ratio and blood glucose

Table 1. Numbers of relationship

Sex	Bound case	correlation	GX-matrix
Male	Upper	176	293
	Lower		330
Female	Upper	274	397
	Lower		601

95.0910. We applied the method to all combinations of 117 examination items and obtain a relation matrix. We call it GX-matrix and it consists of 117×117 elements. Each element is 0 when above two conditions are not satisfied and is a linear regression's coefficient when satisfying them. Our method uses the reference range that changes according to both male/female, and a reference range has upper/lower bound. So, there are four kinds of relation matrices in total. We compare proposed methods results with a result of a method that is based on correlation coefficients. The number of found relationships by both methods is shown in table 1. If the correlation coefficient is greater than 0.6 or smaller than -0.6, the relationship is useful for estimating each other value. If the coefficient is greater than 0.4 or smaller than -0.4, the relationship has possibility to be useful for estimating each other value. In table 1, the number with a correlation method shows found the number of combinations between examination items that have greater than 0.4 or smaller than -0.4 correlation coefficients. GX-matrix's number is the number of combinations that satisfy above two conditions. The table shows that our method can find more relationships than a correlation method does in all of four cases.

4   Discussion

Proposed relation extraction method's result with actual data shows that it is effective on finding hidden relationships that ordinary correlation method cannot reveal. Medical research findings show that there is relationship between states of a kidney disease and BMG [5], and it is known that there is a relationship between blood glucose and a kidney disease. However, few research shows quantitative relationship between BMG and blood glucose. So, proposed method's result is consistent with ordinary medical knowledge, and it can provide quantitative relationship between them. Upper and lower dotted lines in figure 3 show reference range of blood glucose. In the figure, an average of blood glucose increases as an outside-state ratio of BMG urine increases. Even when a client takes only BMG urine examination, the result can provide a caution that blood glucose might be outside reference range if all BMG urine values are outside reference range in all three examination times. It shows that measuring BMG urine can substitute blood glucose's state. In addition, the result shows that blood glucose's average value is important to catch a BMG urine's tendency to be getting worse in spite of blood glucose's average value is inside reference range . A measurement cost of blood glucose is much cheaper than the cost of BMG urine. By clarifying quantitative relationship between an average of one item and frequency of the other item, we can use average values of blood glucose instead of BMG urine's state. Thus, our method's result has possibility to reduce examination's cost by using inexpensive and usual health check examination items instead of using expensive and special examination items that can be only inspected in a big hospital. In this paper, we showed one example that is a relationship between BMG urine and blood glucose. This analysis can also provide the other relationships with BMG urine's outside-state ratios. So, for precisely estimating examination item's value, we need utilize multiple linear regression analysis.

5   Conclusion

We proposed a data mining method for finding hidden relationship in blood and urinalysis data. The analysis result with actual client's data shows that the method could provide hidden relationship between items that cannot be found with a correlation method. In addition, it also showed that this method might reduce examination's cost, and the method would provide new screening strategy for medical diagnosis. In the future, we are planing to evaluate estimation of examination item values with multiple regression linear analysis of this result and then provide cost efficient test strategy for medical diagnosis.

Acknowledgement

This research was supported in part by the Program for Promoting the Establishment of Strategic Research Centers, Special Coordination Funds for Promoting

Science and Technology, Ministry of Education, Culture, Sports, Science and Technology (Japan).

References

1. Abe, A., et al.: Possibility of integrated data mining of clinical data. Data Science Journal 6(suppl.) (March 9, 2007)
2. Bayes-Genis, A., et al.: Pregnancy-Associated Plasma Protein A as a Marker of Acute Coronary Syndromes. The New England Journal of Medicine 345(14) (October 2001)
3. Nissen, S.E., et al.: Statin Therapy, LDL Cholesterol, C-Reactive Protein, and Coronary Artery Disease. The New Englang Journal of Medicine 352(1), 29–38 (2005)
4. Shlipak, M.G., et al.: Cystatin C and the Risk of Death and Cardiovascular Events among Elderly. Persons, The New Englang Journal of Medicine 352(20), 2049–2060 (2005)
5. Michelis, R., Sela, S., Ben-Zvi, I., Nagler, R.M.: Salivary b2-Microglobulin Analysis in Chronic Kidney Disease and Hemodialyzed Patients. Blood Purif 25, 505–509 (2007)

Application of Classification Association Rule Mining for Mammalian Mesenchymal Stem Cell Differentiation

Weiqi Wang[1], Yanbo J. Wang[2], René Bañares-Alcántara[1], Zhanfeng Cui[1], and Frans Coenen[3]

[1] Department of Engineering Science, Parks Road, University of Oxford, OX1 3PJ, UK
{weiqi.wang,rene.banares,zhanfeng.cui}@eng.ox.ac.uk
[2] Information Management Center, China Minsheng Banking Corp., Ltd.
Room 606, Building No. 8, 1 Zhongguancun Nandajie, Beijing, 100873, China
wangyanbo@cmbc.com.cn
[3] Department of Computer Science, University of Liverpool
Ashton Building, Ashton Street, Liverpool, L69 3BX, UK
Coenen@liverpool.ac.uk

Abstract. In this paper, data mining is used to analyze the differentiation of mammalian Mesenchymal Stem Cells (MSCs). A database comprising the key parameters which, we believe, influence the destiny of mammalian MSCs has been constructed. This paper introduces Classification Association Rule Mining (CARM) as a data mining technique in the domain of tissue engineering and initiates a new promising research field. The experimental results show that the proposed approach performs well with respect to the accuracy of (classification) prediction. Moreover, it was found that some rules mined from the constructed MSC database are meaningful and useful.

Keywords: Classification Association Rule Mining, Data Mining, Differentiation, Mesenchymal Stem Cells, Tissue Engineering.

1 Introduction

Mesenchymal Stem Cells (MSCs) have been claimed to be an integral part of tissue engineering due to their proliferation and differentiation potential both in vivo and in vitro [5, 12, 36], and have become one of the most significant research topics in the past several decades. MSCs are able to differentiate along the osteogenic, chondrogenic, adipogenic, myogenic, tendonogenic, neurogenic lineages under appropriate stimuli [28, 31, 33] generating bone, cartilage, fat, muscle, tendon, neuron cells respectively (Fig. 1). The significance of the application of MSCs in clinical therapy has triggered an urgent need for computational prediction of MSC differentiation [14].

A large amount of studies have been carried out with the aim of understanding the mechanisms involved in MSCs' proliferation and differentiation both in vivo and in vitro [4, 17, 21, 25, 26, 27]. However, little has been achieved so far due to the enormous complexity of the intracellular pathways in MSCs, especially during their differentiation process [6].

P. Perner (Ed.): ICDM 2009, LNAI 5633, pp. 51–61, 2009.
© Springer-Verlag Berlin Heidelberg 2009

On the other hand, the experiments and studies which have been executed were not interrelated with each other, i.e. different experiments focused on different combinations of parameters affecting MSC differentiation, including animal species, in vitro vs. in vivo cultures, MSC culture medium, supplements to the culture medium and growth factors, culture type (monolayer vs. 3D culture), cell attaching substrate (for monolayer culture) vs. scaffold (for 3D culture) and, in particular, the differentiation fates of MSCs in terms of the different lineages to which the cells committed [17, 18, 20, 21, 27]. The scattered experimental data results in a large amount of noise in the database and a discrete data structure, which cannot take advantage of traditional mathematical modelling methods.

For this reason, we aim to classify the data according to the different cell fates and predict mammalian MSC differentiation. A series of computational techniques have been compared, and data mining has been found to be a promising method due to its ability of processing discrete and noisy data. In particular, the data mining classification approach known as Classification Association Rule Mining (CARM) [10] is used in this study.

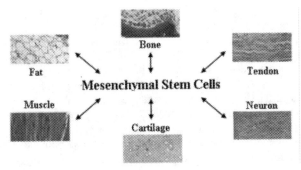

Fig. 1. Differentiation potential of MSCs (modified from [33])

The rest of this paper is organized as follows. In section 2 we describe some related work relevant to this study. Section 3 describes the application of CARM for mammalian MSC differentiation, and the construction of a domain-specific class-transactional database for this study. Experimental results, which demonstrate that the proposed approach performs well with respect to the accuracy of (classification) prediction, are presented in section 4. In section 5, we present our conclusions and discuss open issues for further research.

2   Related Work

2.1   Classification Rule Mining

Classification Rule Mining (CRM) [23, 29] is a technique for identifying Classification Rules (CRs) from a given class database D_c, the objective being to build a classifier to categorize "unseen" data records. Generally D_c is described by a relational database table that includes a class attribute – whose values are a set of predefined class labels $C = \{c_1, c_2, ..., c_{|C|-1}, c_{|C|}\}$. The process of CRM consists of two stages: (i) a training phase where CRs are generated from a set of training data instances $D_R \subset D_C$; and (ii) a test phase where "unseen" instances in a test data set $D_E \subset D_C$ are assigned into predefined class groups. A D_C is established as $D_R \cup D_E$, where $D_R \cap D_E = \varnothing$. Both D_R and D_E share the same database attributes except the class attribute. By convention the last attribute in each D_R record usually indicates the predefined class of this record, noted as the class attribute, while the class attribute is missing in D_E.

Mechanisms on which CRM algorithms have been based include: decision trees [29], naive Bayes [24], K-Nearest Neighbor (K-NN) [19], Support Vector Machine (SVM) [7], etc.

- **Decision Tree Induction:** In this approach CRs are mined based on a greedy algorithm. The approach can be separated into two stages. In the first stage the tree is constructed from D_R and this is followed by a tree pruning phase. In the second stage the pruned tree is used in CR generation. C4.5 [29] is the best known decision tree based CRM method and operates by recursively splitting D_R on the attribute that produces the *maximum information gain* to generate the decision tree. This tree is then pruned according to an error estimate. The result is used to classify "unseen" data records.

- **Naive Bayes:** The typical mechanism found in Bayesian CRM approaches such as [13] is naive Bayes [24], which has been widely applied in machine learning. The general idea of naive Bayes is to make use of knowledge of the probabilities involving attribute values and classes in the training dataset to produce a model of a machine learning application that can then be applied to "unseen" data. The term naive is used to refer to the assumption that the conditional probability of a database attribute value given a class is independent of the conditional probability of other attribute values given that class. A naive Bayes classifier [30] is built using D_R, and comprises a set of conditional probabilities for each database attribute and each class $c_i \in C$, so that there are $n \times |C|$ conditional probabilities, where n represents the number of attributes in D_R and $|C|$ is the size function (cardinality) of C. A naive Bayes classifier also comprises a set of prior class probabilities, one for each class. All these probabilities are then used to classify "unseen" data records in D_E according to Bayes' theorem.

- **K-Nearest Neighbor:** K-NN [19] is a well-known statistical approach used in CRM, and classifies an "unseen" data record $d'_{j'} \in D_E$, by assigning to that record the most frequent class in the set of the K most similar instances to $d'_{j'}$, identified in D_R. To identify the K most similar training-instances for $d'_{j'}$, calculation of the Euclidean distance value between each training data record $d_j \in D_R$ and $d'_{j'}$ is commonly used:

$$distance(d_j, d'_{j'}) = \sqrt{(\Sigma_{\{k = 1...n\}} (d_{j.k} - d'_{j'.k})^2)} , \qquad (1)$$

where $d_{j.k}$ and $d'_{j'.k}$ are the values of the k-th data attribute in D_C for d_j and $d'_{j'}$.

- **Support Vector Machine:** The objective of using SVM [7] is to find a hypothesis $\hat{h}$ which minimizes the *true error* defined as the probability that $\hat{h}$ produces an erroneous result. SVM makes use of linear functions of the form:

$$f(x) = w^T x + b, \tag{2}$$

where w is the weight vector, x is the input vector, and $w^T x$ is the inner product between w and x. The main concept of SVM is to select a *hyperplane* that separates the positive and negative examples while maximizing the smallest margin. Standard SVM techniques produce binary classifiers as opposed to multi-classifiers. Two common approaches to support the application of SVM techniques to the multi-class problem are One Against All (OAA) and One Against One (OAO).

2.2 Association Rule Mining

Association Rule Mining (ARM), first introduced in [1], aims to extract a set of Association Rules (ARs) from a given transactional database D_T. An AR describes an implicative co-occurring relationship between two sets of binary-valued transactional database attributes (items), expressed in the form of an "antecedent ⇒ consequent" rule. Cornelis et al. [11] suggest that the concept of mining ARs can be dated back to the work in 1960's [15].

More generally, we define ARM as follows. Let $I = \{a_1, a_2, ..., a_{n-1}, a_n\}$ be a set of items, and $F = \{T_1, T_2, ..., T_{m-1}, T_m\}$ be a set of transactions (data records), a transactional database D_T is described by F, where each $T_j \in F$ comprises a set of items $I' \subseteq I$. In ARM, two threshold values are usually used to determine the significance of an AR:

1. **Support:** A set of items S is called an itemset. The support of S is the proportion of transactions T in F for which $S \subseteq T$. If the support of S exceeds a user-supplied support threshold σ, S is defined to be a frequent itemset.

2. **Confidence:** Represents how "strongly" an itemset X implies another itemset Y, where $X, Y \subseteq I$ and $X \cap Y = \varnothing$. A confidence threshold α, supplied by the user, is used to distinguish high confidence ARs from low confidence ARs.

An AR $X \Rightarrow Y$ is said to be *valid* when the support for the co-occurrence of X and Y exceeds σ, and the confidence of this AR exceeds α. The computation of support is:

$$support(X \cup Y) = count(X \cup Y) / |F|, \tag{3}$$

where $count(X \cup Y)$ is the number of transactions containing the set $X \cup Y$ in F, and $|F|$ is the size function (cardinality) of the set F. The computation of confidence is:

$$confidence(X \Rightarrow Y) = support(X \cup Y) / support(X). \tag{4}$$

Informally, "$X \Rightarrow Y$" can be interpreted as: if X is found in a transaction, it is likely that Y also will be found.

2.3 Classification Association Rule Mining

An overlap between ARM and CRM is CARM (Classification Association Rule Mining), which strategically solves the traditional CRM problem by applying ARM techniques. The idea of CARM, first introduced in [2], aims to extract a set of

Classification Association Rules (CARs) from a class-transactional database D_{C-T}. Let D_T be a transactional database, and $C = \{c_1, c_2, ..., c_{|C|-1}, c_{|C|}\}$ be a set of predefined class labels, D_{C-T} is described by $D_T \times C$. D_{C-T} can also be defined as a special class database D_C, where all database attributes and the class attribute are valued in a binary manner – "*Boolean attributes can be considered a special case of categorical attributes*" [32]. A CAR is a special AR that describes an implicative co-occurring relationship between a set of binary-valued data attributes and a predefined class, expressed in the form of an "$X \Rightarrow c_i$" rule, where X is an itemset found in D_T (as "$D_{C-T} - C$") and c_i is a predefined class in C.

2.4 Advantages of CARM

CARM offers the following advantages with respect to the CRM techniques mentioned above [3, 35]:

1. The approach is efficient during both the training and categorization phases, especially when handling a large volume of data.

2. The classifier built in this approach can be read, understood and modified by humans.

Furthermore, CARM is relatively insensitive to noise data. CARM builds a classifier by extracting a set of CARs from a given set of training instances. Possible CARs are determined by: (i) a large enough support, and (ii) a large enough confidence. Usually, rules derived from noise in the data will fail to reach these thresholds and will be discarded.

In comparison, CRM approaches other than CARM, i.e. naive Bayes, K-NN, SVM, etc., do not present the classifier in a human readable fashion, so that users do not see why the (classification) predictions have been made. Rules generated by a decision tree based classifier can be read and understood by human; however, Coenen et al. [9] suggest that results presented in the studies of [22] and [23] show that in many cases CARM offers higher classification accuracy than decision tree based classification.

For these reasons it is proposed to use a CARM approach to address the prediction of mammalian MSC differentiation. One of the existing CARM frameworks is the CMAR (Classification based on Multiple Association Rules) algorithm [22]. CMAR generates CARs (from a given set of training instances) through an FP-tree [16] based approach. Experimental results using this algorithm reported in [22] show that it could achieve high classification accuracy for a range of data sets.

3 Database Construction for Mammalian MSC Differentiation

In order to make a (classification) prediction of mammalian MSC differentiation using CARM, a domain-dependent database containing 375 parameters that are believed to influence the MSC differentiation has been built and can be accessed online[1]. The parameters in this database include the most significant ones, such as donor species, in vitro vs. in vivo culture, culture medium, supplements and growth factors, culture

---

[1] http://www.oxford-tissue-engineering.org/forum/plugin.php?identifier=publish&module=publish

type (monolayer vs. 3D culture), substrate (for monolayer culture) vs. scaffold (for 3D culture), MSC differentiation fate, as well as other potentially important parameters including age of donor, cell passage number, concentrations of chemicals and buffer, cell seeding density, incubation duration, pH value, expression of cell markers, expansion fold of cell number, etc. All the records are abstracted from previously published papers and each is stated clearly with the corresponding reference.

The current size of this database is 203 records, each containing attributes including experimental conditions and results, i.e. rule antecedents and rule consequent-classes respectively. There are four types of attributes in the database:

- For the qualitative parameters which have only two possible values, such as the presence/absence of insulin in the culture medium, the corresponding attributes in the database are binary, i.e. '1' refers to 'presence' and '0' refers to 'absence'.
- For the qualitative parameters which have a set of possible categorical values, such as animal species, the attributes are stored as integer symbols.
- For the quantitative parameters, such as cell seeding density, the attributes are stored as real numbers.
- For the descriptive parameters, such as additional conditions, the attributes are stored as text/characters.

3.1 Data Filtering

The database contains a large variety of information, among which some parameters are previously known to be more important and effective to the differentiation fates than others. Those parameters and their significance are listed in Table 1.

Other parameters, such as buffer, gas condition, pH value, etc., seem to be just for facilitating the differentiation. We are also aware that procedures such as radiation,

Table 1. The most significant parameters in the MSC database

Parameters	Significance/Description
Donor Species	Same culture conditions on MSCs from different species of mammal may lead to different results.
In Vitro / In Vivo	MSC differentiation varies significantly from in vivo to in vitro environment.
Culture Medium	Most essential environment where MSCs grow, proliferate and differentiate.
Supplements & Growth Factors	Chemicals that maintain MSC differentiation potential or influence their differentiation fate.
Culture Type (2D vs. 3D)	MSC differentiation sometimes differs significantly from monolayer to 3D culture, even under the same culture medium and supplements.
Substrate (for 2D) / Scaffold (for 3D)	Influences cell viability. A chemically modified substrate can even change MSCs' differentiation fate.
Differentiation Fate	The most significant and obvious result after cell culture. Used as the classes in the database and the prediction in this study.

centrifugation and mechanical stimulation sometimes affect MSCs, and that MSCs from bone marrow are sometimes different from those from other sources, e.g. umbilical cord blood. However, in order to identify plausible MSC differentiation rules the essential parameters considered were those in Table 1; other parameters could be excluded if the purpose were to predict the MSC differentiation fate, according to the "unseen" rules.

3.2 Data Normalization and Cleaning

In this study, only the parameters listed in Table 1 were extracted from the MSC database and used for the (classification) prediction. Consequently, the number of attributes in the abstracted database was reduced from 375 to 105. The database was then discretised and normalised using the LUCS-KDD Discretised Normalised (DN) software[2], so that data are presented in a binary format suitable for use by CARM applications. It should be noted that the database was re-arranged so that occurrences of classes (i.e. osteogenesis, chondrogenesis, etc.) were distributed evenly throughout the database. This then allowed CMAR to be applied to (90% – training set, 10% – test set) divisions of the database with Ten-fold Cross Validation (TCV) accuracy setting. In this study, the discretisation and normalisation process results in a data file with the number of attributes increased to 173.

This discretised and normalised data file contains a number of noisy data, generally caused by the absence of culture conditions such as some growth factors. For example, if the growth factor insulin is absent in a record, this record will have an attribute representing "absence of insulin" after the discretisation and normalisation process. However, the rules that we are looking for in this study are those without the information of "absence", i.e. those only containing the information of "presence". Thus, all the attributes representing "absence" were then eliminated from the data file, resulting in the final input data file for CARM. In this final input file, all the records cover five classes in total, i.e. five kinds of MSC fates: "osteogenesis", "chondrogenesis", "adipogenesis", "neurogenesis", and "proliferation without differentiation".

4 Results

In this section, we aim to evaluate the usage of CARM for mammalian MSC differentiation. The evaluation was performed using the CMAR algorithm although any other CARM classifier generator could equally well have been used. Experiments were run on a 2.00 GHz Intel(R) Core(TM)2 CPU with 2.00 GB of RAM running under Windows Command Processor. The evaluation undertaken used a confidence threshold value (α) of 50% and a support threshold value (σ) of 1% (as previously used in published CARM evaluations [8, 9, 22, 34]). The overall (classification) prediction accuracy is 77.04%.

There were 163 CMAR rules generated from the input data file, among which many are found to be meaningful and useful. Two rules are described as an example, with the actual confidence values presented in square brackets:

---

[2] http://www.csc.liv.ac.uk/~frans/KDD/Software/LUCS-KDD-DN/lucs-kdd_DN.html

1. **Rule # 49:** {in vitro + monolayer + human donor + DMEM + TGFβ1 + plastic substrate} ⇒ {chondrogenesis} [100.0%], which can be interpreted as: in monolayer culture in vitro, human MSCs are most likely to undergo chondrogenesis in the presence of cell culture medium DMEM (Dulbecco's Modified Eagle's Medium) and growth factor TGFβ1 (Transforming Growth Factor β1), on plastic substrate (Fig. 2).

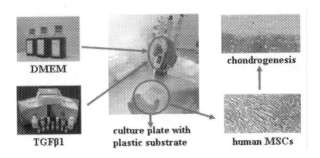

Fig. 2. Description of rule # 49

2. **Rule # 86:** {DMEM + FBS + ascorbate-2-phosphate + Dex} ⇒ {osteogenesis} [93.33%], which can be interpreted as: in DMEM medium supplemented with FBS (Fetal Bovine Serum), MSCs are very likely to be induced to osteogenesis under the stimuli of ascorbate-2-phosphate and Dex (Dexamethasone) together (Fig. 3).

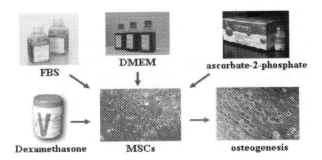

Fig. 3. Description of rule # 86

5 Conclusions

In this study, the (classification) prediction of MSC differentiation has been achieved using CARM, with an accuracy of 77.04%. We have introduced, for the first time, data mining techniques to mesenchymal stem cell, initiating a new promising research field. We are aware, nevertheless, that there is room to increase the (classification) prediction accuracy, as well as to streamline the pruning of generated rules by improving the CARM technique or using alternative mechanisms. Additionally, many of the identified rules were found to be already known in the laboratory. However, the identification of completely original rules is possible and expected if the size and contents of the MSC database are properly expanded in the future.

Acknowledgments. The authors would like to thank the following colleagues from the Department of Engineering Science at the University of Oxford for their valuable suggestions to this study: Paul Raju, Norazharuddin Shah Abdullah, Prof. James Triffitt, Nuala Trainor, Clarence Yapp, Dr. Cathy Ye, Dr. Xia Xu, Dr. Uday Tirlapur, Yang Liu, Zhiqiang Zhao, Dr. Shengda Zhou and Dr. Renchen Liu.

References

1. Agrawal, R., Imielinski, T., Swami, A.: Mining Association Rules between Sets of Items in Large Database. In: Proceedings of the 1993 ACM SIGMOD International Conference on Management of Data, Washington, D.C., USA, pp. 207–216. ACM Press, New York (1993)
2. Ali, K., Manganaris, S., Srikant, R.: Partial Classification using Association Rules. In: Proceedings of the 3rd International Conference on Knowledge Discovery and Data Mining, Newport Beach, CA, USA, pp. 115–118. AAAI Press, Menlo Park (1997)
3. Antonie, M.L., Zaiane, O.R.: Text Document Categorization by Term Association. In: Proceedings of the 2002 IEEE International Conference on Data Mining, Maebashi City, Japan, pp. 19–26. IEEE Computer Society Press, Los Alamitos (2002)
4. Battula, V.L., Bareiss, P.M., Treml, S., Conrad, S., Albert, I., Hojak, S., Abele, H., Schewe, B., Just, L., Skutella, T., Buhring, H.J.: Human Placenta and Bone Marrow derived MSC Cultured in Serum-free, b-FGF-containing Medium Express Cell Surface Frizzled-9 and SSEA-4 and Give Rise to Multilineage Differentiation. Differentiation 75, 279–291 (2007)
5. Beeres, S.L., Atsma, D.E., van der Laarse, A., Pijnappels, D.A., van Tuyn, J., Fibbe, W.E., de Vries, A.A.F., Ypey, D.L., van der Wall, E.E., Schalij, M.J.: Human Adult Bone Marrow Mesenchymal Stem Cells Repair Experimental Conduction Block in Rat Cardiomyocyte Cultures. American College of Cardiology 46(10), 1943–1952 (2005)
6. Bianco, P., Riminucci, M., Gronthos, S., Robey, P.G.: Bone Marrow Stromal Stem Cells: Nature, Biology, and Potential Applications. Stem Cells 19, 180–192 (2001)
7. Boser, B.E., Guyon, I.M., Vapnik, V.N.: A Training Algorithm for Optimal Margin Classifiers. In: Proceedings of the 5th ACM Annual Workshop on Computational Learning Theory, Pittsburgh, PA, USA, pp. 144–152. ACM Press, New York (1992)
8. Coenen, F., Leng, P.: An Evaluation of Approaches to Classification Rule Selection. In: Proceedings of the 4th IEEE International Conference on Data Mining, Brighton, UK, pp. 359–362. IEEE Computer Society Press, Los Alamitos (2004)
9. Coenen, F., Leng, P., Zhang, L.: Threshold Tuning for improved Classification Association Rule Mining. In: Ho, T.-B., Cheung, D., Liu, H. (eds.) PAKDD 2005. LNCS, vol. 3518, pp. 216–225. Springer, Heidelberg (2005)
10. Coenen, F., Leng, P.: The Effect of Threshold Values on Association Rule based Classification Accuracy. Journal of Data and Knowledge Engineering 60(2), 345–360 (2007)
11. Comelis, C., Yan, P., Zhang, X., Chen, G.: Mining Positive and Negative Association Rules from Large Databases. In: Proceedings of the 2006 IEEE International Conference on Cybernetics and Intelligent Systems, Bangkok, Thailand, pp. 613–618. IEEE Computer Society Press, Los Alamitos (2006)

12. Derubeis, A.R., Cancedda, R.: Bone Marrow Stromal Cells (BMSCs) in Bone Engineering: Limitations and Recent Advances. Annals of Biomedical Engineering 32(1), 160–165 (2004)
13. Domingos, P., Pazzani, M.: On the Optimality of the Simple Bayesian Classifier under Zero-one Loss. Machine Learning 29(2/3), 103–130 (1997)
14. Griffith, L.G., Swartz, M.A.: Capturing Complex 3D Tissue Physiology in Vitro. Nature Reviews Molecular Cell Biology 7, 211–224 (2006)
15. Hajek, P., Havel, I., Chytil, M.: The GUHA Method of Automatic Hypotheses Determination. Computing 1, 293–308 (1966)
16. Han, J., Pei, J., Yin, Y.: Mining Frequent Patterns without Candidate Generation. In: Proceedings of the 2000 ACM SIGMOD International Conference on Management of Data, Dallas, TX, USA, pp. 1–12. ACM Press, New York (2000)
17. Hanada, K., Dennis, J.E., Caplan, A.I.: Stimulatory Effects of Basic Fibroblast Growth Factor and Bone Morphogenetic Protein-2 on Osteogenic Differentiation of Rat Bone Marrow-derived Mesenchymal Stem Cells. Journal of Bone and Mineral Research 12, 1606 (1997)
18. Haynesworth, S.E., Baber, M.A., Caplan, A.I.: Cytokine Expression by Human Marrow-derived Mesenchymal Progenitor Cells in Vitro: Effects of Dexamethasone and IL-1a. Journal of Cell Physiology 166(3), 585–592 (1996)
19. James, M.: Classification Algorithm. Wiley Interscience, New York (1985)
20. Kuznetsov, S.A., Friedenstein, A.J., Robey, P.G.: Factors Required for Bone Marrow Stromal Fibroblast Colony Formation in Vitro. British Journal of Haematology 97, 561–570 (1997)
21. Lennon, D.P., Haynesworth, S.E., Young, R.G., Dennis, J.E., Caplan, A.I.: A Chemically defined Medium Supports in Vitro Proliferation and Maintains the Osteochondral Potential of Rat Marrow-derived Mesenchymal Stem Cells. Experimental Cell Research 219, 211–222 (1995)
22. Li, W., Han, J., Pei, J.: CMAR: Accurate and Efficient Classification based on Multiple Class-association Rules. In: Proceedings of the 2001 IEEE International Conference on Data Mining, San Jose, CA, USA, pp. 369–376. IEEE Computer Society Press, Los Alamitos (2001)
23. Liu, B., Hsu, W., Ma, Y.: Integrating Classification and Association Rule Mining. In: Proceedings of the 4th International Conference on Knowledge Discovery and Data Mining, New York, USA, pp. 80–86. AAAI Press, Menlo Park (1998)
24. Lowd, D., Domingos, P.: Naive Bayes Models for Probability Estimation. In: Proceedings of the 22nd International Conference on Machine Learning, Bonn, Germany, pp. 529–536. ACM Press, New York (2005)
25. Magaki, T., Kurisu, K., Okazaki, T.: Generation of Bone Marrow-derived Neural Cells in Serum-free Monolayer Culture. Neuroscience Letters 384, 282–287 (2005)
26. Meuleman, N., Tondreau, T., Delforge, A., Dejeneffe, M., Massy, M., Libertalis, M., Bron, D., Lagneaux, L.: Human Marrow Mesenchymal Stem Cell Culture: Serum-free Medium Allows Better Expansion than Classical α-MEM Medium. European Journal of Haematology 76(4), 309–316 (2006)
27. Muller, I., Kordowich, S., Holzwarth, C., Spano, C., Isensee, G., Staiber, A., Viebahn, S., Gieseke, F., Langer, H., Gawaz, M.P., Horwitz, E.M., Conte, P., Handgretinger, R., Dominici, M.: Animal Serum-free Culture Conditions for Isolation and Expansion of Multipotent Mesenchymal Stromal Cells from Human BM. Cytotherapy 8, 437–444 (2006)
28. Pittenger, M.F., Mackay, A.M., Beck, S.C., Jaiswal, R.K., Douglas, R., Mosca, J.D., Moorman, M.A., Simonetti, D.W., Craig, S., Marshak, D.R.: Multilineage Potential of Adult Human Mesenchymal Stem Cells. Science 284(5411), 143–147 (1999)

29. Quinlan, J.R.: C4.5: Programs for Machine Learning. Morgan Kaufmann Publishers, San Mateo (1993)
30. Rish, I.: An Empirical Study of the Naive Bayes Classifier. In: Proceedings of the 2001 IJCAI Workshop on Empirical Methods in Artificial Intelligence, Seattle, WA, USA (2001)
31. Roelen, B.A., Dijke, P.: Controlling Mesenchymal Stem Cell Differentiation by TGFbeta Family Members. Journal of Orthopaedic Science 8, 740–748 (2003)
32. Srikant, R., Agrawal, R.: Mining Quantitative Association Rules in Large Relational Tables. In: Proceedings of the 1996 ACM SIGMOD International Conference on Management of Data, Montreal, Quebec, Canada, pp. 1–12. ACM Press, New York (1996)
33. Tuan, R.S., Boland, G., Tuli, R.: Adult Mesenchymal Stem Cell and Cell-based Tissue Engineering. Arthritis Research & Therapy 5, 32–45 (2003)
34. Wang, Y.J., Xin, Q., Coenen, F.: A Novel Rule Ordering Approach in Classification Association Rule Mining. In: Perner, P. (ed.) MLDM 2007. LNCS, vol. 4571, pp. 339–348. Springer, Heidelberg (2007)
35. Yoon, Y., Lee, G.G.: Practical Application of Associative Classifier for Document Classification. In: Lee, G.G., Yamada, A., Meng, H., Myaeng, S.-H. (eds.) AIRS 2005. LNCS, vol. 3689, pp. 467–478. Springer, Heidelberg (2005)
36. Zhang, Y., Li, C., Jiang, X., Zhang, S., Wu, Y., Liu, B., Tang, P., Mao, N.: Human Placenta-derived Mesenchymal Progenitor Cells Support Culture Expansion of Long-term Culture-initiating Cells from Cord Blood CD34+ Cells. Experimental Hematology 32, 657–664 (2004)

Computer-Aided Diagnosis in Brain Computed Tomography Screening

Hugo Peixoto[1] and Victor Alves[2]

[1] Hospital Center Tâmega e Sousa, Penafiel Portugal
[2] University of Minho, Braga, Portugal
hugo.peixoto@chts.min-saude.pt
valves@di.uminho.pt

Abstract. Currently, interpretation of medical images is almost exclusively made by specialized physicians. Although, the next decades will most certainly be of change and computer-aided diagnosis systems will play an important role in the reading process. Assisted interpretation of medical images has become one of the major research subjects in medical imaging and diagnostic radiology. From a methodological point of view, the main attraction for the resolution of this kind of problem arises from the combination of the image reading made by the radiologists, with the results obtained from using Artificial Intelligence based applications that will contribute to the reduction and eventually the elimination of perception errors. This article describes how machine learning algorithms can help distinguish normal readings in brain Computed Tomography from all its variations. The goal is to have a system that is able to detect normal appearing structures, thus identifying normal studies, making the reading by the radiologist unnecessary for a large proportion of the brain Computed Tomography scans.

Keywords: Medical imaging, Computer Aided Detection, Brain Computed Tomography, Artificial Intelligence, Machine Learning.

1 Introduction

Not only the advent of new methodologies for problem solving but also the emergence of new information and communication technologies has strongly influenced our societies, including their health care systems. Due to such a fine-tuning, Medical Informatics (MI) specialists are ever more involved in managing Information Systems (IS) in health care institutions, in particular of the hospital information ones. Indeed, the monitoring of hospital IS is a fundamental task that has to be followed when one intends to inform the physicians and the management on time, in order to provide a better care.

On the other hand an appropriate design of educational programs in MI and an increasing number of well-trained MI specialists will help to pursue the goal of transforming and improving the delivery of health care through innovative use of information and communication technology. This evolution in technology used in the medical image practice, confront the radiology physicians with a new problem: the capacity to

P. Perner (Ed.): ICDM 2009, LNAI 5633, pp. 62–72, 2009.
© Springer-Verlag Berlin Heidelberg 2009

interpret a huge image workload. Indeed, the current workflow reading approaches are becoming inadequate for reviewing the 300 to 500 images of a routine Computed Tomography (CT) of the chest, abdomen, or pelvis, and even less for the 1500 to 2000 images of a CT angiography or functional Magnetic Resonance (MR) study. In fact, image analysis and treatment methods present enormous development and an increasing utilization in the area of medical imaging. Given to the general interest and the impressive growth in Computer Aided Detection and Diagnosis (CADD), the application of Artificial Intelligence (AI) based techniques in the interpretation and diagnosis of medical image became a rapidly growing research field [1].

For this assessment Brain CT studies were chosen due to its importance in the overall imaging market. Indeed, specialists working in Oporto region, in the North of Portugal, showed that its population, of about 1,500,000 inhabitants, present a yearly average of 450,000 CT analysis, 198,000 (44%) of which were brain CT studies[2]. However, from a methodological point of view, the main attraction for the resolution of this kind of problem arises from the combination of the image analysis taken from the radiologists, with those obtained using AI based applications, that will contribute to the selection of the urgent studies and to aid the radiologists in their readings [2].

Currently, specialized physicians almost exclusively make interpretation of medical images. The next decades will most certainly be of change, and computer-aided diagnosis systems, that have become one of the major research subjects in medical imaging and diagnostic radiology, will play an important role in the reading process. From a methodological point of view, the main attraction for the resolution of this kind of problems arises from the combination of the image reading made by the radiologists, with the results obtained from using AI based techniques that will contribute to the reduction and eventually the elimination of perception errors[3]. Undeniably, the machine learning procedures may distinguish normal studies from all its variations. Our goal is to have a system that is able to detect normal appearing structures, thus identifying normal studies, making the reading by the radiologist unnecessary for a large proportion of the brain CT scans.

Therefore, it is necessary to effect a rigorous assessment in order to establish which, are the real clinical contributions of such systems in decision support. It is here that AI appears to open the way for computer supported diagnosis in medical imaging. Since histological images are complex and their differences are quite subtle, sub-symbolic systems (e.g. Artificial Neural Networks (ANNs), Genetic Algorithms and Evolutionary Programming, Particle Swarm Organization) should be used to overcome the drawbacks of pure symbolic systems. A methodology for brain CT feature extraction and automatic diagnostic generation is proposed. Preliminary results are presented and discussed in regard to the selected features and learning algorithms.

Nowadays Computer Aided Detection (CAD) studies are mainly concerned with screening mammography and thoracic CT studies. Tao Chan et. al studied the effect of a CAD system on clinicians performance in detection of small Acute Intracranial Hemorrhage on CT concluding that there were significantly improvements in the performance of emergency physicians when they make the diagnosis with the support of CAD [4]. Timothy et. al proposed to assess the effect of CAD on the interpretation of screening mammograms in a community breast center. Their conclusions revealed that the use of CAD in the interpretation of screening mammograms can increase the detection of early-stage malignancies without undue effect on the recall rate or positive predictive value for biopsy [5]. Several commercial CAD products are starting to proliferate and their market is also beginning to develop. Shih-Ping Wang was able to get a US Patent for his CAD system that works over the probability likelihood and predicted values for mammography examinations, aiding the physicians in achieving a correct diagnostic.

The reduction and eventually the elimination of perception errors can be achieved by using neural-network computers taught what is normal with all its variations. The computer should eventually be able to identify normal appearing structures making the reading by the radiologist unnecessary for a large proportion of images.

2 Computer Aided Detection

Automatic segmentation and machine learning are techniques that come from the computer graphics and AI field and are closely related with CAD [3]. Segmentation is used in feature extraction from the images and in many cases this results in excellent clinical information [4]-[5]. In this work, machine learning is concerned with the automatic extraction of information from data, using Knowledge Discovery and Data Mining (KDD) techniques, in our case from the study images. CAD systems should be able to return an adequate answer to the problem of automatic study reading (Fig.1). Here, we were inspired by the work of Kanazawa and Katsuragawa, and the studies follow a similar practical workflow [7]-[8].

Fig. 1. Stages of CAD

Concerning the studies analyses that where performed, the most relevant features from the images had to be determined, i.e., which aspects of these images, reveals or not the existence of a pathology. From an exhaustive analysis of our case set, we determined that the image's histogram and some shape aspects were key features in determining the existence of pathology. Thus it was decided to consider the histogram and a shape function as the most appropriate features. The case set was composed by 265 studies, where the complete DICOM study, the radiologist report and some additional clinical information (e.g., age, gender) was available.

2.1 General Characteristics Vector

Using the ImageJ framework [9] a plug-in was developed to extract a General Characteristics Vector $\overrightarrow{(GCV)}$ in a csv (comma separated values) format. The global General Characteristics Vector, $\overrightarrow{GCV}_g$, is therefore, given by:

$$\overrightarrow{GCV}_g = (SI, A, G, fc(SH), fs(S), xxVar, yyVar, xyVar, D)$$

where SI, A, G, fc(SH), SH, fs(S), S, xxVar, yyVar, xyVar, and D denote, respectively, the study identification, the patient age, the gender of the patient, the content function, the Study Histogram, the shape function, the Study images, the x variance, the y variance, the xy variance, and the diagnostic. The first parameter, identification of the study, is a value of reference that will identify the study. This dimension as well as the age and gender were extracted from the DICOM images meta-data.

On the other hand, Hounsfield stated that the most relevant information of a CT belongs to the interval [-150;150], whose values are given in Hounsfield Units (HU). In a CT image each voxel corresponds to the x-ray linear attenuation coefficient measurement of the tissue in question. HU of x is given by:

$$HUx = 1000 \times \frac{(\mu_x - \mu_{H2O})}{(\mu_{H2O} - \mu_{air})}$$

where x denotes a given voxel, $\grave{\imath}_{H2O}$ denotes the linear attenuation coefficients of water and $\grave{\imath}_{air}$ denotes the linear attenuation coefficients of air, considered at a standard temperature and pressure [10].

The standards presented in Table I were chosen as they are universally available, and suited to the key application for which computed axial tomography was developed, i.e., imaging the internal anatomy of living creatures based on organized water structures and mostly living in air, e.g. humans [11].

Table 1. Hounsfield values

Tissue	Interval
Air & sponge	$(-\infty;-430]$
Air	$[-100;-50]$
Water	$[-5;5]$
Fluid	$[10; 20]$
Muscle	$[10;40]$
Blood	$[38; 50]$
Grey Matter	$[38; 50]$
White Matter	$[38; 50]$
Bone	$[150; \infty)$

The function $fc(SH)$ denotes the content function for the total study; for a single image, one has the function called $fci(IH)$, where i denotes a single image of the study. $fci(IH)$ is now given by:

$$fci(IH) = \left(\frac{x_{HU1}}{t}, \frac{x_{HU2}}{t}, \frac{x_{HU3}}{t}, ..., \frac{x_{HUn}}{t} \right)$$

where x_{HU} denotes the total number of voxels, in image i, for each HU from -2000 to 2000 and t denotes the total number of voxels in the image i. Therefore, $fc(SH)$ will be given by:

$$fc(SH) = fci\left(\sum_{i=1}^{l}\frac{x_{HU1}}{t}, \sum_{i=1}^{l}\frac{x_{HU2}}{t}, \sum_{i=1}^{l}\frac{x_{HU3}}{t}, ..., \sum_{i=1}^{l}\frac{x_{HUn}}{t}\right)$$

where l denotes the last image of the study. Fig. 2 illustrates the resulting $fc(SH)$ from -2000 to 2000 HU.

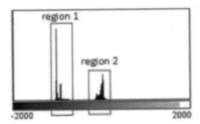

Fig. 2. Study Histogram

After looking at Table 1 and Fig. 2 a threshold from -150 to 150 was performed, since it is where the relevant information for the study lies. This way, the final $fc(SH)$ denotes the *region 2*. *Region 1* corresponds to the ambient air and the sponge support for the patients head. On the other hand, since representing the $fc(SH)$ for each HU leads to a huge amount of data, it was decided to sum up the number of pixels dividing the HU in intervals of five. This way we achieved 60 uniform intervals of 5 HU, being represented by $fc_l(SH)$.

From the shape function $f(S)$ the values of *xSkew*, *ySkew* indicate the symmetry of the series, while the values of *xKurt*, *yKurt*, represent the kurtosis values in the two axes. The remaining parameters, *xxVar*, *yyVar* and *xyVar*, were added to the study and represent the variance in *x*, *y* and *xy* directions, respectively. Finally, and only in the case of the trainings set, the diagnostic value are needed. This parameter stands for a set of values that are given in the form:

- Diagnostic = Ø or
- Diagnostic = {Pathology 1, Pathology 2, ..., Pathology n}

where Ø denotes the empty set. With respect to the first evaluation or assessment, it was decided to consider that if D = Ø its value is set to 0 and is judged as non pathological. If the diagnostic is a non empty set, i.e., if D ≠ Ø, the value of D is set to 1 and is taken as pathological.

3   Methodology

The first step for building a good methodology at the meta-level lead us to define a functional overall workflow of the problem solving process, where one is able to manage with all the variants (Fig. 3).

The clinical practice is at the root of the process, once it is the target as well as the basis for obtaining good knowledge. Going a step further, we obtain the image studies

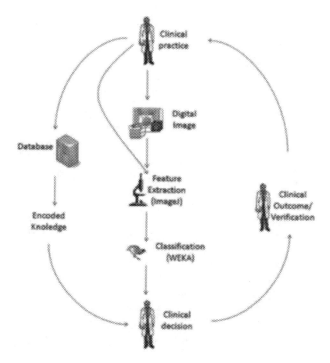

Fig. 3. Workflow Overview

of brain CT's as well as the correspondent diagnostic for the training set. Afterward, ImageJ is used to extract the image features, i.e., content function and shape function.

The Waikato Environment for Knowledge Analysis (WEKA) [12] is then used to perform the computational analysis of the supplied data, applying KDD techniques. The $\overrightarrow{GCV}_g$ vector for WEKA is at that time imported, being therefore possible to observe the diverse attributes, distinctly enumerated.

4 Assessment

The $\overrightarrow{GCV}_g$ vector was converted for different evaluation purposes, having different dimensions with different attributes, so that the best attribute set may be obtained. Indeed, after feeding WEKA with the first output $\overrightarrow{GCV}$ several parameters were removed, since they turn out to be irrelevant to the study (e.g., patient ID). Then, $\overrightarrow{GCV}_g$ move towards $\overrightarrow{GCV}_1$, which is given by:

$$\overrightarrow{GCV}_1 = (\,A, G, f_{c1}(SH), fs(S), \text{xxVar}, \text{yyVar}, \text{xyVar}, D)$$

On a second run, after importing the $\overrightarrow{GCV}_g$, other parameters were removed, in addition to the patient ID. The gender was removed and the shape function $f_s(S)$ was modified. *xKurt*, *yKurt* and *ySkew* were deleted and a new shape function, $f_{s1}(S)$ was built. This move was considered since the turnover of keeping such parameters was unknown, and its real impact had to be evaluated. Subsequently, $\overrightarrow{GCV}_g$ move towards $\overrightarrow{GCV}_2$, which is given by:

$$\overrightarrow{GCV}_2 = (\text{A}, f_{c1}(SH), f_{s1}(S), \text{xxVar}, \text{yyVar}, \text{xyVar}, \text{D})$$

where $f_{s1}(S)$ is built by *xSkew*.

Two more attempts were considered. Firstly, it was removed only the gender, then, the gender was kept in and it was removed the xKurt, yKurt and ySkew. In this way we got $\overrightarrow{GCV}_3, \overrightarrow{GCV}_4$, which are given by:

$$\overrightarrow{GCV}_3 = (\text{A}, f_{c1}(S \neq), f_s(S), \text{xxVar}, \text{yyVar}, \text{xyVar}, \text{D})$$

$$\overrightarrow{GCV}_4 = (\text{A}, \text{G}, f_{c1}(SH), f_{s1}(S), \text{xxVar}, \text{yyVar}, \text{xyVar}, \text{D})$$

All the tasks referred to above were performed using only the tools provided by WEKA; however, little variations in the generation of the $\overrightarrow{GCV}_g$ were also tried to guarantee that all the possible scenarios would be covered. Firstly, the distribution of the intervals in the histogram was changed. One of the adjustments was to introduce the notion of weight to the histogram intervals. So, the intervals were multiplied by the sin function. This was made since most of the relevant information is in the middle of the histogram, and as Fig.4 shows, the sin function attributes more weight to its middle locations.

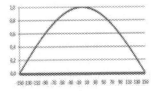

Fig. 4. Sin() function

$\overrightarrow{GCV}_5$ is therefore given by:

$$\overrightarrow{GCV}_5 = (\text{A}, \text{G}, \sin(f_{c1}(SH)), f_s(S), \text{xxVar}, \text{yyVar}, \text{xyVar}, \text{D})$$

Another assessment consisted in grouping the age of the patients into several groups, as it is depicted in Table 2.

The resulting $\overrightarrow{GCV}$, i.e., $\overrightarrow{GCV}_6$ is therefore given by:

$$\overrightarrow{GCV}_6 = (\text{GROUPED\_A}, \text{G}, f_{c1}(SH), f_s(S), \text{xxVar}, \text{yyVar}, \text{xyVar}, \text{D})$$

Finally, the last interaction was to convert the diagnostic parameter in a non 0 or 1 values, giving, instead, a probability of having pathology. This result in the last assessed $\overrightarrow{GCV}$, leads to $\overrightarrow{GCV}_7$, which takes the form:

$$\overrightarrow{GCV}_7 = (\text{GROUPED}_\text{A}, \text{G}, f_{c1}(SH), f_s(S), \text{xxVar}, \text{yyVar}, \text{xyVar}, \text{D}_1)$$

Table 2. Grouping by age

Description (Group)	Age (years)
Baby	0 – 2
Child	2 – 6
Infant	6 – 10
Starter	10 – 14
Juvenile	14 – 16
Junior	16 – 18
Young Adult	18 – 30
Junior Adult	30 – 45
Senior Adult	45 – 60
Old Adult	+ 60

5   Results

According to the results so far obtained, it is feasible to conclude that grouping the age, giving different weights to the histogram intervals, and passing a non 0 or greater than 0 the values in the diagnostic, such changes are not so relevant as initially had been thought, i.e., the results for $\overrightarrow{GCV}_5$, $\overrightarrow{GCV}_6$ and $\overrightarrow{GCV}_7$ do not get close to the other $\overrightarrow{GCV}$. Table 3 presents the attained results for some algorithms using quite a few of $\overrightarrow{GCV}$'s, where the best algorithms are the Simple Logistic and the LMT (Logistic Model Trees), being the most stable one the VFI (Voting Feature Intervals classification algorithm), since it presents good results not only to the best $\overrightarrow{GCV}$, but also for the remaining ones. The best $\overrightarrow{GCV}$ has the following characteristics:

- It is divided in sixty uniform intervals;
- It presents a numerical value for age;
- It presents a Boolean value for the Diagnostic; and
- It comes without gender, xKurt, yKurt and ySkew as parameters.

Table 3. Algorithm scores, false negatives and false positives

GCV	ALGORITHM	SCORE (%)	FALSE POSITIVE (%)	FALSE NEGATIVE (%)
$\overrightarrow{GCV}_1$	VFI	76.92	30.8	15.4
$\overrightarrow{GCV}_1$	Random Committee	76.92	15.4	38.5
$\overrightarrow{GCV}_2$	VFI	80,77	23,1	15,4
$\overrightarrow{GCV}_2$	Simple Logistic	80,77	30,8	7,7
$\overrightarrow{GCV}_2$	LMT	80,77	30,8	7,7
$\overrightarrow{GCV}_3$	VFI	73,08	30,8	23,1
$\overrightarrow{GCV}_4$	VFI	80,77	23,1	15,4

Once the parameters for gender, *xKurt*, *yKurt* and *ySkew* are discharged, the results improved considerably. The same happens when the gender parameter is kept in $\overrightarrow{GCV}_4$. On the other hand, removing the gender parameter and maintaining the other ones, did not score as high as in the previous dataset.

Fig. 5 gives an idea of the form as the results are provided by WEKA. It is possible to distinguish between the false positives and negatives, and the results that are marked as pathological or non-pathological.

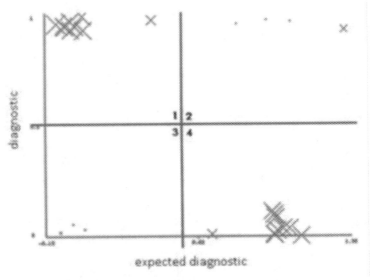

Fig. 5. Expected Diagnostic vs Diagnostic

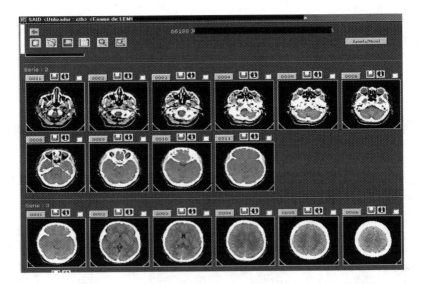

Fig. 6. Diagnostic interface (study selection for diagnostic)

It is interesting to notice that since the results are presented in terms of a percentage, the system not only suggests all diagnostic possibilities but also indicates a measure of accuracy for each suggestion. The final result is a system that provides the user with diagnostic suggestions for the selected brain CT. The prototype built for real-world assessment in a medical environment is WEB based. In Fig. 6 we can see a screen shot of the web page where the study is selected for automatic generation of diagnostics. In Fig. 7 we can see the generated diagnostics where the bars stand for the probability of each diagnostic. On the left screenshot we have two big bars standing for a big probability of the diagnostic "Atrophy" and "Stroke". On the right screenshot of Fig. 7 all bars is very small, meaning that we are in the presence of a normal study, i.e. no pathology is identified.

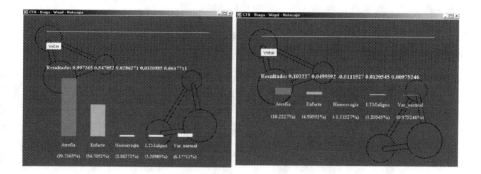

Fig. 7. Diagnostic interface (generated diagnostics)

5 Conclusions

The primary goal of the presented work was reached. We believe that the relevant parameters for the knowledge extraction were identified. The parameters which significantly influenced the studies, as are the xKurt, yKurt, gender, ySkew were also identified.

From the analysis of the learning algorithms, we concluded that the algorithms that generated the best results, considering the presented dataset, were the Simple Logistic and the LMT fed by a vector without the parameters gender, xKurt, yKurt and ySkew, in which the rightness percentage was of 80,77 % and for false negatives was of 7,7 %. However the steadiest algorithm was the VFI that kept a more steady behavior along all studies. It is imperative to refer that a higher number of studies could lead to better results. Though, due to some reluctance in providing medical imaging studies with its corresponding diagnostic report by the radiologists, this will always be a problem for future investigation.

This work could certainly evolve to better results, since it has a great potential of development, based on a safe foundation. The learning process, provided with a bigger number of cases will have to be performed to allow a better assessment of the learning algorithms in the knowledge extraction necessary to identify the existence or not of pathology in examinations of brain CT.

Acknowledgment

We are indebted to *CIT- Centro de Imagiologia da Trindade*, for providing the anonymized dataset and for their help in terms of experts, technicians and machine time.

References

1. Doi, K.: Computer-aided diagnosis in medical imaging: Historical review, current status and future potential. Computerized Medical Imaging and Graphics 31, 198–211 (2007)
2. Alves, V.: PhD Thesis: Distributed Problem Solving – A breakthrough into the areas of artificial Intelligence and Health (In Portuguese), Departamento de Informática, Escola de Engenharia, Universidade do Minho, Braga, Portugal (2002)
3. Castellino, R.A.: Computer aided detection (CAD): an overview. Cancer Imaging (2005)
4. Chan, T., Huang, H.K.: Effect of a Computer-aided Diagnosis System on Clinicians' Performance in Detection of Small Acute Intracranial Hemorrhage on Computed Tomography. Acad Radiol, 290–299 (2008)
5. Freer, T.W., Ulissey, M.J.: Screening Mammography with Computer-aided Detection: Prospective Study of 12,860 Patients in a Community Breast Center. Radiology 220, 781–786 (2001)
6. Wang, S.-P.: Computer-aided diagnosis system and method. Patent nr: 96193822 (1996), http://www.ipexl.com/patents/others/SIPO/Wang_ShihPing/Wang_ShihPing/96193822.html
7. Katsuragawa, S.: Computerized Scheme for Automated Detection of Lung Nodules in Low-Dose Computed Tomography Images for Lung Cancer Screening. Academic Radiology (2004)
8. Kanazawa, K.: Computer-Aided Diagnosis for Pulmonary Nodules Based on Helical CT Images. In: Nuclear Science Symposium – IEEE (1997)
9. Abramoff, M.D., Magelhaes, P.J., Ram, S.J.: Image Processing with ImageJ. Biophotonics International 11(7), 36–42 (2004)
10. Hounsfield, G.: Computerized transverse axial scanning (tomography): Part I. Description of system. Br. J. Radiology 46, 1016–1022 (1973)
11. Geise, R.A., McCullough, E.C.: The use of CT scanners in megavoltage photon-beam therapy planning. Radiology, 133–141 (1977)
12. Witten, I.H., Frank, E.: Data Mining: Practical machine learning tools and techniques, 2nd edn. Morgan Kaufmann, San Francisco (2005)

Knowledge Representation in Difficult Medical Diagnosis

Ana Aguilera and Alberto Subero

Faculty of Science and Technology, Universidad de Carabobo, Valencia, Venezuela
{aaguilef,asubero}@uc.edu.ve

Abstract. This article is based on medical knowledge produced thought collaborative problem solving by a group of experts, in the field of medical diagnosis. In this work, we propose a representation format for a medical case base into a traditional RDBMS representation, which is queryable using standard SQL. We are concerned in difficult medical cases which imply a solution in several steps with several expert solvers (medical specialists). Some queries on this case base are proposed. A case base was implemented and validated in real time with experts on the real scenarios.

1 Introduction

Artificial Intelligence has been applied in numerous applications in the domain of health sciences. In the late 1980's, followed by ground laying work done by Koton [17], and Bareiss [11], Case-Based Reasoning (CBR) appeared as an interesting alternative for building medical AI applications, and then it has been used in the field.

Especially in the medical field, the knowledge of experts does not only consist of rules, but rather of a mixture of textbooks knowledge and practical experience. The latter consists of typical or exceptional cases and the reasoning applied to them by physicians. In medical knowledge based systems there are two sorts of knowledge, objective knowledge, which can be found in textbooks, and subjective knowledge, which is limited in space and time and changes frequently. The problem of updating the changeable subjective knowledge can partly be solved by incrementally incorporating new up-to-date cases. Both sorts of knowledge may clearly be separated: objective textbook knowledge may be represented in form of rules or functions while subjective knowledge is contained in case reports. Thus, the arguments for case-oriented methods are as follows: Case reasoning corresponds with the decision making process of physicians; Incorporation of new cases means automatically updating parts of the changeable knowledge; Objective and subjective knowledge can be clearly separated; As cases are routinely stored, integration into clinic communication systems is easy.

Case-Based Reasoning (CBR) is a recognised and well established method for building medical systems. Many works have been developed. FM-Ultranet [9, 10] is a medical CBR project implemented with CBR-Works. FM-Ultranet detects malformations and abnormalities of foetuses through ultrasonographical examinations. Perner (Perner 1999) proposes a system that uses CBR to optimize image segmentation at the low level unit. This optimization is done in changing image acquisition conditions and image qualities. Jaulent et al. (Jaulent et al. 1997) is applying CBR to histopathology in the breast cancer domain. Their system uses cases that are derived from written

P. Perner (Ed.): ICDM 2009, LNAI 5633, pp. 73–87, 2009.
© Springer-Verlag Berlin Heidelberg 2009

medical reports. CARE-PARTNER [12, 14] is a decision support system for the long term follow-up of stem cell transplanted patients at Fred Hutchinson Cancer Research Center (FHCRC) in Seattle. Montani et al. has focused on CBR in hemodialysis treatments for end stage renal disease [19]. Their system is applied to the efficiency assessments of hemodialysis sessions. Montani et al. [8] attempt to integrate different methodologies into a Multi-Modal Reasoning (MMR) system, used in therapy support for diabetic patients. Costello and Wilson [15] are focusing on the classification of mammalian DNA sequences, and are using a case library of nucleotide (A,T,G,C) segments. Nilsson et al. [20] address the domain of psychophysiological dysfunctions, a form of stress. The system is classifying physiological measurements from sensors. TeCoMED [21], Schmidt and Gierl attempt to use a prognostic model to forecast waves of influenza epidemics, based on earlier observations done in previous years. TeCoMED combines CBR with Temporal Abstraction to handle the problem of the cyclic but irregular behaviour of epidemics. The Auguste project [18], is an effort to provide decision support for planning the ongoing care of Alzheimer's Disease (AD) patients. WHAT [16] is a tutoring medical CBR system for the education of sports medicine students. WHAT is designed to give better matching exercise prescriptions than the conservative rule-based approach taught by most books. Bichindaritz et al. [13] have evolved CARE-PARTNER into a medical training system on the Internet. The intention is to help medical students improve their knowledge by solving practical cases.

The intentional databases (IDB) are an interesting topic of research in the data mining. Theses databases integrate not only data but also generalizations (patterns and models) [4]. An approach to the integration of decision trees into IDB was developed successfully in [4]. This approach consisted in storing the patterns and models in a straightforward way, using the usual relational database tables provided by any Relational Database Management System (RDBMS) and the standard SQL language to represent, store and query the new generalizations made on the data.

In this work, we propose a representation format for a medical case base into a traditional RDBMS representation. The system is queryable using standard SQL. We are focused on difficult medical cases which imply a solution in several steps with several expert solvers (medical specialist).

2   The Nature of Medical Knowledge

We can distinguish three aspects within the medical knowledge [5]: 1) the corpus of medical knowledge. It constitutes the current state of the knowledge cumulated in the various fields of medical science (clinical, complementary examinations, physiopathology, biology and all fundamental sciences applied to medicine). This is normally the knowledge. 2) the medical exercise itself and its evolution. It relates to the application of knowledge to particular cases. This is normally the know-how. 3) knowledge related to the medical innovation [2], since it involves new knowledge and consequently new ways of acting.

The acquisition of the knowledge corpus by the future doctors is the subject of initial formation delivered by faculties of medicine, in addition, this corpus updates during years. On the other hand, there exist several ways for the acquisition of the know-how. The principal one is the trade-guild [5], nonetheless it has the disadvantage that : it is restricted for reasons of time, availability of experts, etc.; it is necessary to give oneself the means of gathering the knowledge and medical know-how, the protocols, the guides of good practice, the bibliographical references of evidence-based medicine [Charlet 02].

3   A Clinical Scenario of Distributed Care

Physicians treat different diagnostic cases, some of them are considered as obvious (classical cases). This happens because the nature of the illness in question is simple or because the physician's experience is significant. There are other less obvious cases for which it is not easy to establish a satisfactory diagnosis directly (complex cases). There are many reasons for this: an insufficient number of available judgment elements; the illness has not evolved enough for certain signs to appear; or for a circumstantial non-schematizable cause. The diagnosis in these cases can then involve great difficulties and requires a complex procedure. This procedure can implement several clinical maneuvers (exploration of various organs or systems, complementary tests and sometimes a further patient observation) which are repeated several times [5].

In the daily practice, the care of a particular patient is often shared by several different clinicians who are located in a variety of physically distributed health care settings. For instance, a general practitioner (GP) may suspect that his patient has breast cancer. However, as he neither has the knowledge nor the resources to confirm this hypothesis, he must refer the patient to a hospital specialist who can make a firm diagnosis. Having confirmed the presence of breast cancer, the specialist must devise a care program for treating the patient - this typically involves the hospital, the patient's GP, and a home care organization jointly executing a series of interrelated tasks. In addition to this inter-organization coordination, there is also a need to ensure that the activities within an organization are effectively and coherently managed. In a hospital, for instance, care typically involves execution of interrelated tasks by doctors, nurses, pharmacy, laboratories, and resource management departments. Shared care is a situation in which clinicians (GPs and specialists) jointly treat the same patient. Patients requiring shared care are, for example, patients suffering from chronic disorders, such as diabetes mellitus, obstructive pulmonary diseases, or cardiological diseases, or patients who receive palliative care at home, patients with multi-system illnesses where physiopathology and the nature of origin of the illness make it necessary to examine the results of several diagnostic procedures.

A clinical case illustrating this type of work is shown in fig. 1 (A more detailed explication is found in [6]). In this case, we can observe the interaction between a GP, a radiologist and a pathologist.

In the analysis of this case we can observe the following elements: *When*: It is introduced by the concept of time. We can note that time is an essential element in medicine. It is very related to : medical specialists, patient, etc. *What*: It is the action executed by the actor: diagnosis, treatments, tests, test results, medical reports, etc.) *How*: It indicates the means of communication between the actors: the medical history itself. All this analysis indicates to us that, there is a flow of information and control which determines the protocol used (circuit) to arrive at the diagnosis and the treatment of the patient. These circuits of the communications involve also the following

Clinical case
A 40 years old male patient, non-smoker, without any obvious particular antecedents in his past medical history. He went to his GP with a non-productive cough of three months. His physical tests are normal. GP recommends him a palliative treatment and somes laboratory and paraclinical tests (Postero-anterior chest x-ray).

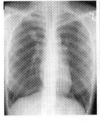

x-ray result : circumscribed 2-3 cm nodule located in the right upper lobe of the lung with the presence of interior calcifications of non-specified type. Heart and rest of study without obvious modifications. Scanner recommended.

CAT Lung scanner: It indicates a 2 x 3 cm mass with non-epimacular aspect located in the right upper lobe of the lung with non-specified calcifications. There is no affectation of Mediastinum lymphatic ganglia. There are no other masses in the thorax.

Pulmonary biopsy: macro and microscopic analysis of post-operative piece.

Diagnosis: ENDOBRONCHIAL HAMARTHOMA.
The patient leaves the hospital and considering the benign origin of the pathology, the doctor recommends an annual check up with his general practitioner.

Fig.1. A clinical case

characteristics: These circuits define a collaborative work between different agents with the total objective directed towards to the patient care; Collaboration in this level includes: inter dependence, division, responsibility and participation. There are two types of collaboration: between doctors with different specialists or within the same speciality; Each agent has subs objectives implied by its speciality. Each one carries out one or more tasks related to the diagnosis, a treatment or a selection of tests; The medical circuits are not fixed, they depend on the case, the patient, the conditions of the disease, the resources employed. This study has guided us in the modeling and the knowledge representation.

3   Knowledge Representation

3.1 Data Source

The data source is stored in the physical database, represented by an entity-relationship scheme (Fig 2). Conceptually, the database has tables that contain all itemsets and all association between data. This source keeps medical information as a set of well defined facts which constitute the medical history of patient (Fig. 2 - right side). Normally, outside of images, the medical history is a set of data which include reports, follow-up notes, complementary results of tests, etc Being by definition the "place" of convergence of flows of information and of knowledge which circulates in the hospital about a patient, the medical history is the point of obliged passage and, by its function of filing, the receptacle of all personal information about the patient [2]. Starting from the medical history, we can note a flow of knowledge which constitutes the knowledge and the know-how of the medical practices. Thus, the medical history is the principal source of knowledge. In our case, we remained in a relational representation such as it is used by the systems of relational databases. These databases provide a powerful and flexible mechanism to store knowledge; this is why they almost completely ensured the storage of information in the commercial systems of business. Moreover, there are recent researches [7, 4] which support this type of idea.

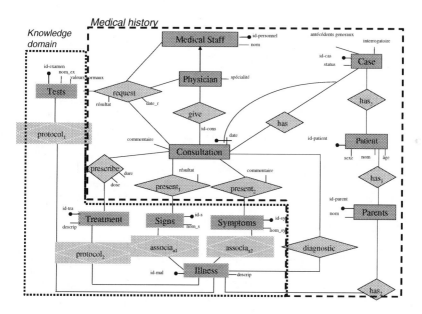

Fig. 2. Entity-relationship scheme of database

The itemsets are updated by each consultation of patients (Fig. 3 shows an instantiation of the database scheme for a particular patient). Many cases can be associated to the same patient and a single case can take several consultations with different doctors. This approach of representation has a number of advantages over other approaches with respect to extensibility and flexibility. The storage and querying of data is easier.

3.2 Knowledge Bases

The knowledge base is also represented and stored in a relational representation (Fig. 2 – left side). Thus, having data and knowledge in the same place facilitates their actualization and their querying.

We have two types of knowledge represented in the relational structure: a facts base and a case base. For the implementation of the relational tables, we use the well-known database management system (DBMS) PostgreSQL version 8.2.

Présente₁

id-sym	id-cas	date	commentaire
Sy001	C001	t1	
Sy002	C001	t1	avec expectoration verdâtre
Sy003	C001	t1	d'une semaine d'évolution

Présente₂

id-sig	id-cas	date	résultat
S001	C001	t1	37,5°C
S002	C001	t1	32 resp./m
S003	C001	t1	crépitations dans les deux bases pulmonaires

Prescrire

id-tra	id-cas	date	dose	dure
T001	C001	t1		
T002	C001	t1	deux fois par jour	Pendant 2 semaines
T003	C001	t1	8h-16h-24h	5 jours

Diagnostic

id-med	id-cas	id-mal	date
M001	C001	Ma001	t1
M001	C001	Ma002	t1
M002	C001	Ma003	t1

Demande

id-med-Demandeur	id-med-processeur	id-cas	date	id-exa	Date-r	résultat
M001	M002	C001	t1	E003	t2	On observe une lésion au poumon droit arrondie, de bord flou, de 3 cm
M001	M003	C001	t1	E001	t3	Une malformation utérine

Fig. 3. View of a medical history

- The Facts base: It consists of elements known in medicine: Medical protocols, tests required by diseases, signs and symptoms by diseases, treatments for the diseases (Fig 4.). Fig. 5 shows an example of this type of relationships. The ICPC (International Classification of Primary Care) [22] and other medical texts have been taken as data source.

ILLNESS	SIGNS
"M0001";"Anemie hypocromique"	"S0004";"Aphasie";
"M0002";"Angioectasies"	"S0005";"Aphasie de Broca"
"M0003";"Cancer colo-rectal"	"S0006";"Aphasie de Wernicke"
"M0007";"Cholangite"	"S0007";"Apnée";
"M0008";"Cholangite purulente"	**SYMPTOMS**
TREATMENTS	"Sy0001";"Adynamie";
"T0001";"Alfa Bloqueants"	"Sy0002";"Amaigrissement";
"T0002";"AlfaMetil"	"Sy0003";"Anémie";
"T0003";"Amiodarona"	"Sy0004";"Asthénie";
"T0004";"Anticonvulsivantes"	"Sy0005";"Céphalée ";
"T0005";"Beta Bloqueantes"	**TESTS**
	"E0001";"Coloscopie"
	"E0002";"Cytologie"
	"E0004";"Électrocardiogramm

Fig. 4. Illness, Treatments, Signs, Symptoms, Medical Tests

- The case base: It is an organized trace of resolutions of the medical cases. This trace keeps all the elements of the 5-W analysis method [23] that includes: when, which, what, where and how (Fig. 6). This base is also represented by a relational structure (Fig. 7).

M0001	Hypocromic Anemia	

Hypochromic anemia is a generic term for any type of anemia in which the red blood cells (erythrocytes) are paler than normal. (Hypo- refers to less, and chromic means colour.) A normal red blood cell will have an area of pallor in the center of it; in hypochromic cells, this area of central pallor is increased. This decrease in redness is due to a disproportionate reduction of red cell hemoglobin (the pigment that imparts the red color) in proportion to the volume of the cell. In many cases, the red blood cells will also be small (microcytic), leading to substantial overlap with the category of microcytic anemia.

id_maladie l id_signe	id_maladie l id_symptome	id_maladie l id_examen
----------+----------+	-----------+-------------+	-----------+-----------+
M0001 l S0041	M0001 l Sy0001	M0001 l E0012
M0001 l S0054	M0001 l Sy0004	M0001 l E0011
M0001 l S0064	M0001 l Sy0005	M0001 l E0023
M0001 l S0065	M0001 l Sy0006	
M0001 l S0070	M0001 l Sy0018	id_maladie lid_trat
M0001 l S0030	M0001 l Sy0008	-------------+---------+
M0001 l S0042	M0001 l Sy0011	M0001 l T0012
	M0001 l Sy0023	

Fig. 5. Presentation of hypocromic anemia with its signs, its symptoms, its tests and its treatments

Quoi : "C0002";
Quand : "2007-08-23 16:55:52.312";
Qui : "practicien hospitalier";
Contexte : "{{S0071,39}}";"{{Sy0013,profuse},{Sy0027,""petit seche et vesperale""}}";
Set d'actions exécutés :
"{{Traitement,antibiotiques},{Demande_ex,radiographie}}"

Fig. 6. Entry in the case base

```
              Table "public.connaissance"
    Column      |         Type          |  Modifiers
 -------------+-----------------------+----------
  id_cas       | character varying(10)  | not null
  date_cons    | timestamp without time zone | not null
  specialite   | character varying(40)  |
  signes       | text[][]               |
  symptomes    | text[][]               |
  accion       | text[][]               |

Indexes: "connaissance_pkey" PRIMARY KEY, btree (id_cas,
date_cons)
Foreign-key constraints: "connaissance_id_cas_fkey"
FOREIGN KEY (id_cas, date_cons) REFERENCES
consulte(id_cas,ddate) ON UPDATE CASCADE ON DELETE CASCADE

CREATE INDEX signes_et_symptomes ON connaissance
   USING btree
   (signes, symptomes);
ALTER TABLE connaissance CLUSTER ON signes_et_symptomes;
```

Fig. 7. Relational structure of case base

We added other elements to the case structure in order to reflect the current state of the case. The detail of description of the columns is made up as follows:

- Id_cas and datetime (*When*): constitutes the primary key for querying and identification of cases. The id_cas enables us to identify all the elements in relation to a particular case and the datetime enables us to keep the chronological trace of the events.
- Physician (*Who*): describes the specialty of the doctor who intervened.
- Signs and Symptoms enable us to describe the current context. We did not consider these elements in our analysis of case, but they are used as a guide to the doctors at the diagnosis step. – the sequence of actions is : diagnosis, result, examination, treatment, tests. It answers the question "how". The values attached to each action show the description of the action carried out. This answers the question "what".

The structure of this base not only stores the problem description and its solution, which is normally represented in a typical case base, but it keeps also the way in which the solution was derived (Fig. 8).

```
"idcas";"datecons";"specialite";"signes";"symptomes";"accion"
```
```
"C0002";"2007-08-23 16:55:52.312";"praticien hospitalier";-
"{{S0071,39}}";"{{Sy0013,profuse},{Sy0027,""petit-seche
vesperale""}}";"{{Traitement,antibiotiques},{Demande_ex,radio
graphie}}"
```
```
"C0002";"2007-08-23 18:44:20.281";"radiologue";-
"{{S0071,""hyperthermie nocturne dans les suites d'un
syndrome grippal""}}";"{{Sy0013}}";"{{resultat,""Accentuation
de la trame broncho-vasculaire des 2 bases, sans foyer
parenchymateux, sans réaction pleurale. Le médiastin
supérieur n'est pas élargi""}}"
```
```
"C0002";"2007-08-23 20:48:00.265";"clinicien
pneumologue";"{{S0071,""hyperthermie nocturne dans les suites
d'un syndrome grippal""}}";"{{Sy0013}}";-
"{{demande_ex,E0026},{demande_ex,E0027},{demande_ex,E0028},{d
emande_ex,E0029}}"
```
```
"C0002";"2007-08-24 13:55:10.515";"clinicien pneumonologue";-
"{{S0011,""avec un syndrome inflammatoire""}}";"";-
"{{diagnostic,M0063},{diagnostic,M0057},{demande_ex,M0030}}"
```
```
"C0002";"2007-08-24 14:33:44.593";"nephrologue";-
"{{S0064}}";"";"{{diagnostic,M0057},{demande_ex,E0031}}"
```
```
"C0002";"2007-08-24 14:42:20.843";"chirurgien
urologue";"";"";"{{realiser_ex,E0031}}"
```
```
"C0002";"2007-08-24 14:50:30.578";"anatomo-
patologue";"";"";"{{diagnostic,M0058}}"
```
```
"C0002";"2007-08-24 15:32:57.703";"praticien hospitalier";-
"";"{{Sy0035,persiste},{Sy0036,""dans le doigts""}}";-
"{{demande_ex,E0006},{demande_ex,E0025},{demande_ex,E0033}}"
```

Fig. 8. Representation of a case in the case base

Certain authors call this type of bases; strategic case bases [3]. For us, there is a context in time (signs and symptoms) and the decision taken (diagnosis, examinations, result, tests or procedure, or treatment); additionally we keep the author of this decision. Fig 9. shows a simplified view of our knowledge base.

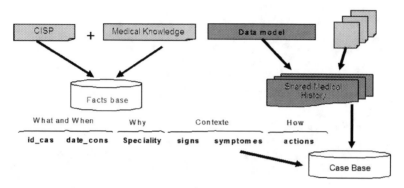

Fig. 9. View of knowledge bases

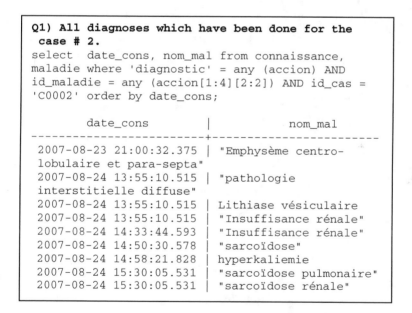

```
Q1) All diagnoses which have been done for the
case # 2.
select  date_cons, nom_mal from connaissance,
maladie where 'diagnostic' = any (accion) AND
id_maladie = any (accion[1:4][2:2]) AND id_cas =
'C0002' order by date_cons;

          date_cons            |          nom_mal
-------------------------------+------------------------------
2007-08-23 21:00:32.375 | "Emphysème centro-
lobulaire et para-septa"
2007-08-24 13:55:10.515 | "pathologie
interstitielle diffuse"
2007-08-24 13:55:10.515 | Lithiase vésiculaire
2007-08-24 13:55:10.515 | "Insuffisance rénale"
2007-08-24 14:33:44.593 | "Insuffisance rénale"
2007-08-24 14:50:30.578 | "sarcoïdose"
2007-08-24 14:58:21.828 | hyperkaliemie
2007-08-24 15:30:05.531 | "sarcoïdose pulmonaire"
2007-08-24 15:30:05.531 | "sarcoïdose rénale"
```

Fig. 10. All illness diagnosed for a particular case

3.3 Construction and Querying of Our Case Base

The construction of our case base is made in real time, in a dynamic way. Each time that a new element appears in the medical history, we record the data concerned and we update our case base, simultaneously. A stored procedure is triggered when the new data appear.

```
Q2) Signs and symptoms present by the case # 2
(select date_cons, nom_sig from connaissance,signe
where id_signe = any(signes) AND id_cas = 'C0002')
UNION
(select date_cons, nom_sym from connaissance,
symptome where id_symptome = any(symptomes) AND
id_cas = 'C0002') order by date_cons;

        date_cons          | nom_sig et nom_sym
------------------------------+--------------------
  2007-08-23 16:55:52.312  | Fièvre
  2007-08-23 16:55:52.312  | Soeur
  2007-08-23 16:55:52.312  | Toux
  2007-08-23 18:44:20.281  | Fièvre
  2007-08-23 18:44:20.281  | Soeur
  2007-08-23 20:48:00.265  | Fièvre
  2007-08-23 20:48:00.265  | Soeur
  2007-08-23 21:00:32.375  | Fièvre
  2007-08-23 21:00:32.375  | Soeur
  2007-08-24 13:55:10.515  | Asthénie
  2007-08-24 14:33:44.593  | Splénomégalie
  2007-08-24 14:58:21.828  | "difficulté a
l'endormissement"
  2007-08-24 15:32:57.703  | "crampes"
  2007-08-24 15:32:57.703  | "dyspnée d'effort"
  2007-08-24 15:44:01.296  | "crampes"
  2007-08-24 15:44:01.296  | Dyspnée
```

Fig. 11. Signs and symptoms presented in a particular case

```
Q3) Which are the treatments applied in the case #
2?
select  date_cons, descripcion from connaissance,
traitement where id_trat = any(accion) AND
'traitement' = any (accion) AND id_cas = 'C0002'
order by date_cons;

   date_cons            | descripcion
--------------------------+---------------
  2007-08-23 16:55:52.312 | antibiotiques
  2007-08-24 14:55:15.703 | corticoids
```

Fig. 12. All treatment prescribed for a particular case

We used the DBMS for sequential storage of data with two indices: one for the primary key (the case and the date), and the other for the context (signs and symptoms). This allows us an effective recovery of the knowledge stored.

From the users point of view, tables with itemsets and rules etc. exist and can be queried like any other table. How these tables are filled (what case base adaptation algorithm is run, with what parameters, etc.) is transparent. The user does not need knowledge about the many different implementations that exist and when to use what implementation, nor does he/she need to familiarize herself with new special-purpose query languages. The whole approach is also much more declarative: the user speci-fies queries on the case that should result from the adaptation process on the case base [4]. Some example queries on the case base are presented. Query 1 (Fig. 10) shows all illness diagnosed for a particular case, the diagnostic process is done in an incremental way for reducing uncertainty in each step. Query 2 (Fig. 11) shows the case context (signs and symptoms) for a particular case , Query 3 (Fig. 12) shows all treatment prescribed for a particular case and query 4 (Fig. 13) shows all actions applied (test, medical procedures) for a particular case.

```
Q4) Which are the executed protocols in the case #
2.
select  date_cons, accion[1:4][1:2] from
connaissance where 'diagnostic' <> all (accion)
AND id_cas = 'C0002' order by date_cons;

          date_cons          |       accion
--------------------------+--------------------------
2007-08-23 16:55:52.312 |
{{Traite ment,antibiotiques},{Demande_ex,radiograph
ie}}

2007-08-23 18:44:20.281 | {{resultat,"Accentuation
de la trame broncho-vasculaire des 2 bases, sans
foyer parenchymateux, sans réaction pleurale. Le
médiastin supérieur n'est pas élargi"}}

2007-08-23 20:48:00.265 |
{{demande_ex,E0026},{demande_ex,E0027},{demande_ex
,E0028},{demande_ex,E0029}}

2007-08-24 14:42:20.843 | {{realiser_ex,E0031}}

2007-08-24 14:55:15.703 |
{{traitement,T0019},{demande_ex,E0032}}

2007-08-24 15:32:57.703 |
{{demande_ex,E0006},{demande_ex,E0025},{demande_ex
,E0033}}

2007-08-24 15:44:01.296 | {{resultat,"'une nette
diminution du syndrome
interstitiel'"},{resultat,"'retrouvent un trouble
important de la diffusion alvéolo-capillaire'"}}
```

Fig. 13. All actions applied for a particular case

4   Validation and Evaluation of Our Case Base

We defined the following questions for validation and evaluation purposes: is the case selected the correct case? : The real cases have been used in the test. The validation of knowledge extracted was made with the assistance of a physician, who interpreted the results. Is the case correct?: certain tests have been done thanks to SQL language of DBMS. We also implemented certain restrictions of integrity which guarantee the unicity of keys, the correct values when loading data, the checking of crossed references, etc.

To test our case base, we worked on eleven cases from the hospital "Ciudad Hospitalaria Enrique Tejeras", Venezuela and the Cavale Blanche hospital, France. The data were collected from the specialists implied in the diagnostic resolution. The way of working was made in an asynchronous way, i.e. the data were initially collected then transcribed using the specialists who worked on the case. For the validation of our case base, we presented it together with the validation criteria to the specialists. We filled the various criteria used in the validation, thanks to the appreciation of the specialists concerned. Finally, we based on certain formal criteria specified for the V&V of the knowledge bases [1] to test our case base. A nonexhaustive list of criteria, considered as attributes of quality, makes it possible to evaluate the quality of the knowledge base and was used to check or validate it according to the target of the user.

These criteria have been answered affirmatively:

Suitability: capacity of the knowledge base to build a set of knowledge suitable for a specified case and rules defined by the user.

Interoperability: capacity of the knowledge base to be interacted by one or more softwares.

Reliability: capacity of the knowledge base to provide a certain level of knowledge from a specific domain.

Fault Tolerance: capacity of the knowledge base to maintain a level of performance specified at the time of faults in the base.

Availability: capacity of the knowledge base to provide knowledge always.

Changeability: capacity of the knowledge base to being modified easily.

Analysability: capacity of the knowledge base to being diagnosed for deficiencies or gaps in the base, and thereafter to being modified.

Stability: capacity of the knowledge base to prevent the unexpected effects of the modifications on the basis of knowledge.

Testability: capacity of the knowledge base to being modified for validation.

Adaptability: capacity of the knowledge base to being adapted to various environments, specified without other modifications pulled by the knowledge base.

Installability: capacity of the knowledge base to being installed in a specified environment.

Coexistence: capacity of the knowledge base to coexist with independent software in an environment sharing the same resources.

Remplaceability: capacity of the knowledge base to being used in the place of another knowledge base for the same field in the same environment.

The negative criteria are :

> *Coherence*: capacity of the knowledge base to provide noncontradictory knowledge.
> *Maturity*: capacity of the knowledge base to prevent failures.
> *Complétude*: capacity of the knowledge base to have all knowledge necessary for the comprehension of the field and all the interconnections between various knowledge elements.
> *Redundancy*: capacity of the knowledge base not to have knowledge duplicated unnecessarily.

4.1 Usefulness of the Case Base

Know-how knowledge: The first utility of the case base is related to reuse. The cases stored will serve as a guide in the solving of other cases with similar contexts, and even as a subject of study among young physicians. We are interested in solving difficult cases, thus the context of application remains broad. The case base serves as a scheme to the knowledge representation from past experiences in describing situations of patient monitoring, as well as his/her context and his/her treatments. Each case represents a prototypical patient with a context of diagnostic problem solving and medical follow-up in the time space.

Communication: Our base could aid to the communication between humans. We emphasize that the role of medical case base facilitates interaction and sharing of knowledge between the group of medical experts.

Aid in the specification of knowledge base system: The case base serves as a basis to build diagnosis aid systems, where diagnosis can be suggested for the treatment of new cases.

Data mining : The construction of the data source (with an important data volume) will also permit to implement other knowledge extraction tools for discovering new patterns and models of knowledge.

Interoperability: Interoperability is a specialty in communications, in this case for two computers. The case base identifies diagnostic problem solving experiences that could be used by other applications even if they are distant and were developed on different bases.

The indexing and retrieval of information. The implementation of the case base and the individual indexes for each item (signs, symptoms, treatments, diagnostics, etc.) provide the conceptual indexes that describe the case and the subsequent retrieval of information on these cases.

5   Conclusion

This article showed the complexity of the medical field dealt with. This field is not obvious and this can be confirmed in the diagnoses that unfold from the real cases considered. We distinguished two types of knowledge: a fact base, which keeps the knowledge, and a case base, which keeps the medical know-how of the cases treated. A case is represented by several temporal entries in the case base. These entries are derived from consultations and contributions of the medical specialists in the diagnostic resolution. Our data source remained in the medical history, and the implementation in the information processing system rested on a relational system. This system showed its adequacy to the problem arising for the representation and recovery of data and knowledge. The temporal aspect is very significant in the medical field and gave it a significant place in all the treated elements. Finally, certain criteria for validation and usability of the case base were studied. The latter constitutes a very vast topic of research in social sciences and software engineering, even though our goal was to present only some criteria for validation and evaluation that were actually applied to our approach.

Acknowledgments. This work is sponsored by FONACIT under No G-2005000278.

References

1. Althoff, K.D.: Validation of case base reasoning system. Leipziger Informatik-Tage (1997)
2. Charlet, J.: L'ingénierie des Connaissances Développements, Résultats et Perspecti-ves pour la Gestion des Connaissances Médicales. Mémoire d'Habilitation à diriger des recherches présenté à l'Université Pierre et Marie Curie Spécialité: Informatique (2002)
3. Christodoulou, E., Keravnou, E.T.: Metareasoning and meta-level learning in a hybrid knowledge-based architecture. Artificial Intelligence in Medicine 14, 53–81 (1998)
4. Fromont, E., Blockeel, H., Struyf, J.: Integrating decision tree learning into inductive databases. In: Džeroski, S., Struyf, J. (eds.) KDID 2006. LNCS, vol. 4747, pp. 81–96. Springer, Heidelberg (2007)
5. Paolaggi, J.B., Coste, J.: Le raisonnement médical, de la science à la pratique clinique. Editions ESTEM (2001)
6. Quintero, J.: Collaborative decision in medicine. Analysis of a case diagnosis. In: Proceedings of the Sciences of Electronic, Technology of Information and Telecommu-nications, SETIT, Sousse, Tunisia, paper R228137 (2003)
7. Szydlo, T., Sniezynski, B., Michalski, R.S.: A Rules-to-Trees Conversion in the Inductive Database System VINLEN (2005),
   http://www.mli.gmu.edu/papers/2005/05-4.pdf
8. Montani, S., Magni, P., Roudsari, A.V., Carson, E.R., Bellazzi, R.: Integrating different methodologies for insulin therapy support in type 1 diabetic patients. In: Quaglini, S., Barahona, P., Andreassen, S. (eds.) AIME 2001. LNCS, vol. 2101, pp. 121–130. Springer, Heidelberg (2001)
9. Balaa, Z.E., Traphoner, R.: Case-based decision support and experience management for ultrasonography. In: German Workshop on Experience Management (GWEM 2003) (2003)
10. Balaa, Z.E., Strauss, A., Uziel, P., Maximini, K., Traphner, R.: Fm-ultranet: a decision support system using case-based reasoning, applied to ultrasono-graphy. In: Workshop on CBR in the Health Sciences. ICCBR 2003, pp. 37–44 (2003)
11. Bareiss, E.R.: Exemplar-Based Knowledge Acquisition: A unified Approach to Concept Representation, Classification and Learning. 300 North Zeeb road, Ann Arbor, MI 48106-1346: UMI (1989)

12. Bichindaritz, I.: Solving safety implications in a case based decision-support system in medicine. In: Workshop on CBR in the Health Sciences. ICCBR 2003, pp. 9–18 (2003)
13. Bichindaritz, I., Sullivan, K.: Generating practice cases for medical training from a knowledge-based decision-support system. In: Workshop Proceedings. ECCBR 2002, pp. 3–14 (2002)
14. Bichindaritz, I., Kansu, E., Sullivan, K.M.: Case-based reasoning in care-partner: Gathering evidence for evidence-based medical practice. In: Advances in CBR: 4th European Workshop. ECCBR 1998, pp. 334–345 (1998)
15. Costello, E., Wilson, D.C.: A case-based approach to gene finding. In: Workshop on CBR in the Health Sciences. ICCBR 2003, pp. 19–28 (2003)
16. Evans-Romaine, K., Marling, C.: Prescribing exercise regimens for cardiac and pulmonary disease patients with cbr. In: Workshop on CBR in the Health Sciences. ICCBR 2003, pp. 45–52 (2003)
17. Koton, P.A.: Using Experience in Learning and Problem Solving. MIT Press, Cambridge (1988)
18. Marling, C., Whitehouse, P.: Case-based reasoning in the care of alz-heimer's disease patients. In: Case- Based Research and Development. ICCBR 2001, pp. 702–715 (2001)
19. Montani, S., Portinale, L., Leonardi, G., Bellazi, R.: Applying case-based retrieval to hemodialysis treatment. In: Workshop on CBR in the Health Sciences. ICCBR 2003, pp. 53–62 (2003)
20. Nilsson, M., Funk, P., Sollenborn, M.: Complex measurement classification in medical applications using a case-based approach. In: Workshop on CBR in the Health Sciences. ICCBR 2003, pp. 63–72 (2003)
21. Schmidt, R., Gierl, L.: Prognostic model for early warning of threatening influenza waves. In: German Workshop on Experience Management. GWEM 2002, pp. 39–46 (2002)
22. Wonca, International Classification of Primary Care, 2nd edn. (ICPC-2), http://www.who.int/classifications/icd/adaptations/icpc2/en/index.html
23. Where Went The 5 Ws?, http://marcusletter.com/Five%20Ws.htm

Forecasting Product Life Cycle Phase Transition Points with Modular Neural Networks Based System

Serge Parshutin, Ludmila Aleksejeva, and Arkady Borisov

Riga Technical University, Institute of Information Technology, 1 Kalku Str., Riga,
Latvia, LV-1658
serge.parshutin@rtu.lv, ludmila.aleksejeva@cs.rtu.lv,
arkadijs.borisovs@cs.rtu.lv

Abstract. Management of the product life cycle and of the corresponding supply network largely depends on information in which specific phase of the life cycle one or another product currently is and when the phase will be changed. Finding a phase of the product life cycle can be interpreted as forecasting transition points between phases of life cycle of these products. This paper provides a formulation of the above mentioned task of forecasting the transition points and presents the structured data mining system for solving that task. The developed system is based on the analysis of historical demand for products and on information about transitions between phases in life cycles of those products.

The experimental results with real data display information about the potential of the created system.

Keywords: Modular Neural Networks, Self-Organizing Maps, Product Life Cycle, Forecasting Transition Points.

1   Introduction

Constantly evolving computer technologies are becoming more and more inherent part of successful enterprises management and keeping its activity at a high level. Different institutions are trying to reduce their costs by fully automatising certain stages of manufacturing process as well as introducing various techniques intended for forecasting certain market indicators that impact general manufacturing process. Different statistical methods are employed as well, though an increasing interest in artificial intelligence technologies and their practical application can be observed ever more.

For quite a long time neural networks have been one of the most popular research areas in the field of various processes forecasting including non-linear ones. The number of publications, books and monographs published within the last few years gives apparent evidence of that. A special place among neural networks is occupied by self-organising maps whose primary goal is to transform

P. Perner (Ed.): ICDM 2009, LNAI 5633, pp. 88–102, 2009.

the incoming vectors of signals that are of deliberate dimensionality into single-
or two-dimensional discrete map.

This paper focuses on studying self-organising map's ability to process discrete
time series of different duration. A task of product life cycle phase transition
point forecasting can serve as an example of analysis of different duration time-
series. From the viewpoint of the management it is important to know, in which
particular phase the product is. One of applications of that knowledge is selection
of the production planning policy for the particular phase [10]. For example, for
the maturity phase in case of determined demand changing boundaries it is
possible to apply cyclic planning [2], whereas for the introduction and decline
phase an individual planning is usually employed. This paper proposes a model
of modular multi-network system that ensures the solving of the aforementioned
task as well as provides an analysis of system testing results.

The paper is organised as follows: Section 2 formulates the task of forecasting a product life cycle phase transition point, followed by the Section 3 with
structure of the created model presented and with functional aspects of the system elements described. The gathered experimental results are presented and
analysed in Section 4 followed by conclusions.

2   Problem Statement

Any created product has a certain life cycle. The term "life cycle" is used to
describe a period of product life from its introduction on the market to its
withdrawal from the market. Life cycle can be described by different phases:
traditional division assumes such phases like introduction, growth, maturity and
decline [9]. For products with conditionally long life cycle, it is possible to make
some simplification, merging introduction and growth phases into one phase -
introduction.

An assumption that three different phases, namely, introduction, maturity and
end-of-life are possible in the product life cycle, gives us two possible transitions.
The first transition is between introduction and maturity phases and the second -
between maturity and product's end-of-life.

From the side of data mining [3,4,5] information about the demand for a
particular product is a discrete time series, in which demand value is, as a rule,
represented by the month. A task of forecasting a transition points between life
cycle phases may be formulated as follows. Assume that $D = \{d_1, \ldots, d_i, \ldots, d_n\}$
is a dataset and $d = \{a_1, \ldots, a_j, \ldots, a_l\}$ is a discrete time series whose duration
equals to l periods, where $l \in L = \{l_1, \ldots, l_h, \ldots, l_s\}$ and varies from record to
record in the dataset D. For simplification, the index of d is omited. Time series
d represents a particular phase of a product life cycle, say introduction. Assume
that for a particular transition, like introduction to maturity, a set of possible
transition points $P = \{p_1, \ldots, p_k, \ldots, p_m\}$ is available. Having such assumptions
the forecasting of a transition point for a new product, represented by a time
series $d' \notin D$, will start with finding an implication between historical data sets
D and P, $f : D \to P$; followed by application of found model to new data.

3   Model of the Modular Neural Networks Based System

This section contains a description of the general structure of the proposed system as well as provides information about the functions and importance of each block of the system.

3.1   Structure of the System

The main distinctive feature of the developed modular multi-network system (multiSOM system) is its ability to process discrete time series of different duration l, that represent information about the changes in certain indicator over time, e.g. previously mentioned demand for a product.

In general, the format of historical input data that could be processed by the system has to comply with these conditions:

- Each record displays the demand for a product, collected within known period of time, the length of which is set by the system - day, week, month, etc. In other words, each record is a demand time series.
- Each record has one or both markers - transition indicators:

  - marker *M1* indicates the period when product switched from Introduction phase to Maturity phase;
  - marker *M2* indicates the period when product switched from Maturity phase to End-of-Life phase.

- Each record has a marker indicating the moment of the actual beginning of the Introduction phase (ABI).

The last condition is based on the fact that in many databases records are kept from the defined moment in time. It is evident that not all historical products were introduced on the market at the same moment in time. Marks on transitions can guarantee that a model will be build; if we have patterns of transitions in historical data, then, theoretically, in presence of a model for generalisation, we are able to recognise those patterns in new data. To lessen the impact of noisiness and dominance of data, the dataset should be preprocessed. Data should be normalized and obvious outliers should be excluded [13,14].

Figure 1 shows a structure of the proposed system consisting of three main blocks: Data Management Block, Neural Block and Decision Making Block.

Data Management Block. The Data Management Block (DMB) performs tasks of processing input data and their distribution over modules in Neural Block. Data distribution over neural modules occurs in accordance with load distribution policy specified by the user. Let us define total system load as the number of elements in set L. At the given distribution, load distribution policy shows which part of general load will be taken by each neural module. The chosen distribution directly affects the number of neural modules in the system.

Let us illustrate load distribution. Assume that the duration of discrete time series in dataset D varies from 4 to 11 periods, thus $l \in L = \{4, 5, \ldots, 11\}$. In this

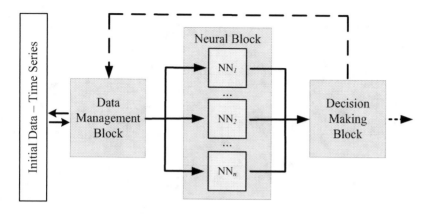

Fig. 1. Structure of the system

case, total system load will consist of eight values that time series duration can take. Let us assume that the load has to be distributed uniformly over modules at the condition that an individual load on separate module should not exceed three values that is $q = 3$. Under such conditions, three neural modules will be created in the Neural Block by the moment of system initialisation. The first neural module will process time series of duration $l \in \{4, 5, 6\}$; records with $l \in \{7, 8, 9\}$ will be sent to the second module. The remaining time series with duration of 10 and 11 periods will be processed by the third neural module.

In the real world situation, the information about the demand for a new product is becoming available gradually: after a regular period finishes, new demand data appear. Due to that specifics of system application environment, at the stage of learning it is necessary to use On-line data flowing imitation procedure. This procedure is implemented at the DMB level. The algorithm employed is executed taking into account the following details. Time series d contains demand data within introduction or maturity phase; it has duration l and is connected with the appropriate marker ($M1$ or $M2$ depending on a represented phase of a product life cycle) with value p. Provided that the system is able to process discrete time series with durations equal to l_{min} and greater, the algorithm of On-line data flowing procedure will include these steps:

1. Define $l^* = l_{min}$;
2. Send to the system first l^* periods of record d and a marker value p;
3. If $l^* < l$ then increase value of l^* by one period and return to step 2; else proceed to step 4;
4. End processing of record d.

Figure 2 illustrates execution of On-line data flowing procedure at $q = 3$.

Data processing without imitation of data flowing, *Off-line data processing* is based on this principle: at the moment when record d is ready to be sent to the system, the entire record with marker p is being sent to the corresponding module.

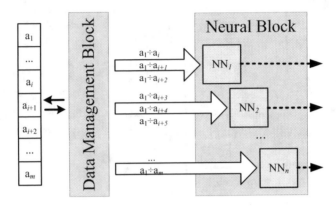

Fig. 2. On-line data flowing

Neural Block. According to the system structure depicted in Figure 1, DMB directs the data further to the corresponding modules of Neural Block (NB). Each module is a self-organizing neural map; according to the chosen policy of system load distribution it processes specific fraction of input data. The number of modules in the Neural Block, as well as the number of synaptic weights neurons will have in each neural network, is also affected by the chosen policy of system load distribution.

Each self-organising map is based on the modified Kohonen map. The necessity to modify the classical Kohonen map appears because of the stated task to use the same SOM for processing discrete time series of different duration. The number of synaptic weights of a neuron in the classical Kohonen map equals to the length of records - the duration of time series, in the input dataset [6,8,11,1]. Due to such limitations, it is possible to use the classical Kohonen map in the developed system only when $q = 1$ for each neural network. For to be able to maintain the system functionality while $q > 1$, it is necessary to apply some modifications to the classical Kohonen map.

In this work a heuristic, based on substituting the distance measure, is considered. We propose substitution of a Euclidean distance measure with a measure based on Dynamic Time Warping. Using DTW allows one to process time series with different duration. Classical Dynamic Time Warping algorithm, that is applied in the system, is fully described in [7]. The Dynamic Time Warping algorithm was experimentally compared with Derivative Dynamic Time Warping algorithm in the scope of the task, defined in Section 2, (see source [12]). The gathered results speak in favour of using the DTW for the specified task.

Functional aspects of NB are as follows: in the NB neural maps are organised at the stage of learning after that clusters are formed in each map; whereas at the stage of system application NB determines object's -record's or time series'- membership in one of previously formed clusters and forwards information from cluster further according to the scheme of transition point determination.

Decision Making Block. Decision Making Block is the one that receives information from the Neural Block on the best matching cluster found for a particular object. Following certain algorithms, whose examples are provided in the next section, a decision is shaped regarding a current object. The decision can contain both a particular value and a percentage distribution among several possible values of transition points. Taken into account that the final decision is made by a decision making person, the second option proves to be most preferable.

3.2   Basic Algorithm of System Functioning

The basic system functioning algorithm consists of the following three main stages: System training; System testing and validation and Application of the system to the real world problem. These stages are graphically displayed in Figure 3 and as can be seen from the scheme, the system functioning algorithm is cyclic.

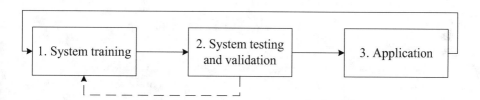

Fig. 3. General steps of the algorithm

As the system is employed, new data are accumulated an increase in whose amount will inevitably lead to the necessity to update the system. This process is initiated by feedback from step 3 to step 1. In its turn, feedback from step 2 to step 1 enables entering necessary corrections into the system even at the testing stage, which is more convenient from the viewpoint of management as compared to error correction after the application of the system to the real world problem. The following current subsection considers each of the three steps in more detail.

Step 1. System Training. The initial phase of system training process is determination and setting of basic system's parameters: the number of neural networks (modules) in the system, dimensionality and topology of networks and the number of synapses each neuron will have in each network. The number of modules in a neural block, n, is calculated empirically. Given a policy assuming uniform distribution of general load among the networks, formula (1) can be used to calculate the number of modules in the system:

$$n = \left\lceil \frac{|L|}{q} \right\rceil , \tag{1}$$

where q - each network's individual load; $\lceil \cdot \rceil$ - symbol of rounding up.

In parallel with calculating the number of modules in the system, it is necessary to determine a topology of neuron dislocation in the network. Several frequently used topologies can be applied [8,1] - tetragonal, hexagonal and a combination of the first two mentioned (see Figure 4).

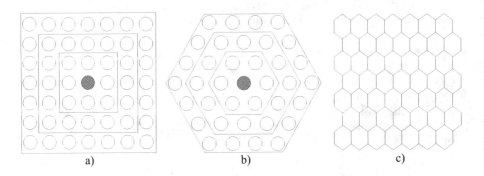

a)   b)   c)

Fig. 4. Network topologies: a) - tetragonal; b) - hexagonal; c) - a) and b) combination

After the number of modules, n, is determined, for each module m_i an interval of time series durations $[l_{i,min}; l_{i,max}]$ is set. The records with duration $l \in [l_{i,min}; l_{i,max}]$ will be processed by module m_i. Given a uniform load distribution, equation (2) can be used:

$$\begin{cases} i = 1, & l_{i,min} = l_{min} \; , \\ i > 1, & l_{i,min} = l_{i-1,max} + 1 \; ; \end{cases}$$
$$l_{i,max} = l_{i,min} + q - 1 \; . \tag{2}$$

The number of neurons of each network is determined empirically depending on the task stated. The number of synapses of a neuron in each network can be calculated using formula (3) below:

$$b_{j,i} = \left\lceil \frac{l_{i,min} + l_{i,max}}{2} \right\rceil \; , \tag{3}$$

where $b_{i,j}$ is the number of synapses of a neuron j in the network m_i.

As an alternative, a median calculation can be used for determining the number of synapses of a neuron. Assuming that module m_i can process discrete time series of duration $l \in [l_1, l_2, \ldots, l_i, \ldots, l_k]$, let us denote for each value l the number of records in the training set having duration equal to l, as $f \in [f_1, f_2, \ldots, f_i, \ldots, f_k]$. By having such an assumption, a median of time series durations a module m_i can process, may be calculated with formula (4):

$$Median = \frac{\sum_{i=1}^{k}(l_i \cdot f_i)}{\sum_{i=1}^{k} f_i} \; . \tag{4}$$

The calculated median must be rounded to the integer thus obtaining the number of synaptic connections of a neuron. Will the median be rounded to a smaller or a greater value, to a large extent depends on the task to be solved.

As the main parameters are set, the initialisation of the system begins. Synaptic weights of each neuron in each network are assigned initial values - usually small values produced by random number generator. At this moment networks in the system have no organization at all. Then the following main processes are launched: Competition, Cooperation and Synaptic adaptation.

According to the selected process of data input, Data Management Block forwards each record of the training set to the corresponding network. Then the process of neuron competition for the right to become the winner or best-matching neuron for the arrived record begins. Discriminant function - the distance between the vector of the synaptic weights and discrete time series - is calculated using the DTW algorithm. Thus the neuron with the least total distance becomes the winner or best-matching neuron.

The winner neuron is located in the centre of the topological neighbourhood of co-operating neurons. Let us define lateral distance between the winner neuron (i) and and re-excited neuron (j), as $ld_{j,i}$. Topological neighbourhood $h_{j,i}$ is symmetric with regard to the point of maximum defined at $ld_{j,i} = 0$. The amplitude of the topological neighbourhood $h_{j,i}$ decreases monotonically with the increase of lateral distance $ld_{j,i}$, which is the necessary condition of neural network convergence [6]. Usually a Gaussian function if used for $h_{j,i}$ calculation (formula 5).

$$h_{j,i(d)} = \exp\left(-\frac{ld_{j,i}^2}{2 \cdot \sigma^2(n)}\right) . \tag{5}$$

A decrease in the topological neighbourhood is gained at the expense of subsequent lessening the width of σ function of the topological neighbourhood $h_{j,i}$. One of possible kinds of value σ dependence on discrete time n is an exponential decline (formula 6).

$$\sigma(n) = \sigma_0 \cdot \exp\left(-\frac{n}{\tau_1}\right) \quad n = 0, 1, 2, \dots , \tag{6}$$

where σ_0 is the beginning value of σ; τ_1 - some time constant, such as the number of learning cycles.

To ensure the process of self-organisation, the synaptic weights of a neuron has to change in accordance with the input data, i.e. adapt to the input space. Let us assume that $w_j(n)$ is the vector of synaptic weights of neuron j at time moment (iteration, cycle) n. In this case, at time instant $n+1$ the renewed vector $w_j(n+1)$ is calculated by formula (7):

$$w_j(n+1) = w_j(n) + \eta(n) \cdot h_{j,i(d)}(n) \cdot (d - w_j(n)) , \tag{7}$$

where η - learning rate parameter; d - discrete time series from learning dataset.

Note how the difference between discrete time series and the vector of synaptic weights is calculated in expression (7). When the load is $q = 1$, that is when

each neural network is processing discrete time series with a certain fixed duration, and DTW is not used, the difference between d and $w_j(n)$ is calculated
as the difference between vectors of equal length. In other cases when DTW
is employed, the fact of time warping has to be taken into account. For this
purpose, it is necessary to fix in the memory a warping path on whose basis
the distance between the vector of synaptic weights of the winner neuron and
discrete time series was calculated. Thus the following information is becoming
available: according to which value of the discrete time series the corresponding
synaptic weight of the neuron has to be adjusted.

The network organization contains two main processes - initial organization
followed by convergence process. The initial organisation takes about 1000 iterations. During this process each network gets an initial organization, the learning
rate parameter decreases from 0.1, but remains above 0.01.

The number of iterations the convergence process takes is at least 500 times
larger than the number of neurons in the network. The main difference from
the initial organization is that during the convergence process the topological
neighbourhood of a winner neuron contains only the closest neighbours or just
the winner neuron.

As soon as the process of self-organising neural network training is finished,
the process of cluster formation begins. Each record d of the training set is passed
to the system. The same principle of data input (On-line/Off-line) is used as in
organizing neural networks. Algorithms for module m_i and winner neuron n_j^*
determination fully coincide with those used in network organization.

In parallel, for each neuron n_j^* these statistics are kept: records with which
value of the key parameter p have got to neuron n_j^* and how many times. Cluster
c_i is a neuron n_j that at least once became the winner neuron during cluster
formation.

Once the last record of the training set is processed, for each cluster $c \in$
$C, C = \{c_1, c_2, \ldots, c_f\}$ a base value of the key parameter p^* is defined which will
be produced by the system as an output for record d that has got to the given
cluster during system testing.

The simplest way to determine the base value of the key parameter p^* for
cluster c_i is to select value $p = p'$ that has the highest frequency according to
the previously collected statistics for values p in cluster c_i.

Situations might occur when a cluster has several possible values p'. As of
today, in the system developed by the authors, if the above situation occurs, the
least by module value p' is assumed as a base value of the key parameter. As
applied to the task of product life cycle phase transition period forecasting, it
means that the value will be chosen that forecasts transition in earlier period as
compared to others. Also variants are possible, when out of several p' the largest
by module value p' is chosen or the value p' closest to the median of distribution
of values p fixed in the cluster.

Step 2. System Testing and Validation. To test the system, the same data input
principle (On-line/Off-line) is used as in organizing neural networks. Two criteria
are employed to evaluate the effectiveness of the system: Mean Absolute Error

- MAE, to evaluate the accuracy of the system and Logical Error to evaluate whether decisions made by the system are logically correct. The Mean Absolute Error (MAE) is calculated using formula (8):

$$MAE = \frac{\sum_{i=1}^{k} |p_i - r|}{k} \quad i = [1, 2, \ldots, k] , \tag{8}$$

where k - the number of records used for testing; p_i - real value of the key parameter for record d_i; r - the value of the key parameter forecasted by the system.

Logical error provides information about the logical potential of the system. To calculate the logical error, it is necessary to define logically correct and logically incorrect decisions. As applied to the task of forecasting product life cycle phase transition period, logically correct and logically incorrect decisions could be described as follows:

1. Assume that discrete time series d, entering the system has a duration equal to l_d but the value of the key parameter - the period of product life cycle phase transition, is $p = p_d$, where $p_d > l_d$. This statement means that a real time of transition between the phases of the product life cycle has not come yet. Accordingly, logically correct decision is to forecast transition period r_d, where $r_d > l_d$. Logically incorrect decision in this case will be if $r_d \leq l_d$.
2. Assume that discrete time series d, entering the system has a duration equal to l_d but the value of the key parameter - the period of product life cycle phase transition, is $p = p_d$, where $p_d \leq l_d$. This statement gives evidence that real transition moment has already come. Accordingly, logically correct decision could be forecasting transition period r_d, where $r_d \leq l_d$. In its turn, logically incorrect decision will take place if $r_d > l_d$.

The statement that at $r_d = l_d$ transition has occurred can be considered correct as the availability of data about some period in record d shows that the period is logically completed and, consequently, the transition - if any was assumed in this period - is also completed.

Step 3. System Application. Application of the system means not only using a model, but also monitoring the performance of the system and collecting the new data. As the new data is collected, the steps of the main algorithm should be repeated (see Figure 3), and the system (neural modules) must be reorganized.

4   Experiments and Gathered Results

The fact that the data describes real life process and marks of transitions were putted by experts implies that some noisiness in data is present.

The obtained dataset contains 312 real product demand time series with minimal duration equal to 4 and maximal - to 24 periods. Each time series contains the demand during the introduction phase of a specific product plus one period of the maturity phase, and is marked with *M1* marker. Table 1 presents an example of data after the normalization process is finished. The first 11 periods are given. To normalize the data, the Z-score with standard deviation normalization method was applied. As the true bounds of the demand data in the dataset are unknown and the difference between values of various time series is high, the chosen normalization method is one of the most suitable ones.

Table 1. Example of normalized data (first 11 periods)

ID	*M1*	P1	P2	P3	P4	P5	P6	P7	P8	P9	P10	P11
1	3	-4.620	0.060	-0.127	0.247							
2	8	-0.493	-0.549	-0.361	-0.607	-0.623	-0.596	1.592	1.637	-2.817		
3	23	-1.790	-1.664	-1.352	-1.070	-0.893	-0.807	-0.886	-0.547	-0.683	-0.149	-0.133
4	17	-0.969	-0.262	-0.800	0.044	-0.545	-0.169	-0.491	-0.329	-1.078	-0.188	-1.147
5	8	1.625	1.582	1.512	1.872	1.723	1.785	-0.623	-0.581	-0.403		
6	23	0.454	-0.092	-0.492	1.009	-2.080	-0.741	0.908	-2.454	-0.549	0.870	0.309

Figure 5 displays all periods for the same time series as presented in Table 1. As can be seen, the time series differs not only in duration, but also in amplitude and its pattern.

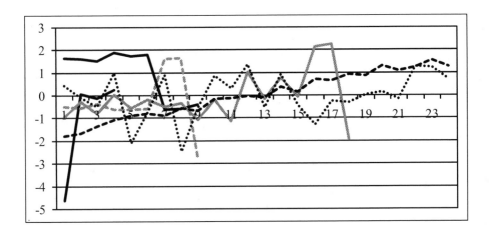

Fig. 5. Example of normalized data

The main target of the performed experiments was to analyse comparatively the precision of the system with different neural network topologies applied while using different network load q. The system was tested by sequentially applying each of three mentioned neural network topologies (see Figure 4) while network load q was changing incrementally from one to five. To calculate the system errors - Mean Absolute Error (MAE) and Logical Error (LE), a 10-fold cross validation method was applied, totally giving 150 system runs.

Table 2 contains the number of neurons in a neural network for each of three topologies, as also supplies the number of training and convergence cycles.

Table 2. Network and learning parameters

Topology	Neurons	Training cycles	Convergence cycles
a	25	1000	12500
b	37	1000	18500
c	25	1000	12500

The learning parameters, used for network organisation in each run, are given in Table 3. For each learning parameter the starting value and the minimal (last) value are supplied, as also the type of a function used for managing the parameter decline process.

Table 3. Learning parameters

Parameter	Starts with	Ends with	Function
Learning coefficient - η	0.9	0.01	Exponential
σ for Gaussian neighbourhood	0.5	0.01	Exponential

While testing the system in Online mode, for each neural network topology and for each of five defined values of q a Mean Absolute Error and a Logical Error were obtained. The gathered results are accumulated in Table 4 and Table 5, and graphically displayed in Figures 6 and 7 respectively.

Table 4. On-line Mean Absolute Error - MAE

Topology	$q = 1$	$q = 2$	$q = 3$	$q = 4$	$q = 5$
a	2.917	2.950	2.940	3.186	3.042
b	3.061	2.912	3.151	2.989	3.152
c	3.066	3.148	3.166	3.267	3.102

The obtained results show that created system with certain precision is able to predict transition points for new products, using a model, built on a historical data. In On-line mode System was able to make a logically correct (refer to "System Testing and Validation" step in subsection 3.2) decision in at least 82.7% and at most in 86.5% of times. Thus the Mean Absolute Error lies close to three periods. Such an error may be unacceptable for a system, a forecast of

Table 5. On-line Logical Error - LE

Topology	$q = 1$	$q = 2$	$q = 3$	$q = 4$	$q = 5$
a	16.1%	13.5%	13.9%	15.3%	14.2%
b	17.3%	14.7%	15.3%	16.2%	15.4%
c	16.6%	15.4%	15.6%	16.0%	14.0%

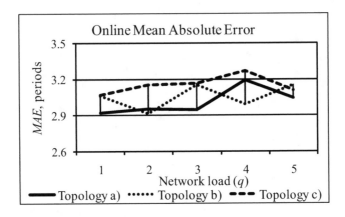

Fig. 6. On-line Mean Absolute Error

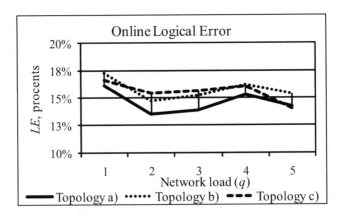

Fig. 7. On-line Logical Error

which will be just accepted without any further processing. Again mentioning the specificity of the dataset obtained and also bringing the point that the final decision is made by a decision making person, it is possible to conclude that the created system can be used as a data mining tool to gain an additional knowledge for solving a product life cycle phase transition points forecasting task, as well as other tasks, connected with forecasting a value of a target parameter for a time dependent variable.

5   Conclusions

For the practitioners of management of the product life cycle the knowledge, which describes in which phase the product currently is and when the transition between phases will occur, is topical. Such knowledge, in particular, helps to select between the cyclic and non-cyclic policy of planning supply chain operation.

In this paper, the task of forecasting the transition points between different phases of product life cycle is stated, and the structure of data mining system, which helps to solve this task, is shown. On the basis of the analysis of historical demand data for products it is possible to learn the modular neural network based system, which will be able to forecast the transition points in life cycle of new products. Experimentally gathered results show that the created system has its potential and can process real demand data, returning a forecast with a certain precision.

One aspect is that in the future it is necessary to examine the developed system on the data from different production fields, and, which is also important, to have a response from practitioners of supply chain management who will use these systems. Also relevant is to consider and analyse the possibility of using other Data Mining techniques in in the place of the Self-Organising maps.

Another aspect, modest data volume that was used for practical experiments, is related to the fact, that it is necessary to have transition marks in historical data from experts and practitioners. The more products, the more complicated for human to make all these marks - in practice the amount of marked data will always be restricted. As a result, possible direction of future research is treatment of forecasting the transition points in the context of a semi-supervised learning [15]. In this case, there is a small set with marked transitions and also a large dataset in which transitions are not marked. In such a situation it is necessary to create a model, which will be able to apply the knowledge, gathered on the small set of marked data, to the new (test) data.

Acknowledgements

The present research was partly supported by the ECLIPS project of the European Commission "Extended collaborative integrated life cycle supply chain planning system", and also supported by ESF Grant 2004/0002/VPD1/ESF/ PIAA/04/NP/3.2.3.1/0001/0002/0007.

References

1. Agarwal, P., Skupin, A. (eds.): Self-Organising Maps: Application in Geographic Information Science. John Wiley & Sons Ltd., Chichester (2008)
2. Campbell, G.M., Mabert, V.A.: Cyclical schedules for capacitated lot sizing with dynamic demands. Management Science 37(4), 409–427 (1991)

3. Dunham, M.: Data Mining Introductory and Advanced Topics. Prentice-Hall, Englewood Cliffs (2003)
4. Han, J., Kamber, M.: Data Mining: Concepts and Techniques, 2nd edn. Morgan Kaufman, San Francisco (2006)
5. Hand, D.J., Mannila, H., Smyth, P.: Principles of Data Mining. MIT Press, Cambridge (2001)
6. Haykin, S.: Neural Networks, 2nd edn. Prentice-Hall, Englewood Cliffs (1999)
7. Keogh, E., Pazzani, M.: Derivative dynamic time warping. In: Proceedings of the First SIAM International Conference on Data Mining, Chicago, USA (2001)
8. Kohonen, T.: Self-Organizing Maps, 3rd edn. Springer, Heidelberg (2001)
9. Kotler, P., Armstrong, G.: Principles of Marketing, 11th edn. Prentice-Hall, Englewood Cliffs (2006)
10. Merkuryev, Y., Merkuryeva, G., Desmet, B., Jacquet-Lagreze, E.: Integrating analytical and simulation techniques in multi-echelon cyclic planning. In: Proceedings of the First Asia International Conference on Modelling and Simulation, pp. 460–464. IEEE Computer Society Press, Los Alamitos (2007)
11. Obermayer, K., Sejnowski, T. (eds.): Self-Organising Map Formation. MIT Press, Cambridge (2001)
12. Parshutin, S., Kuleshova, G.: Time warping techniques in clustering time series. In: Proceedings of 14th International Conference on Soft Computing MENDEL 2008, pp. 175–180. Brno University of Technology (2008)
13. Pyle, D.: Data Preparation for Data Mining. Morgan Kaufmann Publishers, an imprint of Elsevier (1999)
14. Tan, P.-N., Steinbach, M., Kumar, V.: Introduction to Data Mining. Pearson Education, London (2006)
15. Zhu, X.: Semi-supervised learning literature survey. Technical Report 1530, Department of Computer Sciences, University of Wisconsin (2008)

Visualizing the Competitive Structure of Online Auctions

Stephen France[1] and Douglas Carroll[2]

[1] Lubar School of Business, UW – Milwaukee, 3202 N. Maryland Avenue., Milwaukee, Wisconsin, 53201-0742
[2] Rutgers University, Graduate School of Management, Newark, New Jersey, 07102-3027
france@uwm.edu, dcarroll@rci.rutgers.edu

Abstract. Visualizations of product competition are common in marketing research. Competitive product relationships can be modeled using data from a variety of sources, including questionnaires, surveys and brand switching data. Product competition applications based on brand switching data are usually restricted to high volume, frequent purchase products such as coffee and frozen foods. Analysis of competitive product structure requires data for multiple purchases from a single consumer, data that are not usually available for large value, rare purchase items such as cars and computers. We use bid information from online auctions as a source of competitive product structure information for these items. We develop a simple algorithm for creating a distance matrix representing market structure between brands and brand features from online auction data. We take data from eBay mobile phone auctions in the USA and based upon the auction data develop visualizations of product competition for brands and brand features.

Keywords: Auctions, Bidding, MDS, Marketing, Visualization.

1 Introduction

Visualizations of product competition can be created from a variety of sources, including perceptual data, behavioral data, and product attributes. Perceptual data is typically gathered using questionnaires or a user panel. Information can be elicited and recorded in different ways, including direct preferences between brands, comparing brands on different attributes, and using a Likert scale questionnaire for measuring latent attributes. Behavioral data is typically gathered from Point Of Sale (POS) purchase data, from which one can derive brand loyalty and brand switching data.

Obtaining behavioral data or perceptual data is expensive in terms of time and money. A retailer may have data from POS terminals available and have some method (e.g., a loyalty card scheme) of tracking customer purchases. A manufacturer may not have available purchase information across competing brands, and may have to purchase such information from a data provider such as AcNielsen, who gathers data from multiple retailers. The use of behavioral data to determine product competition is usually limited to high volume, frequent purchase items; examples used in the academic literature include coffee and frozen foods. Given a high value, rare purchase

P. Perner (Ed.): ICDM 2009, LNAI 5633, pp. 103–116, 2009.

item such as a car, it is not feasible to use purchase information to model product competition. Years' worth of historical purchase information would be required to get information on multiple purchases for a single consumer, and a consumer's purchase intensions and behavior may have changed drastically in the intervening period between purchases. Obtaining perceptual data, usually via surveys of consumer panels, requires great effort and expense, and only a small number of consumers can participate, possibly leading to a small, unrepresentative sample of consumers.

Research on product competition typically concentrates on the sale of new products, i.e., the primary market. For many products, typically high value, infrequent purchase items, there is an important resale or secondary market. Online auctions provide an important part of the secondary market.

In this paper we propose a method of using online auction data to determine product competition. We concentrate on the visualization of product competition; we also briefly discuss how clustering techniques can be used to segment products. By using freely available auction data the cost of analyzing product competition is much lower than when perceptual or behavioral data are used. As previously described, it is difficult to analyze primary data for infrequent purchase items. Even if primary data were available (e.g., new product purchase data or a questionnaire on product preferences), the secondary data provides valuable extra information. In this paper we concentrate on auctions processed through one particular website, that of eBay.com. eBay.com has the largest auction volume of any online auction website. We choose a particular high volume product class, that of mobile phones, on which to demonstrate the analysis of product competition.

2 Literature

Behavioral product competition/market structure analysis is the study of the structure of market information from behavioral data, usually customer purchases. Two common representations of behavioral data are brand switching matrices and market share matrices. Brand switching data is usually derived from supermarket POS data. To create a brand switching matrix, the sales data are split into discrete purchase occasions and a brand switching matrix $\mathbf{S}$ is calculated, where s_{ij} is the probability of purchasing brand j in the next period given that brand i was purchased in the current period. The matrix $\mathbf{S}$ is a stochastic probability matrix, with entries in each row adding up to one. [3] describes brand switching behavior in such a matrix, assuming a heterogeneous zero order of population of consumers. This work is extended by [12] and [13], which introduces the idea of loyal segments and brand switching segments and develops a latent class procedure to model these segments. Market Share Analysis[6][7] uses the idea of a market share matrix, which can be calculated for any set of purchase data. The market share matrix is decomposed using an overlapping clustering methodology.

We propose using online auction data as a source of behavioral market structure information for products, particularly infrequent purchase consumer items such as mobile phones, computers, and TVs. These data can then be used to produce visualizations showing product competition. We concentrate in particular on eBay auctions. In an eBay auction, one or multiple products are sold as a lot. An item is put up for auction for a certain time period; the auction closes at the end of that time period. Bidders do not bid an actual price as in a standard English auction, but bid the amount that they are willing to pay for an item. Given an existing user with the current high bid p and willingness to pay price q, and a new bid by a different user with a willing to pay price r at least one price increment above p, then there are two possible price scenarios. If r is less than or equal to q then the existing user is still the highest bidder and if r is greater than q then the new bidder becomes the highest bidder. In both cases the bid price is raised to one price increment above r. [2] describes the mechanism design for eBay auctions in great detail and build a structural model of eBay auctions that helps describe concepts such as the winners curse and shilling.

There have been various attempts to use exploratory statistical techniques to help explore auction data sets. In [15][21][22], a series of exploratory data analysis (EDA) tools are developed for exploring auction data. Further EDA work [16] uses Functional data analysis (FDA) to model auction prices over time. In modeling the effect of concurrent events on auction output, [17] introduces the idea of spatial distances between products sold in eBay auctions. Distances are defined between products, based upon the differences in features between the products. A Hamming[14] distance or matching distance between each pair of products is calculated for the discrete features and a Euclidean distance is calculated for the continuous features. The distance between any two products is taken as the weighted sum of the Hamming and Euclidean distances between the products.

In this paper, we build upon the EDA methods developed in the previous described papers, but we concentrate on finding an interpretation of product competition from the bidding data. As per [17], we calculate a set of pairwise distances between products, but we create an algorithm for using bidding information to derive distances between the products, with closely competing products having smaller relative distances between them than pairs of products not competing closely. From these distances we create spatial visualizations of product competition.

3   Methodology

Market structure analysis is a method of association. The associations between purchases are used to build the structure. This structure is developed from buying patterns, e.g., a brand switching matrix measures the aggregate probabilities of shoppers choosing a certain brand, given that they purchased a different brand in a previous period.

For our analysis of auction data, we explore the idea of bidding patterns for brands. A similarity relation is defined between each pair of brands. The relation is symmetric and same-brand similarity is calculated, so if there are brands $1, \cdots, k$ then there are $k \times (k + 1) / 2$ relations. $\mathbf{S}$ is defined as a $k \times (k + 1) / 2$ lower triangular similarity matrix, where s_{ij} is the similarity relation between brand i and brand j. If a consumer bids on two items then these items are both in his or her consideration set. We consider this to be a measure of similarity between the items. This is analogous to the idea of similarity used when analyzing brand switching matrices. Rather than similarity between two brands being defined by a consumer purchasing these brands in subsequent periods, similarity is defined by a consumer bidding on these different brands within a certain time period. If a consumer bids on brand i and brand j then 1 point of similarity is added to s_{ij}. If the consumer bids for more than two items, then for each pair of items in the bidding consideration set a point of similarity is added.

One problem with this approach is that bidders may bid for different items during the considered time period. For example, a consumer may bid for an item for his/her self at the beginning of the month and bid for someone else towards the end of the month. EBay has a number of commercial bidders, who may bid for multiple items at any one time. Such situations lead to "noise" in the distances derived for competitive product structure. In fact, during initial experimentation, this "noise" led to visualizations with null structure[4].

It is not possible to completely eliminate this "noise" but it can be reduced in two possible ways. Firstly, bidders who win large numbers (say $N > 10$) of items are discounted in order to minimize the effect of commercial bidders. Secondly, a weighting scheme is defined based upon the time between bids. A consumer's period of interest for a brand is defined to be the time period between the consumer's first bid for a product and either the time at which the consumer's last bid is outbid (in the case of the consumer not winning the auction) or the end of the auction (in the case of the consumer winning the auction). A consumer's time difference between products i and j is defined as TD_{ij}, where TD_{ij} is 0 if the consumer's time windows for products overlap and is the difference between the end of the first time window and the start of the second time window if the time windows do not overlap. A weight is assigned, based upon a monotone decreasing function of TD_{ij}.

The algorithm is outlined below:

1. Initialize $s_{ij} = 0$ and $w_{ij} = 0$ for $i = 1, \cdots, k$ and $j \leq i$.
2. Select a consumer who has bid on at least two items. If a consumer has bid on only one item then no meaningful product similarity information can be obtained.
3. For each pair of items in that consumer's bidding consideration set, record the brands i and j, and add 1 to s_{ij}.
4. Set $w_{ij} = w_{ij} + F(TD_{ij})$ where $w_{ij} \geq 0$ and if $a > b$ then $F(a) \leq F(b)$.
5. Repeat steps 2 to 4 until all consumers who have bid on at least 2 items in the bidding period have been processed.

There is a major problem in using the raw similarity matrix calculated in steps 2 to 3 in that the similarity is biased towards brands that have large numbers of bidders. To solve this we calculate the expected similarity given independent heterogeneous probabilities of bidding for different brands (i.e., there is no association pattern).

$$E\left(s_{ij}\right) = \frac{N_i N_j}{N_T} \quad \forall i > j \tag{1}$$

Where N_i is the number of bids for brand i, N_j is the number of bids for brand j, and N_T is the number of bids for all brands. A normalized similarity measure is defined in (2).

$$ns_{ij} = \frac{s_{ij}}{E\left(s_{ij}\right)} = \frac{s_{ij} N_T}{N_i N_j} \tag{2}$$

The weighted normalized similarity measure, given in (3), is calculated by multi-plying s_{ij} by the average weight value. As 1 is added to s_{ij} for every weight added to w_{ij}, dividing w_{ij} by s_{ij} gives the average weight.

$$ws_{ij} = ns_{ij} \times \frac{w_{ij}}{s_{ij}} = \frac{w_{ij}N_T}{N_i N_j} \tag{3}$$

To calculate dissimilarities between brands i and j for $i < j$, the similarities are sub-tracted from the largest non-diagonal similarity.

$$d_{ij} = \max_{i \neq j} \left(ws_{ij} \right) - ws_{ij} \tag{4}$$

One can ensure that the values defined by (4) are distances by adding the following positive additive constant to each value of d_{ij}. This ensures that the triangle inequality holds for all possible sets of three points.

$$k = \max \left\{ \max_{i,j,k} \left\{ d_{ik} - d_{ij} - d_{jk} \right\}, 0 \right\} \tag{5}$$

The resulting distance matrix can be used as input to a multidimensional scaling procedure, or any other procedure that requires a distance matrix (e.g., clustering, latent class analysis etc).

Our methodology gives great flexibility in that one can choose the granularity for the product. For example, the product can be defined as an overarching brand name (e.g., Motorola or Samsung) or the product can be defined at the individual model level. If the level of aggregation is too high then there is a risk of missing out on im-portant information; for example, a brand name company could have products in many different segments. If the level of aggregation is too low then there may not be sufficient information to accurately estimate distances between individual items and visual representations of structure may be hard to interpret.

4 Experimentation and Results

We experimented with data from a month's worth of mobile phone auctions in eBay. We took all mobile phone auctions in the USA which were for the phone only; i.e., no contract or service agreement was included. In order to collect the data we developed a web-bot application using Excel and Visual Basic for Applications (VBA). For each auction, we saved both the details of the auction and the details of all bids for the auc-tion. For a single auction, the information we collected included the auction end date, the closing price, the camera size, the brand, the model, the geographical location and the various features of the mobile phones. For bids, the information we collected in-cluded the user name, the bid level, and the bid time.

In total we collected data for approximately 180,000 auctions, with approximately 1,840,000 bids for these auctions. Not all of the auctions and bids were useable. For example, some data were left incomplete by auction owners and some of the free form entered data were inconsistent and hard to interpret. EBay hides the user names for all bids over $200, so in our analyses we were unable to capture the high end of the mobile phone market. We performed some data cleaning procedures (e.g., matching text values close to one another) in the brand field in order to maximize the number of auctions from which we could get useable information. Users range from those with only one bid, to one user who had made nearly 3000 bids over the monthly period.

In order to develop representations of product competition, the concept of a brand needs to be defined. The brand name, i.e., Motorola, Samsung, LG, etc, can be used, but there is a large degree of heterogeneity within a single brand defined by a brand name. For example, a manufacturer may have a luxury highly featured mobile phone, and also have a value mobile phone, with very few features.

Each mobile phone has a model, and there are many models for each brand name. However, there were a large number of models recorded and the data quality was poor; in fact it was obvious that many sellers either did not know or were confused about the model number for their phone.

To account for within brand heterogeneity, rather than using models, we used a combination of brand names and features to define brand granularity; i.e., each brand defined by a brand name has a set of sub brands, each with different feature sets. The features in eBay auctions are selected from a fixed set of possible features, rather than being entered free form; this results in good data quality. Some features have nominal values and either exist or do not exist. Other features can be ordinal; e.g., the newness feature is split into new/unopened, new, refurbished, and used, which are discrete categories, but are ordered. Splitting brands using a nominal category doubles the number of brands in the analysis. For example, starting with 10 brands and splitting the brands by two different features gives 40 brands.

For our example analysis we examined market structure for all brand names with over 100 auctions. There were 18 brand names. We selected three features on which to split the brands. These were Camera, Bluetooth, and MP3 player. We picked these three features as they give a good spread of auctions with different combinations of these features. Using these features gives $18 \times 2 \times 2 \times 2 = 144$ different brands. We ran the algorithm described in the methodology section and thus created a (144×143) / 2 distance matrix of relative structure between the brands. We used the weighting scheme in (6), which experimentation showed gave good visualizations, avoiding the problem of null structure.

$$F\left(TD_{ij}\right) = \begin{cases} 1 & \text{for } TD_{ij} \leq 10 \\ \dfrac{10-\left(TD_{ij}-10\right)}{10} & \text{for } 10 < TD_{ij} \leq 20 \\ 0 & \text{for } TD_{ij} > 20 \end{cases} \tag{6}$$

We visualized the structure of the auctions using the technique of distance based metric multidimensional scaling (DMMDS). The distances generated by our algorithm use count data and thus are essentially metric. Past empirical work [11] has shown that DMMDS produces strong visualizations of metric data, with stronger distance recovery and neighborhood agreement than for visualizations produced by singular value decomposition (SVD) based methods such as classical MDS[24],[25].

DMMDS (e.g., [18],[19]) performs a gradient based optimization procedure in order to minimize the STRESS function given in (7).

$$STRESS = \sqrt{\frac{\sum_i \sum_j \left(d_{ij} - \hat{d}_{ij}\right)^2}{\sum_i \sum_j d_{ij}^2}} \qquad (7)$$

Where $\hat{d}$ is the source distance/dissimilarity value between i and j, and d_{ij} is the Euclidean distance between i and j for the derived solution, which is of a pre-chosen dimensionality. With DMMDS the distances can be transformed with a scaling factor and an additive constant, leading to a linear regression based optimization.

Initial experimentation revealed cases where there were some distances based on a large number of bids and some distances based on few or no bids (typically between niche products catering to very different niches). In order to minimize "noise" due to lack of information for some distances, we used the weighted version of STRESS given in (8), weighting on the amount of information that we have available. The concept of weight here is completely separate from the weights used to model the time differences between bidding windows. The expected number of bids, based upon marginal probabilities, is used as a proxy for the amount of available bidding information.

$$STRESS = \sqrt{\frac{\sum_i \sum_j \left(v_{ij} \left(d_{ij} - \hat{d}_{ij}\right)^2 \right)}{\sum_i \sum_j \left(v_{ij} \cdot d_{ij}^2 \right)}} \quad \text{where } v_{ij} = E\left(S_{ij}\right) = \frac{N_i N_j}{N_T} \qquad (8)$$

The STRESS function is non-convex and thus any solution is not guaranteed to converge to a globally optimal solution. We produced solutions in one, two, three, and four dimensions. We used the recommendation of Arabie[1] of running multiple

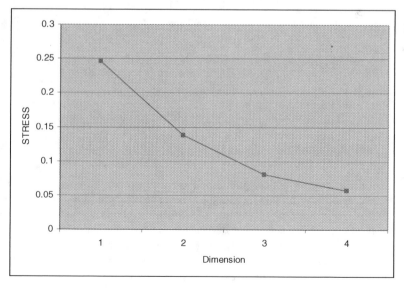

Fig. 1. STRESS values for optimal solution

optimizations in order to have a good chance of obtaining the optimal solution; we performed 100 runs for each dimensionality. For each dimensionality, we selected the run that produced the lowest value of STRESS. Fig. 1 gives the plot of STRESS values on the abscissa against solution dimensionality on the ordinate. With DMMDS, one can obtain the approximate dimensionality of the data by finding the "elbow" of the curve on the graph of STRESS. From Fig. 1, the dimensionality of the data is approximately two or three.

We used the GGobi[4] package to produce visualizations from the two and three dimensional DMMDS solutions. Fig. 2 and Fig. 3 both show visualizations of the optimal two dimensional DMMDS solution. In Fig. 2, the features of a product are shown as "F=ijk", where i, j, and k are binary variables indicating a 1 if a product has a feature and a 0 if the product does not have a feature. The features shown are "i=Has BlueTooth", "j=Has Camera", and "k=Has MP3".

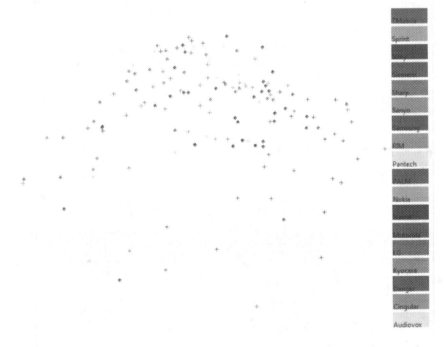

Fig. 2. 2D solution colored by brand: Dots for first set of colors and crosses for second set

One can see from the visualizations that there is greater homogeneity in brand names than in feature sets; i.e., a consumer is more liable to bid for a phone with the same brand name and a different feature set than for a phone with a different brand name and the same feature set. Fig. 4 shows a competitive subset of the three leading brands (brands with the highest number of bids). One can see that LG and Samsung phones compete quite closely with one another, while Motorola phones do not compete strongly with either of the other two leading brands.

Fig. 3. 2D solution by colored by features: Dots for first set of colors and crosses for second set

Fig. 4. Three major brands

Other visualizations are possible. Fig. 5 shows a visualization for the optimal three dimensional solution. Each combination of two dimensions is plotted separately, with a different combination of color and glyph used for each brand. The key is the same as that for Fig. 4. Any one of the sub-visualizations could be expanded in order to help further understand product competition. Profile plots can be used to further explore the structure of the solution and relate features to the dimensions of the visualization. Fig. 6 shows a profile plot of the two dimensional solution, mapping "Brand name" against each of the two dimensions. Given the visualizations in Fig. 2 and Fig. 3, we infer that the "Brand Name" is more important than the brand features with respect to product position (i.e., items with the same brand name are relatively homogeneous with respect to their position in the visualization). This homogeneity results in tightly clustered lines from the brand to dimensions, rather than the more random pattern that would be generated by using brand features in the profile plot. For example, one can see that Sony phones are clustered approximately one third of the way down from the top of both dimension 1 and dimension 2.

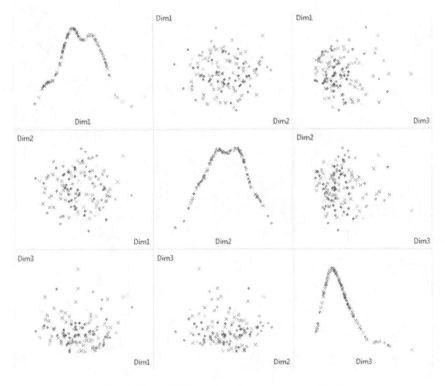

Fig. 5. Plots of 3D solution: Color scheme as Fig. 2

We are not restricted to MDS based visualization techniques for exploring competitive structure. Our algorithm produces a standard distance matrix and this distance matrix can be used as input to a wide variety of statistical and psychometric techniques.

For example, hierarchical clustering can be used to give a tree structure representation of brands. A partitioning clustering technique such as *k*-means clustering could be used to segment the brands. An overlapping clustering methodology such as INDCLUS[5] or extensions[6][7], could be used to decompose the normed similarity matrix and assign brands to latent clusters, giving market shares for the latent classes of brands.

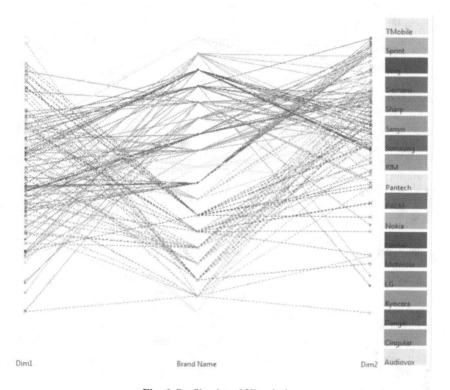

Fig. 6. Profile plot of 2D solution

5 Discussion

In this paper we have introduced a technique for using online auction data as a resource for analyzing brand competition/market structure. We have developed an algorithm for creating a measure of similarity between brands, based upon users who bid for items across different brands. This measure of similarity is normalized using the expected similarity given an independent vector of choice probabilities for the expected number of bids. The normalized measure of similarity can be converted to dissimilarities, and by ensuring that the data conform to the metric axioms, can be converted to distances.

We created a test data set containing a month's worth of mobile phone auctions from eBay. We created a market structure distance matrix for 18 brands and 144 sub-brands, based upon the values for three different features. We created visualizations of market structure from the distance matrix using multidimensional scaling (MDS).

Traditionally, visualization of competition between brands has used brand switching data, usually derived from supermarket POS data. More recently, techniques have been developed that use the idea of market share and the decomposition of a market share matrix. We consider the use of auction data for obtaining market structure to be complementary to these approaches. The knowledge of market structure for the secondary auction market can be used to complement the knowledge of market structure for the primary market. This secondary market structure is particularly useful for infrequent purchase items, where it is difficult to derive primary market structure from purchase data.

Auction structure analysis could be used in conjunction with perceptual mapping techniques. As it is expensive to convene panels and conduct surveys, a product manufacturer could use auction structure analysis to find a subset of brands that compete closely with its own, and then use perceptual market structure analysis and perceptual mapping techniques on this subset of brands. A good overview of perceptual market structure analysis is given in [8].

Our measure of similarity/dissimilarity uses a simple heuristic for association between bids. We make some basic assumptions in order to simplify the heuristic. We assume that the bidding structure is symmetric. That is if someone bids on phone A and then bids on phone B then this is equivalent to someone bidding on phone B and then bidding on phone A. We could remove this assumption and allow for an asymmetric similarity/dissimilarity matrix. This will give additional structural information, but many techniques for analyzing market structure rely on symmetric input, restricting the array of techniques that can be used. Standard distance scaling assumes a symmetric matrix; asymmetric versions of MDS have been created, for example, [9], but useable software is difficult to obtain. Still, for future work we should experiment with an asymmetric version of our algorithm, and test the degree of the asymmetry produced.

In selecting brands to be visualized, we used the rather ad-hoc approach of selecting brand names with more than 100 auctions. Some sort of analysis of the stability of the similarities, relative to the number of bids for a pair of brands, would be very useful. Similarities/dissimilarities could be calculated over different time periods, and one could infer how much of the change in competitive structure is due to an actual change in structure and how much is due to measurement error, due to insufficient data.

We experimented with a data set with a month's worth of mobile phone auctions on eBay. We built a web-bot and then wrote scripts to clean and restructure the data. The resulting data set is much larger and richer than most data sets used for eBay related empirical research. However, we only experimented on one product class. We could attempt to explore more than one product class and also to generalize the idea of "features" across different product classes. EBay has begun to realize the commercial potential from auction day and gives/sells access to auction data and user demographics via a web services programmable interface. In particular, the addition of demographic data allows the building of more complex models for segmentation and latent class analysis. Analyzing the user bid information using the eBay supplied web services should hopefully remove the $200 limit on identifying user bids, thus allowing the analysis of higher value product categories.

References

1. Arabie, P.: Concerning Monte Carlo Evaluations of Nonmetric Multidimensional Scaling Algorithms. Psychometrika 38, 607–608 (1973)
2. Bajari, P., Hortacsu, A.: The Winner's Curse, Reserve Prices, and Endogenous Entry: Empirical Insights from eBay Auctions. The RAND Journal of Economics 34, 329–355 (2003)
3. Bass, F.M.: The Theory of Stochastic Preference and Brand Switching. Journal of Marketing Research 11, 1–20 (1974)
4. Buja, A., Logan, B.F., Reeds, J.A., Shepp, L.A.: Inequalities and Positive-Definite Functions Arising from a Problem in Multidimensional Scaling. The Annals of Statistics 22, 406–438 (1994)
5. Carroll, J.D., Arabie, P.: INDCLUS: An Individual Differences Generalization of the ADCLUS Model and the MAPCLUS Algorithm. Psychometrika 48, 157–169 (1983)
6. Chaturvedi, A., Carroll, J.D.: Deriving Market Structures Via Additive Decomposition of Market Shares (Application of Three-Way Generalized SINDCLUS). In: DIMACS Workshop on Algorithms for Multidimensional Scaling (2001)
7. Chaturvedi, A., Carroll, J.D., France, S.L.: A Methodology for Deriving Market Structure Via Additive Decomposition of Market Shares. Working Paper, Rutgers Business School (2007)
8. Cooper, L.G., Inoue, A.: Building Market Structures from Consumer Preferences. Journal of Marketing Research 33, 293–306 (1996)
9. DeSarbo, W.S., Manrai, A.K.: A New Multidimensional Scaling Methodology for the Analysis of Asymmetric Proximity Data in Marketing Research. Marketing Science 11, 1–20 (1992)
10. Elrod, T., Russell, G.J., Shocker, A.D., Andrews, R.L., Bacon, L., Bayus, B.L., Carroll, J.D., Johnson, R.M., Kamakura, W.A., Lenk, P., Mazanec, J.A., Rao, V.R., Shankar, V.: Inferring Market Structure from Customer Response to Competing and Complementary Products. Marketing Letters 13, 219–230 (2002)
11. France, S.L., Carroll, J.D.: Development of an Agreement Metric Based Upon the RAND Index for the Evaluation of Dimensionality Reduction Techniques, with Applications to Mapping Customer Data. In: Perner, P. (ed.) MLDM 2007. LNCS, vol. 4571, pp. 499–517. Springer, Heidelberg (2007)
12. Grover, R., Srinivasan, V.: A Simultanous Approach to Market Segmentation and Market Structure. Journal of Marketing Research 24, 139–153 (1987)
13. Grover, R., Srinivasan, V.: An Approach for Tracking within-Segment Shifts in Market Shares. Journal of Marketing Research 26, 230–236 (1989)
14. Hamming, R.W.: Error Detecting and Error Correcting Codes. Bell System Technical Journal 26, 147–160 (1950)
15. Hyde, V., Jank, W., Shmueli, G.: Investigating Concurrency in Online Auctions through Visualization. The American Statistician 60, 241–250 (2006)
16. Jank, W., Shmueli, G.: Modelling Concurrency of Events in On-Line Auctions Via Spatio-temporal Semiparametric Models. Journal of the Royal Statistical Society: Series C 56, 1–27 (2007)
17. Jank, W., Shmueli, G.: Functional Data Analysis in Electronic Commerce Research. Statistical Science 21, 155–166 (2006)
18. Kruskal, J.B.: Multidimensional Scaling for Optimizing a Goodness of Fit Metric to a Nonmetric Hypothesis. Psychometrika 29, 1–27 (1964)

19. Kruskal, J.B.: Nonmetric Multidimensional Scaling: A Numerical Method. Psychometrika 29, 115–129 (1964)
20. Macqueen, J.B.: Some Methods for Classification and Analysis of Multivariate Observations. In: Le Cam, L.M., Neyman, J. (eds.) Fifth Berkeley Symposium on Mathematical Statistics and Probability. Statistics, vol. 1, pp. 281–297. University of California Press, Berkeley (1967)
21. Shmueli, G., Jank, W.: Visualizing Online Auctions. Journal of Computational & Graphical Statistics 14, 299–319 (2005)
22. Shmueli, G., Jank, W., Aris, A., Plaisant, C., Shneiderman, B.: Exploring Auction Databases through Interactive Visualization. Decision Support Systems 42, 1521–1538 (2006)
23. Swayne, D.F., Lang, D.T., Buja, A., Cook, D.: GGobi: Evolving from XGobi into an Extensible Framework for Interactive Data Visualization. Computational Statistics & Data Analysis 43, 423–444 (2003)
24. Torgerson, W.S.: Theory and methods of scaling. Wiley, New York (1958)
25. Torgerson, W.S.: Multidimensional Scaling, I: Theory and Method. Psychometrika 17, 401–419 (1952)

Credit Risk Handling in Telecommunication Sector

Monika Szczerba and Andrzej Ciemski

Warsaw University of Technology, Institute of Computer Science,
Nowowiejska 15/19, 00-665 Warsaw, Poland
M.Szczerba@stud.elka.pw.edu.pl,
A.Ciemski@ii.pw.edu.pl

Abstract. This article presents an application of data mining methods in telecommunication sector. This sector becomes a new area of research for particular problem solving e.g. churn prediction, cross-up selling marketing campaigns, fraud detection, customer segmentation and profiling, data classification, association rules discovery, data clustering, parameter importance analysis etc. Credit risk prediction became a new research domain in pattern recognition area aimed to find the most risky customers. This article is devoted to assessing credit risk from the moment of opening a customer account to the moment of closing an account due to non-payment. Algorithms are used to identify and insolvency of a debtor. Credit scoring is presented in a form of activation models, which are used to predict customers' debt as well as indicate clients with the highest, medium and smallest credit risk. Practical part of the article is based on the real customer database in a telecommunication company.

1 Introduction

The research issue of risk and risk management is a growing scientific domain. The term of risk has been introduced in many areas of social, economic and science life. First of all the term of risk can be found in finance, banking, insurance and medicine [1][2][3]. Some of the authors are trying to build a common risk theory, but there is still the risk theory placed in the specific context e.g. insurance, banking [4][5][6]. There is a need of risk research for business related issues concerning crucial business processes. Especially the risk research has became valid for companies which the core of its business is a provision of services e.g. telecommunication services.

For telecommunication sector especially crucial became the churn analysis and proper using methods of data analysis [7][8]. In the recent years also significant from the point of view of operational processes for telecommunication companies became the credit risk analysis for the individual and business customers in the activation process. Authors of this article notice a possibility to use machine learning and data mining methods in pattern recognition as a new approach

P. Perner (Ed.): ICDM 2009, LNAI 5633, pp. 117–130, 2009.

in order to find the most risky customers. This article presents (based on the authors business practice) the outcomes of the research work and the approach applied in the commercial projects for telecommunication companies.

Technological development within telecommunications area has significantly progressed over recent years. This is connected with the growth in the competitiveness within this sector. Telecommunication companies are trying to get as many customers as they can by offering them a lot of attractive deals and products. Finding new customers becomes more difficult though when the market gets more saturated. Customers who respond to those offers are very valuable, because they generate profit for the company. Unfortunately among them there are some who fail to pay their bills and put the company at risk of making considerable losses. To minimize this risk companies can take precautions by using data mining methods.

Data mining help identify customers who may possibly fail to pay their bills. To measure the customer's credit risk level it is essential to make analysis of the activation data. The results of analysis are fundamental part of the process aiming to prevent the company from increasing bad debt. Among tools used for data analysis we can distinguish predictive models, which are the most important part in analysis process. Predictive models help to identify customers with higher, lower and the lowest credit risk.

This article discusses issues concerning credit risk as well as methods detecting customers who may fail on their payments after the activation process. Each of the chapters will be described below.

Chapter two introduces credit risk. It begins with explanation of the term and how it is usually comprehended, and it ends with description of credit risk types on which predictive models were based. Discussed issues form fundamental knowledge needed to understand problem of credit risk and they are good introduction into next chapters. Chapter three presents description of classification trees as one of the data mining methods used to analyze credit risk problem. Classification trees are very well known method used in commercial projects. Chapter four introduces population of the data which was prepared for credit risk analysis. Chapter five describes two of seven predictive models detecting customers with the credit risk among the population. The first one is activation model predicting credit risk based on all customer population and the second one is predicting credit risk for individual customers. Chapter six and seven present summary of results and future research plans.

2   Credit Risk

"Credit risk" term is used in both everyday and scientific language. It can also be interpreted differently depending on type of economic activity. Usually the term "credit risk" means the risk of unattainability of goals. The meaning credit risk is not the same across different sectors like for example banking, insurance, telecommunication, energy, industrial and public sector. For telecommunication area credit risk means decrease in potential profits, lack of the cash flow and financial difficulties which can lead the company to the bankruptcy.

Telecommunication sector is one of the most competitive markets, because it changes very quickly by bringing new technologies and creating new business models. The more knowledge the company has got about its customers the more attractive deal it can offer and this way gain competitive advantage over its competitors. Credit risk is sometimes defined as possibility that customer will not keep to contact conditions and will create a financial loss for the company. Telecommunication sector is changing and developing very quickly and it is important to find methods to lower financial risk and protect company's profits. Data mining methods became the best solution for this problem.

Telecommunication companies started to make several attempts to control risk management by for example verifying customers' credit history before signing a contact with them and imposing deposit on customers with lower credit reliability. The deposit is a guarantee for company in case if the customer refuses to pay. Risk management activities are called credit scoring which means capability to fulfill financial contact obligations with a company.

The term "credit scoring" defines customer's credit risk. Credit scoring can be divided into application scoring and behavioral scoring. Application scoring is used when customer is entering in a contract with a service provider. This process includes filing in the application form (address, age, sex). Credit scoring process can be also used after signing the contract and is based on behavioral customer data (for example the history of payments). It is called behavioral scoring.

Special treatment of customers with a high credit risk allows the company to minimize its financial losses. It is also very important to have all the necessary information when setting a credit limit for a new customer. Proper identification of the customer and scoring his risk level (high, medium, low) lets the company to lower its credit risk. Classification trees are used here to model credit risk.

3   Decision Trees

Decision Trees are one of the data mining methods used to classify all observations of population into groups.

Decision Trees, also known as classification trees, consist of nodes and edges called branches. The node is called predecessor if a node has got a few branches connected to other nodes called successors. If successor has not got any outgoing branches, the node is known as a final node (leaf). Successor is created as a result of decision rule in a node, which splits observations in two groups and sends them to successors.

Construction of classification trees is based on a sample of data (population). Decision rule divides observations from the sample and assigns them to new nodes. Each node is characterized by a diversity of observations. All observations create population which is also known as a class or group. This population consists of observation vectors described below:

$$x_{11}, x_{12}, ..., x_{1n_1} \text{ from the class (population) 1}$$
$$x_{21}, x_{22}, ..., x_{2n_2} \text{ from the class (population) 2}$$
$$...$$
$$x_{g1}, x_{g2}, ..., x_{gn_g} \text{ from the class (population) g,}$$

where: $x_{ki} = \left(x_{ki}^1, x_{ki}^2, ..., x_{ki}^p\right)$ is i-the observation of k-th class. Values of the observation are coming from the p-dimensional set, where $x_{ki} \in D$. In other words, this set is a sequence of n-arranged random pairs, which can be noted as: $(x_1, y_1), (x_2, y_2), ..., (x_n, y_n)$, where $n = n_1 + n_2 + ... + n_g$, x_i means i-th observation and y_i is a observation class label. Decision rule is based on sample of data which is also known as learning sample. Learning sample consists of g subsample and each of subsamples consists of observations from one class (group).

When observations are already classified in the node, then split criterion is set up for each node. If all observations are classified within the same final node, the node changes to class label and shows how many observation it contains. The main target of classification trees is to predict classification of new observations, which are based on division rules created on the basis of a learning sample.

Each of classification trees consists of subtrees, which are part of a main tree. Subtrees can be defined as "subtree of a T tree is a tree, which is a part of T tree". The main target of split rules is to divide learning sample in two groups in a way that observations assigned in new groups should be the most similar to each other.

There are tree most popular rules used in automatic creation of classification trees: misclassification rate, Gini Index, Entropy reduction and Chi-Square test [9][10].

4   The Data and Variables

Database of a telecommunication company consists of individual and business customers who signed a contract with a company between 1st of January 2007 and 31st of March 2008. During this period of 15 months customers were observed and their payment behaviour examined.

Database includes 53433 observations which make up customers' population. This population consists of 48663 customers who are willing to pay for invoices (good customers) and 4770 customers who had their contracts canceled because they failed to pay (bad customers). Customers were then divided in two groups - individual and business, where the first one includes 35941 observations and the second one 17492 observations. Among the group of individual customers there are 32486 good customers and 3455 bad ones. By analogy, in a business group there are 16177 good payers and 1315 bad payers.

These two groups of customers were also divided based on a type of services they use, that is Internet and telephone. Table 1 presents a size of the population of customers using different type of service. This division will be used later to construct predictive models.

Table 1. The size of population divided with regards to the customer type and service type

Population	Amount	Good payer	Bad payer	Bad payers per
All population	53433	48663	4770	8,93 %
Individual customers	35941	32486	3455	6,46 %
Individual customers - Internet	15595	14841	754	1,41 %
Individual customers - Telephone	20346	17645	2701	5,05 %
Business customers	17492	16177	1315	2,46 %
Business customers - Internet	2631	2491	140	0,26 %
Business customers - Telephone	14861	13686	1175	2,20 %

5   Credit Risk Models

Seven models were constructed to examine the customers' population. Before they are described in detail, their construction principles will be presented.

Principle of model construction was customer's activation data signing a contract with a telecommunication company. Models were called activation models. Models include customers who signed a contract for 12, 24 and 36 months and decided to choose either telephone or internet services or both. Contracts were signed between 01st January 2007 and 31st March 2008. Models were built for individual and business customers, regardless of the number of services on customer's account.

Modeling includes customers who signed a contract from 01st January 2007 and 31st March 2008, who failed to make their payments but their contract was not terminated. Main target of the model was to predict a risk of payment failure on 75th, which is when a customer's account is closed during debt collection process. Every model has got defined a target function called reason of terminating a contract which divides customers' population into good payers (BWSFE value) and bad payers (WSFE value).

Customers who canceled a contract within 10 days from a signing date (based on special offer conditions), or who did not receive parcel with telephone or Internet device from a courier (which means that activation process did not start) or customers who were validated negatively were removed from database, which was the basis of modeling. The contacts, which were overload notes, were also eliminated from database. The next principle is that not all the invoices had payment corresponding with invoice during period 01st January 2007 and 31st March 2008 in spite of invoices having balance status in the database and corresponding balance identifier (which means there is a payment connected to invoice). Therefore , if as of 2008/03/31 there was no payment recorded, it was assumed that the customer did not make a payment and that he has became a debtor.

There are seven activation models constructed. First one is based on the whole customer population, Two activation models based on a type of customer (individual and business). Four activation models for individual and business customers divided further by service type. Two of them are presented below.

5.1   Credit Scoring Application Model for Whole Population

Model analyses whole population where 8.93 % is a percentage of bad payers. Percentage of bad payers was presented on figure 1. Good payers are marked as BWSFE.

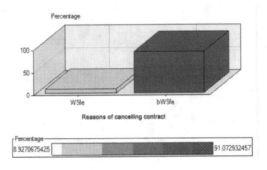

Fig. 1. Percentage of bad payers in relation to good payers in the whole population

Customers who are bad payers cause financial losses to the company. The reason of these losses is customer's failure to pay his bills. This is causing lack of cash flow in the company and impacts the company in a negative way creating bad debt. Bad debt is very dangerous and it might be the reason of financial collapse of the company. Therefore it is crucial for companies to detect and take up preventive actions.

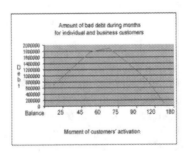

Fig. 2. Amount of bad debt for individual and business customers

Bad debt occurrence was presented on the chart which describes the amount of bad debt during months (Fig. 2). There were taken into consideration following lengths of time: 25, 45, 60, 75, 90, 120 and 180 days from the moment of customers' activation. The chart is based on entire customer population.

The main aim of the model is to identify reasons for bad debt formation which will be used to predict a moment of deactivation caused by failure to pay on 75th day, which is when a customer's account is closed during debt collection process.

Whole population has been classified and arranged according to the importance of attributes showed on figure 3. Figure 3 presents a list of variables with importance measure calculated for each of them, based on training (training column) and validation (validation column) set. The last of columns, called Importance, shows graphic representation of earlier calculations, where dark grey line means variable estimation based on training set and bright grey line - estimation based on validation set.

Variable	Nodes	Training	Validation	Importance
tariff	4	1.000	1.000	
age	1	0.567	0.467	
contract duration	2	0.458	0.442	
non stand. contract agreem.	1	0.281	0.303	
phone restriction	1	0.216	0.161	
CLIR service	1	0.215	0.271	

Fig. 3. Classification of variable importance

According to attributes described above, a classification tree was constructed for entire population. Classification tree was presented on figure 4.

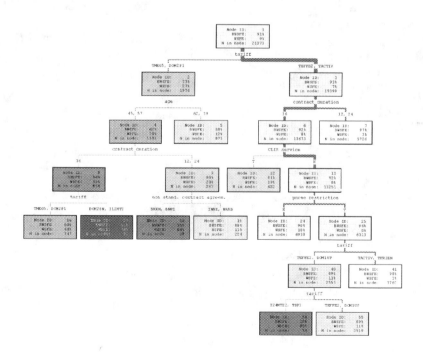

Fig. 4. Classification tree based on whole population

A colour of a node depends on quantity of observations which affect target variable. The more observations determining non-payment there are, the darker the node is. In addition, the thickness of the line can vary as well. It depends on quantity of observations in branches in relation to quantity of observation in the root of a tree. It means that line is thicker when the number of observations is bigger. A tariff, which made the first split in a root of a tree, was chosen as the most important variable. A classification tree is presented in figure 4. Split method was used here based on Chi-square test.

Whole population has been divided into good and bad payers, which is illustrated in leaves of classification tree. Leaves of classification tree have been analyzed and classified from the most risky customers to the least risky customers. Leaves in the figure 5 have been presented in order according to what percentage of target variable had been used. The colours used in the figure describe which type of set was used in data analysis. Dark grey bar means variable estimation on training set and bright grey bar is estimation based on validate set. For every decision rule important results have been presented, which in addition show a more detailed domain in telecommunication company reality.

The constructed model can assessed with regards to matching training data and validation data. Figure 5 presents on the x-axis a model assessment measure for different subtrees, where a subtree is a tree which was built by pruning final nodes. The chart also presented misclassification of observation rate in training and validate dataset. A subtree fits better to the dataset, if misclassification rate of the subtree is closer to zero value. If we look closely at subtree 11 with 11 leaves, we will see it has got the smallest misclassification rate based on training and validate set, which means the subtree fits the most data. Trees tend to fit better to training data rather than to validate data because decision rules are created to fit training data.

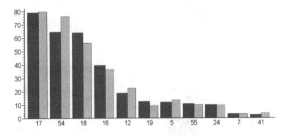

Fig. 5. The most risky customers

Figure 6 presents model assessment which estimates how the model fits the data. The chart compares a set of observations in validation dataset with estimated model results based on training dataset. A Y-axis contains characteristics, which depend on observation frequency in various groups.

The chart (figure 7) illustrates how results (predictions) have improved after applying model in comparison to baseline, when estimation was not been done.

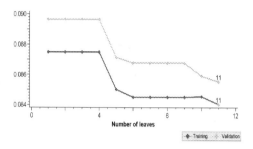

Fig. 6. Misclassification rate chart based on training and validation data

Random model often means making decisions at random. On the basis of this chart it is easy to notice how many times a probability of better results increases when model is applied. Model can be also compared with base model which does not use estimation. The baseline curve on the chart presents results for constant number of successes, which means probability of success on validated dataset. In addition, it has been tested how a node slip criterion effects construction of a tree. Therefore two additional models where constructed where split criterion were Gini Index and Entropy reduction. Results were presented on the figure 7.

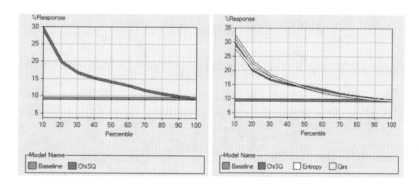

Fig. 7. Results of comparison credit scoring models with various node split criterion

The best result was achieved by model with Gini Index as split criterion. It has received the highest answer in subsequent units. The second best model after Gini Index criterion is a model which used Chi-square test as split criterion. The last one is a model which was using Entropy reduction.

5.2   Credit Scoring Application Model for Individual Customer Population

Credit Scoring model for individual customers population. Model analyses individual customers population where 9.61 % is a percentage of bad payers. Percentage of bad payers was presented on figure 8. Good payers are marked as BWSFE.

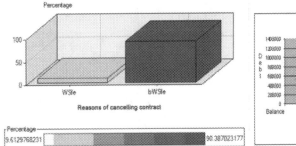
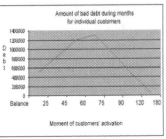

Fig. 8. Percentage of bad payers in relation to good payers in individual customers population

Customers who are bad payers cause financial losses to the company. The reason of these losses is customer's failure to pay his bills. This is causing lack of cash flow in the company and impacts the company in a negative way creating bad debt.

Variable	Nodes	Training	Validation	Importance
tariff	3	1.000	1.000	
age	3	0.803	0.584	
financial credibility	1	0.370	0.241	
contract duraction	1	0.361	0.319	
phone restriction	1	0.242	0.196	
gross discount rate	1	0.232	0.212	
call forwarding	1	0.211	0.008	
CLIR service	1	0.202	0.225	
fax service	1	0.199	0.143	
installation address	1	0.099	0.133	

Fig. 9. Classification of variable importance

The main aim of the model is to identify reasons for bad debt formation which will be used to predict a moment of deactivation caused by failure to pay on 75th day, which is when a customer's account is closed during debt collection process. Whole population has been classified and arranged according to the importance of attributes showed on figure 9.

Figure 9 presents a list of variables with importance measure calculated for each of them, based on training (training column) and validation (validation column) set. The last of columns, called Importance, shows graphic representation of earlier calculations, where dark grey line means variable estimation based on training set and bright grey line - estimation based on validation set.

According to attributes described above, a classification tree was constructed for entire population. Classification tree was presented on figure 10.

A colour of a node depends on quantity of observations which affect target variable. The more observations determining non-payment there are, the darker the node is. In addition, the thickness of the line can vary as well. It depends on quantity of observations in branches in relation to quantity of observation in the root of a tree. It means that line is thicker when the number of observations is

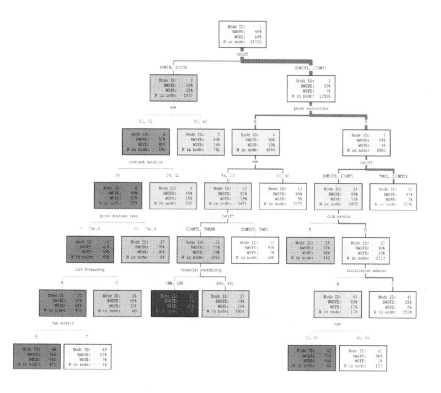

Fig. 10. Classification tree based on individual customer population

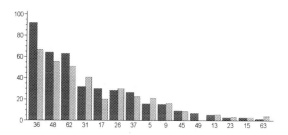

Fig. 11. The most risky customers

bigger. A tariff, which made the first split in a root of a tree, was chosen as the most important variable. Split method was used here based on Chi-square test.

Whole population has been divided into good and bad payers, which is illustrated in leaves of classification tree. Leaves of classification tree have been analyzed and classified from the most risky customers to the least risky customers. Leaves in the figure 11 have been presented in order according to what percentage of target variable had been used. The colours used in the figure describe which type of set was used in data analysis. Dark grey bar means variable estimation on training set and bright grey bar is estimation based on validate set. For every decision rule important results have been presented, which in addition show a more detailed domain in telecommunication company reality.

The constructed model can assessed with regards to matching training data and validation data. Figure 12 presents on the x-axis a model assessment measure for different subtrees, where a subtree is a tree which was built by pruning final nodes.

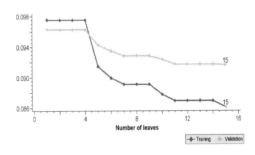

Fig. 12. Misclassification rate chart based on training and validation data

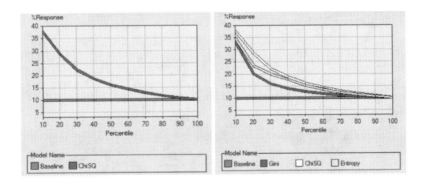

Fig. 13. Lift chart

The chart also presented misclassification of observation rate in training and validate dataset. A subtree fits better to the dataset, if misclassification rate of the subtree is closer to zero value. If we look closely at subtree 15 with 15 leafs, we will see it has got the smallest misclassification rate based on training and validate set, which means the subtree fits the most data. Trees tend to fit better to training data rather than to validate data because decision rules are created to fit training data.

Figure 13 presents model assessment which estimates how the model fits the data. The chart compares a set of observations in validation dataset with estimated model results based on training dataset. A Y-axis contains characteristics, which depend on observation frequency in various groups.

The chart (figure 13) illustrates how results (predictions) have improved after applying model in comparison to baseline, when estimation was not been done. Random model often means making decisions at random. On the basis of this chart it is easy to notice how many times a probability of better results increases when model is applied. Model can be also compared with base model which does not use estimation. The baseline curve on the chart presents results for constant number of successes, which means probability of success on validated dataset. In addition, it has been tested how a node slip criterion effects construction of a tree. Therefore two additional models where constructed where split criterion were Gini Index and Entropy reduction. Results were presented on the figure 13.

The best result was achieved by model with Chi-square test as split criterion. It has received the highest answer in subsequent units. The second best model after Chi-square test is a model which used Entropy reduction as split criterion. The last one is a model which was using Gini Index.

6   Summary

In previous chapter there were models and methods described which support a company in decision making. These models allow calculating profitability for a telecommunication company through measuring customer value and determining level of the risk. By using these models we can divide customers into groups with high, medium, low risk and examine their features are as well as choices they make. In addition these models can be used to prevent financial debt by, for example, implementing deposit policy. The deposit is usually imposed on certain groups where customers were assigned after data analysis.

Activation models presented here were used to predict customer's failure to pay after the debt collection process had finished. The most significant variable is a tariff, chosen by an individual customer. This variable divides whole population of customers into two groups where one is estimated to be ninety percent of good payers and the other one ten percent of bad payers.

The most risky customers share some common features. This group can be divided into certain types of clients. Individuals customers who may fail to pay their bills activate service for phone restriction and they do not apply for phone installation address. Existing business customers on the other hand do not need to show their document of financial reliability when signing a new contract with a telecommunication company. In addition individual customers take advantage of exceptions when signing up for new contract. Exceptions usually include illegible copy of identity document, out of date national judicial register or illegible payment date on the bill. The characteristic quality of a business customer is that they do not agree to disclose full information about their financial obligations. Customers with high probability of failing to pay choose the highest Internet transfer and most expensive tariff with special start price. The contract is usually signed for 36 months. Furthermore, credit limit of 300 polish zlotych increases financial risk for the company.

The lowest risk customers are individual customers, who choose low, medium or no specific tariff at all, or who negotiate better offers, or the ones who do not get any credit limit at all.

7   Future Research

The aim of future research is to develop activation models on the 45 day after debt collection process has begun. If the customer fails to pay within 45 days from date of payment, the debt collection blocks customer's outcoming calls. Therefore it is important to build models which would predict non-payment of an invoice. The next topic for examination could be analysis of other activation models based on different split criterion in decision trees. The aim of this project is to discover new rules which would be able to predict customer ability to pay.

Another aim of future research might be further analysis of activation models based on different split methods than those described earlier in the article. Results of this new analysis could bring very useful information in discovering new rules which identify customers creating financial risk for a telecommunication company.

Telecommunication Company, which was used as a basis for research in this article, is currently growing on the market and is aiming to increase the number of its customers. Once it achieves a satisfactory level of clients, the company will start to analyze customer's data to prevent non-payment of invoices and to improve the process of client verification.

References

1. Bluhm, C., Overbeck, L., Wagner, C.: An Introduction to Credit Risk Modeling. Chapman & Hall/CRC (2002)
2. Keating, C.: Credit Risk Modelling. Palgrave (2003)
3. Gundlach, M., Lehrbass, F.: CreditRisk+ in the Banking Industry. Springer, Heidelberg (2004)
4. Grandell, J.: Aspects of Risk Theory. Springer, Heidelberg (1991)
5. Lando, D.: Credit Risk Modeling: Theory and Applications. Princeton University Press, Princeton (2004)
6. Bühlmann, H.: Mathematical Methods in Risk Theory. Springer, Heidelberg (1996)
7. Lu, J.: Predicting Customer Churn in the Telecommunications Industry – An Application of Survival Analysis Modeling Using SAS. In: SUGI27 Proceedings, Orlando Florida (2002)
8. Hadden, J., Tiwari, A., Roy, R., Ruta, D.: Churn Prediction: Does Technology Matter. International Journal Of Intelligent Technology (2006)
9. Breiman, L., Friedman, J.H., Olsen, R.A., Stone, C.J.: Classification and Regression Trees. Chapman & Hall/CRC (1984)
10. Hastie, T., Tibshirani, R., Friedman, J.H.: The Elements of Statistical Learning: Data Mining, Inference, and Prediction. Springer, Heidelberg (2001)

Sales Intelligence Using Web Mining

Viara Popova[1], Robert John[2], and David Stockton[1]

[1] Centre for Manufacturing
De Montfort University
Leicester, LE1 9BH, UK
[2] Centre for Computational Intelligence
De Montfort University
Leicester, LE1 9BH, UK
{vpopova,rij,stockton}@dmu.ac.uk

Abstract. This paper presents a knowledge extraction system for providing sales intelligence based on information downloaded from the WWW. The information is first located and downloaded from relevant companies' websites and then machine learning is used to find these web pages that contain useful information where useful is defined as containing news about orders for specific products. Several machine learning algorithms were tested from which k-nearest neighbour, support vector machines, multi-layer perceptron and C4.5 decision tree produced best results in one or both experiments however k-nearest neighbour and support vector machines proved to be most robust which is a highly desired characteristic in the particular application. K-nearest neighbour slightly outperformed the support vector machines in both experiments which contradicts the results reported previously in the literature.

Keywords: web content mining, text mining, machine learning, natural language processing.

1 Introduction

In the current information-rich world, useful, well-structured and reliable information is a precious commodity. While sources such as the World Wide Web provide a wealth of information on a wide variety of topics, finding the precise information we need can be a time consuming, frustrating and sometimes unsuccessful process. As a high percentage of the WWW content is in textual form, it is obvious that text mining has the potential to help alleviate the problem.

The case study discussed in this paper looks at web content as a source of sales intelligence for a company in the railway industry. The context of this research is within a larger project (a project on Data Storage, Management, Retrieval and Analysis: Improving customer demand and cost forecasting methods, funded by the Technology Strategy Board in the UK) with main goals of improving demand prediction and resource planning using information from a variety of sources. One potential source of information is the WWW and more specifically customers', competitors' and suppliers' websites. The company in focus produces anti-vibration mountings which are

P. Perner (Ed.): ICDM 2009, LNAI 5633, pp. 131–145, 2009.
© Springer-Verlag Berlin Heidelberg 2009

used for engines in trains, trams and other vehicles in the railway industry. Information about demand for railway rolling stock contains an indication of the demand for anti-vibration mountings. Such information can sometimes be found on the relevant companies' websites in the form of news items and press releases. Locating and extracting the web pages with relevant data is the focus of this paper.

The larger goal of the project is to produce a prediction for the demand for the product which will then be used for intelligent resource planning. The prediction is based on (among other sources) the extracted information on orders for rolling stock as this is a clear indication for demand for anti-vibration mountings. Other factors such as statistics on railway transport, market share information and historical trends will also be considered however this paper concentrates on extracting demand information from news stories. While the main goal is to provide input for the demand prediction module, simply providing the sales representatives with the relevant news items can already be very valuable as they can then take a proactive approach and try to sell their products to manufacturers that they know have received large orders for rolling stock and will need a supply of anti-vibration mountings.

To this end, a variety of techniques and algorithms are used in this paper including machine learning, natural language processing (NLP) methods and web crawling. The company websites can be built in many ways with different structure, using various languages and tools. As a result, the news and press releases can be located in different parts of the web site, even on the main page. The approach taken here is, instead of downloading the whole website, to try to locate the pages as well as possible by locating the news/press section of the website and then downloading content available there at some appropriate depth level. Here depth level refers to how far the program will crawl from the starting page – the starting page is at depth level 0, all pages it links to are at depth level 1, the pages that they link to are at depth level 2 an so on.

However even when the news section is correctly located a large portion of the downloaded pages will not be news items at all. Furthermore most of the news items will not be relevant either as they contain no demand-related information. For this "needle-in-the-hay-stack" problem, we use machine learning methods to produce a model that can filter out the irrelevant documents.

The structure of the paper is as follows. Section 2 presents a brief overview of the relevant literature. Section 3 described the knowledge extraction system developed for this application, the tasks it performs and the approaches taken to realise these tasks. Section 4 presents the result of the experiments on the working of the system for the two main tasks – downloading potentially useful web pages and classification to filter out those that are not news or do not contain the demand-related information that we seek. Section 5 gives a summary and conclusions for the paper and discusses possible directions for future research.

2   Literature Overview

Web content mining [9] is a sub-area of Web mining that is concerned with the discovery of useful information from Web content/data/documents. A special case of web content mining is applied to text such as unstructured data or semi-structured data, for example HTML documents which constitute the most of the content available on the

Web. According to [9], the process of Web content mining can be decomposed into four subtasks: resource finding, information pre-processing (including information selection), generalisation and analysis. This decomposition is consistent with the main processes of the knowledge extraction system presented in this paper: locate and download potentially useful web pages, pre-process the data, use machine learning to classify the documents and present the results of the classification to the user for interpretation and evaluation. The system is discussed further in Section 3.

Most of the research in this area uses a representation of the document as a bag of words or vector representation [20] which considers the words in the document as features ignoring the context in which they appear and other information about the document such as structure, etc. The words are usually represented either as Boolean features which are 0 if the word appears in the document or 1 otherwise or as numeric features corresponding to the frequency of the word in the document. A more elaborate representation of the document can include n-grams, phrases, named entities, etc. [13]. Our research uses named entities as described and motivated in Section 3.

A closely related area is the area of web information extraction [6] which aims at extracting relevant information (facts) from the web rather than relevant documents. The two main approaches to IE are the knowledge engineering approach which employs manually created grammars and rules and the automatic training approach which trains the system using manually annotated training documents. The first approach produces higher quality results but at the expense of more manual effort and domain knowledge.

In the approach reported here a combination of the two approaches is used by: (1) pre-processing the documents using manually built lists of key words and rules to create annotations marking the potentially relevant data, (2) selecting the documents considered relevant by applying machine learning and (3) extracting from them pre-specified bits of information (amount, product and date) using the annotations. For creating the annotations, the GATE annotation engine ANNIE is used which provides means for defining lists of relevant key words and phrases as well as an expressive language to define rules used in the annotation process. Section 3 provides more details on how ANNIE is used and customised for our specific purposes.

One frequent problem with textual data is that it can produce data sets with a large number of features where many of the values will be zero and the percentage of irrelevant features can be high. In addition to various pre-processing techniques, learning algorithms are used which can handle such data. Very good results have been reported when using approaches such as support vector machines (SVM) [22], k-nearest neighbour (kNN) [24], Naïve Bayes [24] and others, however, not much research is done on the comparison of their performance on the same problems. The evaluation in [25] reports best results for kNN, Linear Least Squares Fit (LLSF) (see [26] for details on LLSF) and neural network (NN) (see [24] for details on NNs). SVM has also been shown to perform well on data containing a large number of features [3,8].

One obvious application of Web mining is the automatic classification of news stories collected from the Web (e.g., [1,3,8,10,14,21,23]) that can be used for example to provide personalised news agents [1] or for tracking and discovery of news topics (referred to as new event detection and topic detection and tracking) [10]. Research in these areas usually takes corpora of collected and categorised news stories (e.g. Yahoo sports news in [21], Reuters news corpora in [10,23]) and does not consider the issue of discovering the stories among the other web pages available on the web site.

[21] reports an approach for feature reduction using a combination of principle component analysis (PCA) and class profile-based features (CPBF). As it is recognised that selecting the right features is crucial for the quality of classification, the most regular words in each category were identified manually, weighted and combined with the results from PCA. The resulting features are then used as an input for a NN for classification.

Another approach to reducing the number of features is reported in [23] by representing documents by significance vectors defined by combining the frequencies of words in different topics. In this way each document is represented by n values where n is the number of topics. Furthermore, as a pre-processing step, the hypernym relation in WordNet is used to group similar words which greatly reduces the number of words considered. Finally, the significance vectors are used in self-organising map for classification.

The specific application reported in this paper is not concerned with categorising the stories in a number of predefined topics but rather in discovering news stories among a large number of other web pages and extracting from them only those that contain specific information about orders (which can be used to assess the future demand for the products). The discovered positive examples can be used to extract the specific relevant bits of information such as how much, when, where, etc. or alternatively can be presented directly to the sales representatives.

As the variety of web pages that can be encountered in this application is huge, including pages in several languages other than English, the number of unique words would be higher than in a corpus only containing news stories and only news stories written in one language. Furthermore, among this wide variety of documents, the positive examples are extremely rare, thus extra care needs to be taken to help the learning process build a good model. As a result, the bag-of-words approach was considered not appropriate and the initial experiments (not reported in this paper) did not give satisfactory results. The approach taken was, instead of relying on a general-purpose feature-reduction technique, to design a small number of domain-specific features as described in Section 3.

This application also differs from the area of personal news agents as user profiling is not an issue – it is predefined what is considered an interesting news story and the system does not need to be personalised and adapted to search for different topics of information.

The area of focussed crawling [e.g. 2,5,16] aims at developing efficient means for crawling the web and discovering documents relevant for a pre-defined set of topics without the need for collecting and indexing all available documents in order to answer all possible queries. For example [2] reports an approach that consists of a classifier that evaluates the relevance of a document and a distiller that identifies the most relevant nodes to continue crawling. [5] uses context graph to model the context in which the relevant pages are found. [16], on the other hand, presents a tool for focussed crawling using adaptive intelligent agents.

For the application reported in this paper such an approach would not be necessary as the crawler searches for a very specific pre-specified target (the news section link) on a predefined set of company websites. Therefore a simpler, domain-specific approach would be more efficient and effective. The process is only performed once or at long intervals of time (to refresh the results in case the general structure of the website changes or when a new website is added).

3 The Knowledge Extraction System

The Knowledge Extraction System (KES) (see Fig. 1) needs to perform two general tasks: find potential news stories and classify them to discover which are relevant and which not. For input it is given a list of URLs pointing to the websites of companies representing users and consumers of the products under consideration. In the particular application reported in this paper the focus is on the railway industry and more specifically companies that are potential buyers of locomotives, train carriages, engines, and related products.

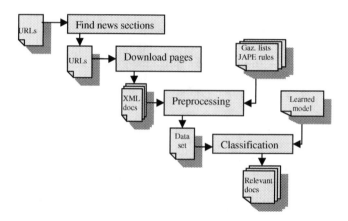

Fig. 1. The knowledge extraction system

The company websites contain various types of information such as company profile, catalogue, recruitment information and so on. Sometimes they also contain news items and press releases describing recent developments related to the company for example big orders, new management appointments, takeovers and mergers, financial information, interviews with senior management, presentation of new products, exhibitions and so on. From all these types of information we are only interested in orders made by these companies for one or more of the relevant products. The percentage of such information on the whole website is extremely low which is why the approach taken is to first find out where the news and press releases are located on the website (the news section) and then download all pages starting from the identified news section main page to a specified depth level.

The news section is located by looking for the key words "news", "press" and "media" in the URLs and the captions of the links found on the main page. The source of the page is used rather than the text which gives access to the URLs of the links and also allows searching for URLs embedded in, for example, JavaScript code rather than just in the HTML code. While not all URLs can be recovered easily from the

JavaScript code (without implementing a full interpreter for the language) still some of the URLs can be extracted which increases the success rate of the algorithm. If such link is found, it is added as a possible news section link. If no such link is found, then the search continues for every frame and every link on the main page pointing to a page on the same website (host). If no results are found after considering all such links then the main page is added as a possible news page with depth level for crawling equal to 0 (only the current page is downloaded) as some company websites do not have a dedicated news section and post news on the main page instead. The following pseudocode shows the algorithm at a high abstraction level. The notation substring(s,r) for two strings s and r indicates that the string r is a substring of s. In the actual code several more checks are performed, for example, to filter out links to image, video and other irrelevant files.

Algorithm: **discover-news-sections**

Input: a webpage link w

Output: the set N of possible new section links

1 N = **discover-news-sections-in-page**(w)

2 If N = ∅ then extract all links F to frames in w

3 If F ≠ ∅ then for each frame f in F:
         N = N ∪ **discover-news-sections-in-page**(f)

4 Otherwise extract all links L in w

5 For each link l_i in L:
         N = N ∪ **discover-news-sections-in-page**(l_i)

6 If N = ∅ then add w to N

7 Return N

discover-news-sections-in-page

Input: a webpage link w

Output: the set N of possible new section links

1 Extract all links L in w and the set C of captions for the links in L

2 For each link l_i in L steps 3-4 are performed

3 If (substring(l_i, "news")) or (substring(l_i, "press") or (substring(l_i, "media")) then add l_i to N

4 Otherwise if for the caption c_i of l_i: (substring(c_i, "news") or (substring(c_i, "press") or (substring(c_i, "media")) then add l_i to N

5 Return N

The approach differs from using a traditional search engine in the fact that it looks for links rather than pages. The keywords should be located either in the caption or in the URL of the link and any other occurrences in the text of the page are ignored. One reason for this choice is that many websites use menus with the most important links which are added to all pages of the website to facilitate the navigation. Therefore if the menu contains a link to the news section with the caption "News" then this word will appear on every page of the site. A second reason for this choice is that, while the link to the news section (or its caption) will most probably contain one of the keywords, the text of the page it points at might not contain any of the keywords or, for example, display them in an image banner (from which the text cannot be easily extracted). On the other hand, it might contain the keyword on every page reporting a news story or a press release which will result in a high percentage of irrelevant pages (as we are only interested in the top page of the news section) and unnecessary search in the deeper sections of the website. In our algorithm, the search is performed top-down and stops when a good candidate is found.

When all news stories need to be downloaded, the necessary depth level can differ quite a lot from one website to another. However the intended use of the system will be at regular intervals (e.g., a month) therefore in many cases only the most recent stories should be downloaded. These are usually located on the main news page and can be downloaded at depth level at most 2. The downloaded documents are stored in XML format. Section 4 reports the results of the performed experiments to test the process of finding news sections and downloading news stories.

The second general task aims at filtering out pages that are not relevant using text classification. In order to convert the set of XML documents to a data set that can be used for classification, they need to be pre-processed at a number of stages. The pre-processing is done using the GATE framework [4,12] and its information extraction engine called ANNIE. The following tools are used:

1. tokeniser which splits the text in simple tokens such as numbers, punctuation, words of different types, etc.
2. gazetteer which uses lists of words and phrases, matches their appearances in the text and annotates them with the appropriate label,
3. sentence splitter which splits the text into sentences,
4. part-of-speech tagger annotates words and symbols with part-of-speech tags,
5. named entities (NE) transducer which uses JAPE rules to annotate different types of information such as locations, dates, etc.

The gazetteer and the NE transducer were customised by adding gazetteer lists and JAPE rules to produce new annotations reflecting the specific application in the following way:

1. added gazetteer lists containing product names and product types for the specific application, e.g. products such as "train set", "locomotive", "tram", and product types such as "high-speed", "freight", "passenger", "low-floor", "two-car", "diesel", etc.,
2. added gazetteer lists containing words and phrases indicating that the page contains information about an order, e.g. "order", "purchase", "new", etc. Similar words (e.g. "order", "orders", "ordered") are combined in one gazetteer list to produce a separate concept,
3. added gazetteer lists containing words and phrases indicating that the topic of the page is different, e.g. "recruit", "exhibition", "roll out", "lease", etc. Again similar

words (e.g. "recruitment", "recruited", "appointed") are combined in one gazetteer list to produce a separate concept,

4. added JAPE rules to the NE transducer to produce new annotations for the added gazetteer lists,
5. for increased performance, all default NE transducer rules were removed except these related to dates and time,
6. added JAPE rules to produce a new annotation indicating the number of products in the following way: product annotations preceded by product type annotations are combined in a new annotation called product group. When a product group is preceded by a number, the number is assumed to indicate the quantity of the products and is assigned a new annotation called amount.

As a result, the following 22 concepts were produced by the pre-processing step: product group, amount, date, six concepts for positive indication, and thirteen concepts for negative indication which were used to produce the data set containing the number of appearances of each concept in each document. Thus we do not use the bag-of-words approach for producing the data set but rather a specific selection of concepts. The motivation for this choice was that the classification task was considered difficult, with limited number of training documents and rare positive class therefore it was necessary to use any available background knowledge. The added difficulty of the particular application stems not only from the fact that a large number of the documents are not at all news stories and should be filtered out but also because for the news stories and press releases it is sometimes difficult (even to a human reader) to distinguish between the useful stories and those that are not useful. One example of the latter is the fact that some companies buy rolling stock which they later lease to other companies. When a story reports on the products being leased it is not always obvious whether these are newly purchased rolling stock that is yet to be manufactures (indication of demand for the purpose of our application) or has been transferred from another lease (no real demand for our application).

In comparison to performing a search with a traditional search engine, our approach is much more flexible and allows discovering and using complex relationships between keywords (represented by concepts). For example just the occurrence of the word "order" does not guarantee that the story contains useful information as it might refer to a product not relevant for our application. If the product is relevant, the order might refer to a lease agreement. However it might still be the case that, in order to realise the agreement, new rolling stock should be purchased from a third party. Not all of these relationships are known in advance and a better approach is to learn them from the data. In addition, the number of words in the gazetteer lists (synonyms, related words and expressions, etc.) makes the formulation of search criteria for the search engine impractical at best if at all possible. Furthermore, the concepts annotated using the gazetteer lists contain useful information that needs to be extracted as the output of the whole process (e.g. product group, amount and date).

Using the produced data set, learning algorithms can be applied to build models that classify the documents as positive or negative examples. For the learning task, the data mining package WEKA [24] was used. Section 4 reports on the selected algorithms and the comparison of the results.

4 Experiments

Experiments were performed in three stages to test the whole process including locating the news sections, downloading the pages and classification. Tables 1 and 2 summarise the results from the first two stages while Tables 5 and 6 summarise the results for the classification process.

As an input to the whole process, a list of 60 URLs of company websites was compiled manually. The software developed for finding news sections and downloading news stories was tested on this set of 60 websites (see Table 1). In 82% (49) of the cases, the news sections were successfully found by the program.

Using the resulting list of URLs (potential news sections detected by the program), the software for downloading the news stories was tested at a number of depth levels for the crawling. Manual evaluation was performed on the results for 17 of these websites (see Table 2). For 77% (13) of them, all available news stories were downloaded when crawling at either depth 1 or depth 2. In the rest of the cases the most recent stories were successfully downloaded which is considered sufficient since the process should be repeated at regular intervals to collect only the new stories and releases for current demand predictions.

The 17 websites were further evaluated on the usefulness of the information they provide in the stories with respect to demand prediction. It was observed that the percentage of useful stories is relatively small and varies between websites. Only 4 of the websites contained more than 4 useful stories considering the full set of stories available from these websites. Eight of the websites did not contain any useful stories at all.

Table 1. The data sets used for the learning process

Input	Finding news sections
60 URLs – a manually compiled list of company URLs	49 (82%) – news section was found
	11 (18%) – news section was not found

Table 2. The experimental results for downloading news stories

Input	Downloading news stories	Usefulness of news stories
The news section URLs found by the program for 17 randomly selected company websites	13 (77%) – all stories found at depth level 2	4 contain >4 useful stories
	4 (23%) – most recent stories found at depth ≤ 2	5 contain between 1 and 4 useful stories
		8 contain no useful stories

For further tests one of these websites was selected since it contained a relatively large number of news stories (145) out of which 38 were considered useful. It was used to test the learning process for classification of the downloaded web pages

Table 3. The data sets used for the learning process

Class	Training set	Test set
Positive	55 documents (25 + 30)	13 documents
Negative	90 documents (60 + 30)	252 documents

(350 pages in total) into useful news stories and other web pages (including not useful stories, news stories in a language other than English, or web pages that are not news stories at all).

In order to build a training set, a set of 25 positive and 60 negative examples were selected at random (see Table 3). Furthermore a set of 30 positive and 30 negative examples of news stories collected from other websites were added to increase the size of the training set and provide more data to the algorithm for separating the useful from not useful news stories which proved to be a more difficult problem than separating the news stories from other web pages. The rest of the data (252 negative and 13 positive examples) were used as a test set.

The WEKA package provides implementations of a large number of algorithms for classification out of which several were selected as they were expected to perform best – they have been reported in the literature as giving good results in text classification or for data with similar characteristics (such as a large number of attributes with sparse values and a rare positive class) (e.g., [8,25]). The following statistical and AI approaches were selected:

- Multilayered Perceptron [24] (referred to as MP in Tables 4 to 6) – a neural network that in this implementation has one hidden layer and uses backpropagation for training,
- Multinomial Naïve Bayes [7,15] (MNB) – a variation of Naïve Bayes used for text classification,
- Complement Naïve Bayes [19] (CNB) – a variation of Multinomial Naïve Bayes for unbalanced data,
- k-Nearest Neighbour [24] (kNN) – an instance-based classifier which bases the classification on the classes of the instance's k neighbours located nearest to it,
- Support Vector Machines [24] (SVM) – an implementation of a sequential minimal optimisation algorithm for training a support vector classifier as in [17],
- Decision Tree [18] (C4.5) – a C4.5 decision tree with or without pruning,
- Logistic Regression [11] (LR) – an implementation of a multinomial logistic regression model,
- RBF Network [24] (RBF Net) – an implementation of a neural network with Gaussian radial basis activation functions; it uses the k-means clustering algorithm to provide the basis functions and learns a logistic regression model.

For each algorithm a number of experiments were conducted with different values of the relevant parameters. Only the best results for each are reported in this paper. Table 4 shows the parameters for which these best results were achieved for each algorithm.

Table 4. The parameters of each algorithm for which the best results (as presented in Tables 5 and 6) were achieved

Algorithm	Parameters, no normalisation	Parameters, with normalisation
MP	hidden neurons = 4	hidden neurons = 4
MNB	Not applicable	Not applicable
CNB	Not applicable	Not applicable
kNN	k = 4, with attribute normalisation	k = 4, k = 2, without attribute normalisation
SVM	RBF with $\gamma = 1.2$	RBF, $\gamma = 1.2$, $\gamma = 0.6$
C4.5	With and without pruning	Without pruning
LR	Not applicable	Not applicable
RBF Net	Number of clusters = 7	Number of clusters = 5

Table 5 presents the results evaluated using the following comparison measures:

1. False positive – the number of negative examples misclassified as positive
2. The percentage of false positive out of all negative examples
3. False negative – the number of positive examples misclassified as negative
4. The percentage of false negative out of all positive examples
5. Precision – the number of true (correctly classified) positive divided by the total number of documents classified as positive
6. Recall – the number of true positive divided by the number of all positive examples
7. F-measure – a frequently used measure in information retrieval which is:

$$2 \times \text{precision} \times \text{recall} / (\text{precision} + \text{recall})$$

In Table 5 the best two results (according to the F-measure and the number of misclassified examples) are highlighted – for kNN, C4.5 and SVM with slightly better results for kNN than the other two algorithms.

Further experiments were performed taking into account the length of the documents by dividing the number of appearances of each concept by the total number of words in the document. The intuition behind this approach is that the number of appearances in a short document should count more than the same number of appearances in a longer document. Table 6 shows the results for all algorithms when using normalisation in this way and the best results are highlighted – for MP and kNN with slightly better results for MP according to the F-measure. SVM also performed well with slightly worse results than kNN.

Table 5. The results from the experiments for the text classification without normalisation with respect to the length of the document

Algorithm	False pos	%	False neg	%	Precision	Recall	F-measure
MP	16	6	4	31	0.36	0.692	0.474
MNB	19	8	4	31	0.321	0.692	0.439
CNB	17	7	5	38	0.32	0.615	0.421
kNN	**8**	**3**	**1**	**8**	**0.6**	**0.923**	**0.727**
SVM	**10**	**4**	**2**	**15**	**0.524**	**0.846**	**0.647**
C4.5	**9**	**4**	**2**	**15**	**0.55**	**0.846**	**0.667**
LR	15	6	3	23	0.4	0.769	0.526
RBF Net	27	11	2	15	0.289	0.846	0.431

Table 6. The results from the experiments for the text classification with normalisation using the length of the document

Algorithm	False pos	%	False neg	%	Precision	Recall	F-measure
MP	**10**	**4**	**1**	**8**	**0.545**	**0.923**	**0.686**
MNB	0	0	12	92	1	0.077	0.143
CNB	0	0	13	100	-	0	-
kNN, k = 2	**7**	**3**	**3**	**23**	**0.588**	**0.769**	**0.667**
kNN, k = 4	**9**	**4**	**2**	**15**	**0.55**	**0.846**	**0.667**
SVM, γ = 1.2	12	5	2	15	0.478	0.846	0.611
SVM, γ = 0.6	14	6	1	8	0.462	0.923	0.615
C4.5	25	10	2	15	0.306	0.846	0.449
LR	19	8	2	15	0.367	0.846	0.512
RBF Net	33	13	3	23	0.233	0.769	0.357

From the results in Table 5 and Table 6 several conclusions can be made:

1. The normalisation did not improve the results for most algorithms indicating that taking into account the length of the document in this particular application does not add useful information that is not already reflected in the data set. The only exception was MP which improved significantly. The other algorithms produced either similar or worse results.
2. kNN and SVM produced good results in both settings and kNN slightly outperformed SVM in both cases. This agrees with the results of further experiments with different subsets of concepts not reported in this paper. It can be concluded that both algorithms are robust and less sensitive to changes in the data which is a feature important for this application due to the high variation in the data that will be used. In [8] and [3] SVM was reported to consistently outperform kNN on data sets consisting of manually collected and classified news stories from a number of different sources. In our application however the results do not support the same conclusion and while the difference in performance is not big, still kNN gave consistently better results.
3. Contrary to the expectations, the two variations of the Naïve Bayes did not produce good results in any of the experimental settings which indicates that they are not suitable for the particular application.
4. C4.5 and MP produced good results in one of the experiments however seemed more sensitive to changes in the data preparation and did not perform well in the other experiments. The decision trees produced by C4.5 were relatively small: with 9 concepts for the experiments in Table 5 and 5 concepts for the experiments in Table 6. In the trees, the concepts for positive indication (as described in Section 3) showed monotone positive behaviour while the concepts for negative indication showed monotone negative behaviour which is the expected outcome and indicates that they indeed discern the two classes in the desired way.
5. In both cases MP outperforms RBF Net and RBF Net gives the worst results among the selected algorithms for both runs (except MNB and CNB in Table 6 which classified all test examples as negative thus clearly did not succeed to learn a usable model).

5   Conclusions and Future Work

This paper presents a knowledge extraction system for providing sales intelligence based on information downloaded from the WWW. The main focus is on locating and extracting news items and press releases of relevant companies and, using classification, discovering these documents that contain relevant information for sales intelligence. The particular application reported here is customised for a company within the railway industry producing anti-vibration mountings, however similar principles can be used to develop systems for other products and areas.

Experiments were conducted to evaluate the performance of the system and to select a machine learning approach for the classification task. A number of algorithms used in the literature were tested and the results point out that the k-nearest neighbour algorithm for instance-based classification and support vector machines give consistently good performance and k-nearest neighbour slightly outperforms the support vector machines on both experimental settings. Other algorithms that gave good results in one of the experiments were the multi-layer perceptron and C4.5 decision tree however they gave significantly worse results in the other experiment indicating that they are more sensitive to changes in the data.

A number of directions for future research can be mentioned. The concepts used in the study were manually selected and designed using gazetteer lists and JAPE rules. However when changing the application area, other concepts might become relevant. An interesting line of research would be to develop tools for automatically extracting information to help design new concepts using, for example, genetic algorithms and/or clustering.

One of the difficulties of this particular application is that the relevant information is sometimes buried among other irrelevant information and can sometimes confuse even a human reader. In fact a degree of uncertainty surely exists in the manual classification. Thus a more natural approach might be to assign degrees of relevance instead of clear-cut labels such as relevant versus irrelevant.

The next phase of this research is to extract the specific data from the relevant pages and, together with data from other sources, use it to generate a prediction for the demand for the product. This prediction can then be used for resource and inventory planning for the supply chain.

Acknowledgements

The research reported here has been funded by the Technology Strategy Board (Grant No. H0254E).

References

1. Billsus, D., Pazzani, M.: A Personal News Agent that Talks, Learns and Explains. In: Proceedings of the Third International Conference on Autonomous Agents (Agents 1999), Seattle, Washington (1999)

2. Chakrabarti, S., van den Berg, M., Dom, B.: Focused crawling: A new approach to topic-specific web resource discovery. Computer Networks 31(11-16), 1623–1640 (1999)
3. Cooley, R.: Classification of News Stories Using Support Vector Machines. In: IJCAI 1999 Workshop on Text Mining (1999)
4. Cunningham, H., Maynard, D., Bontcheva, K., Tablan, V.: GATE: A Framework and Graphical Development Environment for Robust NLP Tools and Applications. In: Proceedings of the 40th Anniversary Meeting of the Association for Computational Linguistics (ACL 2002), Philadelphia (2002)
5. Diligenti, M., Coetzee, F., Lawrence, S., Giles, C.L., Gori, M.: Focused crawling using context graphs. In: Proceedings of the 26th International Conference on Very Large Databases (VLDB), pp. 527–534 (2000)
6. Eikvil, L.: Information Extraction from World Wide Web - A Survey. Technical Report 945 (1999)
7. Frank, E., Bouckaert, R.R.: Naive Bayes for Text Classification with Unbalanced Classes. In: Fürnkranz, J., Scheffer, T., Spiliopoulou, M. (eds.) PKDD 2006. LNCS, vol. 4213, pp. 503–510. Springer, Heidelberg (2006)
8. Joachims, T.: Text Categorization with Support Vector Machines: Learning with Many Relevant Features. In: Nédellec, C., Rouveirol, C. (eds.) ECML 1998. LNCS, vol. 1398. Springer, Heidelberg (1998)
9. Kosala, R., Blockeel, H.: Web Mining Research: A Survey. SIGKDD Explorations 2, 1–15 (2000)
10. Kumaran, G., Allan, J.: Text Classification and Named Entities for New Event Detection. In: Proceedings of SIGIR 2004, pp. 297–304 (2004)
11. le Cessie, S., van Houwelingen, J.C.: Ridge Estimators in Logistic Regression. Applied Statistics 41(1), 191–201 (1992)
12. Li, Y., Bontcheva, K., Cunningham, H.: SVM Based Learning System For Information Extraction. In: Winkler, J.R., Niranjan, M., Lawrence, N.D. (eds.) Deterministic and Statistical Methods in Machine Learning. LNCS (LNAI), vol. 3635, pp. 319–339. Springer, Heidelberg (2005)
13. Manning, C.D., Schütze, H.: Foundations of Statistical Natural Language Processing. MIT Press, Cambridge (1999)
14. Masand, B., Lino, G., Waltz, D.: Classifying News Stories Using Memory Based Reasoning. In: International ACM SIGIR Conference on Research and Development in Information Retrieval, pp. 59–65 (1992)
15. Mccallum, A., Nigam, K.: A Comparison of Event Models for Naive Bayes Text Classification. In: Proceedings of the AAAI 1998 Workshop on Learning for Text Categorization (1998)
16. Menczer, F.: ARACHNID: Adaptive Retrieval Agents Choosing Heuristic Neighborhoods for Information Discovery. In: Fisher, D. (ed.) Proceedings of the 14th International Conference on Machine Learning (ICML 1997). Morgan Kaufmann, San Francisco (1997)
17. Platt, J.: Fast Training of Support Vector Machines using Sequential Minimal Optimization. In: Schoelkopf, B., Burges, C., Smola, A. (eds.) Advances in Kernel Methods - Support Vector Learning. MIT Press, Cambridge (1998)
18. Quinlan, R.: C4.5: Programs for Machine Learning. Morgan Kaufmann Publishers, San Mateo (1993)
19. Rennie, J.D., Shih, L., Teevan, J., Karger, D.R.: Tackling the Poor Assumptions of Naive Bayes Text Classifiers. In: Proceedings of the International Conference on Machine Learning (ICML 2003), pp. 616–623 (2003)

20. Salton, G., McGill, M.: Introduction to Modern Information Retrieval. McGraw-Hill, New York (1983)
21. Selamat, A., Omatu, S.: Web Page Feature Selection and Classification Using Neural Networks. Information Sciences 158, 69–88 (2004)
22. Vapnik, V.: The Nature of Statistical Learning Theory. Springer, New York (1995)
23. Wermter, S.: Hung, Ch.: Selforganizing Classification on the Reuters News Corpus. In: Proceedings of the 19th international conference on Computational linguistics, Taipei, Taiwan, pp. 1–7 (2002)
24. Witten, I.H., Frank, E.: Data Mining: Practical Machine Learning Tools and Techniques, 2nd edn. Morgan Kaufmann, San Francisco (2005)
25. Yang, Y.: An Evaluation of Statistical Approaches to Text Categorization. Information Retrieval 1, 69–90 (1999)
26. Yang, Y., Chute, C.G.: A Linear Least Squares Fit Mapping Method for Information Retrieval from Natural Language Texts. In: Proceedings of the 14th International Conference on Computational Linguistics (COLING 1992), pp. 447–453 (1992)

A Sales Forecast Model for the German Automobile Market Based on Time Series Analysis and Data Mining Methods

Bernhard Brühl[1,3], Marco Hülsmann[1], Detlef Borscheid[2], Christoph M. Friedrich[1], and Dirk Reith[1,*]

[1] Fraunhofer Institute for Algorithms and Scientific Computing (SCAI), Schloss Birlinghoven, 53754 Sankt Augustin, Germany
[2] BDW Automotive, Maybachstr. 35, 51381 Leverkusen, Germany
[3] Present address: Universität zu Köln, Seminar für ABWL, Risikomanagement und Versicherungslehre, Kerpener Str. 30, 50937 Köln, Germany
dirk.reith@scai.franhofer.de

Abstract. In this contribution, various sales forecast models for the German automobile market are developed and tested. Our most important criteria for the assessment of these models are the quality of the prediction as well as an easy explicability. Yearly, quarterly and monthly data for newly registered automobiles from 1992 to 2007 serve as the basis for the tests of these models. The time series model used consists of additive components: trend, seasonal, calendar and error component. The three latter components are estimated univariately while the trend component is estimated multivariately by Multiple Linear Regression as well as by a Support Vector Machine. Possible influences which are considered include macro-economic and market-specific factors. These influences are analysed by a feature selection. We found the non-linear model to be superior. Furthermore, the quarterly data provided the most accurate results.

Keywords: Sales Forecast, Time Series Analysis, Data Mining, Automobile Industry.

1 Introduction

Successful corporate management depends on efficient strategic and operative planning. Errors in planning often lead to enormous costs and in some cases also to a loss of reputation. Reliable forecasts make an important contribution to efficient planning. As the automobile industry is one of the most important sectors of the German economy, its development is of utmost interest.

The introduction and the development of mathematical algorithms, combined with the utilization of computers, have increased the reliability of forecasts enormously. Enhanced methods, e.g. Data Mining, and advanced technology allowing the storage

---

* Corresponding author.

P. Perner (Ed.): ICDM 2009, LNAI 5633, pp. 146–160, 2009.
© Springer-Verlag Berlin Heidelberg 2009

and evaluation of large empirical data sets generate the means of producing more reliable forecasts than ever before. At the same time, the methods have become more complex. However, the explicability of a forecast model is as important as its reliability. Therefore, the main objective of our work is to present a model for sales forecasts which is highly accurate and at the same time easily explicable.

Although Lewandowski [1] [2] investigated sales forecast problems in general and concerning the automobile industry in particular in the 1970s, few studies have focused on forecasts concerning the German automobile industry thereafter. Recent publications have only been presented by Dudenhöffer and Borscheid [3] [4], applying time series methods to their forecasts. Another approach has been chosen by Bäck et al. [5], who used evolutionary algorithms for their forecasts.

The present contribution pursues new routes by including Data Mining methods. But there are also differences in the time series methods which are applied. A detailed analysis of the past is needed for a reliable forecast of the future. Because of that, one focus of our work is the broad collection and analysis of relevant data. Another focus is the construction and the tests of the used model. The data base of our models consists of the main time series (registrations of new automobiles) and the secondary time series, also called exogenous parameters, which should influence the trend of our main time series. To eliminate parameters with insignificant influence on the main time series, a feature selection method is used. These tasks are solved for yearly, monthly and quarterly data. Then the results are analyzed and compared in order to answer the following main questions of this contribution:

1. Is it possible to create a model which is easy to explain and which at the same time provides reliable forecasts?
2. Which exogenous parameters influence the sales market of the German automobile industry?
3. Which collection of data points, yearly, monthly or quarterly data, is the most suitable one?

2 Data

The main time series comprises the number of registrations of new automobiles in Germany for every time period. Hence, the market sales are represented by the number of registrations of new automobiles, provided by the Federal Motor Transport Authority.

The automobile market in Germany increased extraordinary by the reunification of the two German states in 1990. This can only be treated as a massive shock event which caused all data prior to 1992 to be discarded. Therefore, we use the yearly, monthly and quarterly registrations of the years 1992 to 2007. The sales figures of these data are shown in Figures 1-3 and the seasonal pattern of these time series is clearly recognizable in the last two figures.

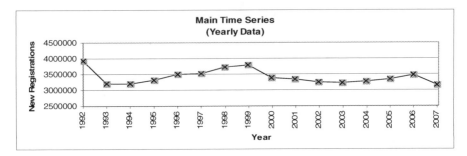

Fig. 1. Registrations of new automobiles from 1992 to 2007: Yearly data

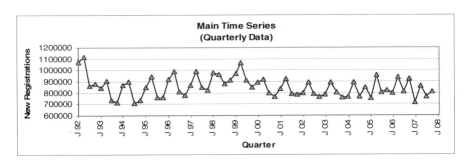

Fig. 2. Registrations of new automobiles from 1992 to 2007: Quarterly data

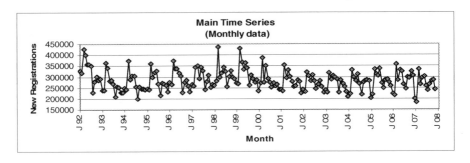

Fig. 3. Registrations of new automobiles from 1992 to 2007: Monthly data

Our choice of the exogenous parameters fits the reference model for the automobile market given by Lewandowski [2]. In this model the following properties are considered:

a) Variables of the global (national) economy
b) Specific variables of the automobile market
c) Variables of the consumer behavior w.r.t. the changing economic cycle
d) Variables that characterize the influences of credit restrictions or other fiscal measures concerning the demand behaviour in the automobile industry.

Based on this model, the following ten market influencing factors, shown in Table 1, are chosen [6] (exogenous parameters)

Table 1. In the first column, the exogenous parameters are indicated. In the second column, the time intervals of the publication of these parameters are given. Yearly published parameters are denoted by a "Y", quarterly by a "Q", and monthly by an "M". The parameters are used by different units. In the case that the relative deviation to the reference year 2000 is used, the parameter will be denoted by a "D". If absolute numbers are used, the parameter will be denoted by an "A". Two parameters are computed by market experts and have no unit. These will be denoted by an "E". This is shown in the third column. In the forth column, the data source is given. The three data sources are the Federal Statistical Office (FSO), the German Federal Bank (GFB), and BDW Automotive (BDW). In the last column, a short explanation of the parameters is provided.

Exogenous Parameters	Publishing Interval	Parameter Unit	Data Source	Explanation
Gross Domestic Product	Y; Q	D	FSO	The Gross Domestic Product (GDP) is a quantity for the economic strength of a national economy in a specific time period. It measures the value of all goods and services within a country.
Available Personal Income	Y; Q	D	FSO	The Available Personal Income of the private households represents the income which could be used to consume and to save. It is calculated by the sum of the primary income, the monetary social benefit and other continuous transfers. This sum is reduced by income and property tax, social security contributions and other continuous transfers.
Consumer Price Index	Y; M	D	FSO	The Consumer Price Index measures the average shift in prices of all goods and services which are bought by the private households for consumption.
Interest Rate	M	A	GFB	The Interest Rate is the rate for credit capital, which is published by the German Federal Bank and the European Central Bank.
Unemployment Rate	M	A	FSO	The Unemployment Rate is the percentage of unemployed persons relative to all available civil manpower.
Industrial Investment Demand	Y; Q	D	FSO	The Industrial Investment Demand includes all material investments in a specific period.
Petrol Charge	Y; M	D	FSO	The mean of all prices of the different petrol forms like unleaded gasoline or diesel is the Petrol Charge.
Private Consumption	Y; Q	D	FSO	The purchase of the goods and services of the private households for consumption is called Private Consumption. It includes non-real purchases like the use of a freehold apartment or consumer spending of private organizations.
Latent Replacement Demand	Y; Q; M	E	BDW	In general, all goods have a mean product lifetime after which it needs to be replaced. This initial purchase could be delayed for various reasons, for example due to financial shortages. In these cases a so-called Latent Replacement Demand [2] arises. In contrast, there are cases which lead to an earlier initial purchase. These variations are considered in the Latent Replacement Demand.
Model Policy	Y; Q; M	E	BDW	The Model Policy considers the influence of the introduction of new vehicles, revisions of existing models and campaigns. A certain base level will be calculated by the data of the past [4]. In periods with increased innovations, the value increases above the base level. In periods with fewer innovations, the value deceases below the base level.

In Table 1, it is shown that not all exogenous parameters used are published on a monthly, quarterly, and yearly base. In cases in which the necessary values are not given directly, the following values are taken:

Yearly data analysis: The averages of the Unemployment and Interest Rate of each year are used.

Quarterly data analysis: The average of the Unemployment and Interest Rate of each quarter is used. For the parameters Consumer Price Index and Petrol Charge the values of the first months of each quarter are taken.

Monthly data analysis: In the case of the quarterly published parameters, a linear interpolation between the values of two sequential quarters is used.

3 Methodology

3.1 Time Series

Time Series Model
In this contribution an additive model with the following components to mimic the time series is applied.

Let x_t, $t = 1,...,T$, be the time series observed in the past. Then x_t can be written as:

$$x_t = m_t + (s_t) + (p_t) + e_t,$$

where m_t is the trend component, s_t the seasonal component (only for monthly and quarterly data), p_t is the calendar component (only for monthly data), and e_t represents the error component.

Seasonal Component
For the estimation of the seasonal component there are many standard methods like exponential smoothing [7], the ASA-II method [8], the Census X-11 method [9], or the method of Box and Jenkins [10]. In this contribution, the Phase Average method [11] is used because it is quite easy to interpret. To get accurate results with this method, the time series must have a constant seasonal pattern over time and it has to be trendless. A constant seasonal pattern is given in our time series. To guarantee the trend freedom, a trend component is estimated univariately and subtracted before the seasonal component is estimated. The latter is done by using a method which is close to the moving average method [12]. Because of the small given data set, differing from the standard method, the following formula is used to compute the mean m of a period:

$$m_t = \frac{1}{t}\sum_{i=1}^{t} x_i \quad t = 1,...,T$$

Trend Component
Although a univariate trend estimation would be easier to explain, this route is not followed in this contribution because the assumption that the registrations of new automobiles in Germany are not influenced by any other parameter is not justified. Hence, the most important component, the trend, is estimated multivariately.

For the linear trend estimation the Multiple Linear Regression (MLR) [12] is used. The Support Vector Machine (SVM) with ε-Regression and Gaussian kernel [13] [14] [15] is chosen as a representative of a non-linear estimation because the SVM has proven to provide suitable results in other industrial projects [16] [17]. However, this choice might be altered in future publications.

Calendar Component
The calendar component considers the number of working days within a single period. For the estimation of the calendar component p_t a method close to a standard method used in [1] is chosen. At first the auxiliary value p^*_t is computed by:

$$p^*_t = \frac{N_t - A_t}{G_t} \quad t = 1,...,T$$

where N_t is the mean of the working days in the past of the according period, A_t the number of working days, and G_t the total number of days of the period t.

Then the absolute values p_t of the calendar component is calculated via Linear Regression using the values x_t, $t = 1,...,T$ of the main time series and p^*_t, $t = 1,...,T$.

Error Component
The error component is estimated with the Autoregressive-Moving-Average-Process of order two [8]. A condition to use this method is the stationarity of the error component. This condition is tested by the *Kwiatkowski-Phillips-Schmidt-Shin* Test (KPSS-Test) [18]. In the cases of non-stationarity, it is set to zero.

3.2 Data Pre-processing

Time lag
In reality, external influencing factors do not always have a direct effect on a time series, but rather this influence is delayed. The method used to assign the time lag is based on a correlation analysis.

Time lag estimation
If the value y_t of a time series Y has its influence on the time series X in $t + s$, the time lag of the time series Y is given by the value s. Then the correlation between the main time series X with its values $x_1,...,x_T$ and all of the k secondary time series Y^i, $i=1,...,k$, with its values $y^i_1,...,y^i_T$, is computed. Afterwards the secondary time series is shifted by one time unit, i.e. the value y^i_t becomes the value y^i_{t+1} and the correlation between the time series $x_2,...,x_T$ and $y^i_1,...y^i_{T-1}$ is assigned. This shifting is repeated up to a pre-defined limit. The number of shifts of every highest correlation between the main and a secondary time series is the value of the time lag of this secondary time series.

Smoothing the exogenous parameters by using the time lag
It is assumed that every value y_t of an exogenous parameter Y is influenced by its past values. The time lag indicates how many past data points influence the current value. This results in the following method:

Let s be the time lag of $Y = y_1,...,y_T$, then the current value y_t is calculated by the weighted sum

$$y_t = \begin{cases} \displaystyle\sum_{j=1}^{s} \lambda(1-\lambda)^{s-j} y_{t-j} & t = s+1,...,T \\ \\ y_t & t = 1,...,s \end{cases}$$

where $\lambda \in (0,1)$ is the weighting factor.

Normalisation

To achieve comparability between factors which are not weighted in the same way, these factors have to be normalized to a similar range. With that step numerical errors can be tremendously reduced. As normalization method, the z-Transformation is applied. It refines the mean value to zero and the standard deviation to one: Let v_t be any factor at a particular time t, $t \in T$, then the z-Transformation is calculated by

$$v_{t,normalized} = \frac{v_t - \mu(v)}{\sigma(v)},$$

where $\mu(v)$ is the mean and $\sigma(v)$ the standard deviation of v.

Feature Selection

Methods which are typically used for Feature Selection are the correlation analysis, the *Principal Component Analysis* (PCA) [19], the *Wrapper Approach* [20], and the *Filter Approach* [21]. Here, the Wrapper Approach with two different regression methods - the Multiple Linear Regression and the Support Vector Machine - is chosen for dimension reduction. Compared with other methods, this method provides more explicable results even for small data sets. Additionally, forecasts with the PCA are calculated as a reference model for our results. The PCA results are not easily explicable, as the PCA-transformed parameters can not be traced back to the original ones. Therefore the results have not been considered for the final solution

4   Evaluation Workflow

The data, i.e. the main time series and the exogenous parameters, are divided into training and test data. The methods introduced in Chapter 3 are used to generate a model on the training data, which is evaluated by applying it to the test data. The complete evaluation workflow is shown in Figure 4.

Step 1: Data Integration: The bundling of all input information to one data source is the first step in the workflow. Thereby, the yearly, quarterly or monthly data ranges from 1992 to 2007. The initial data is assumed to have the following form:

Main Time Series	Secondary Time Series
$x_t = m_t + s_t + p_t + e_t, t = 1, ..., T$	$y^i_t, t = 1, ..., T$ and $i = 1, ..., k$

Step 2: Data Pre-processing: Before the actual analysis, an internal data pre-processing is performed, wherein special effects contaminating the main time series are eliminated. For example, the increase of the German sales tax in 2007 from 16%

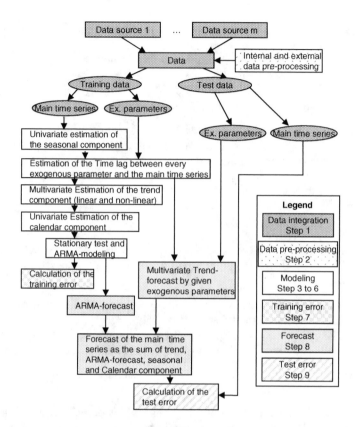

Fig. 4. Evaluation Workflow: First, the data is collected and bundled. After a data pre-processing, it is split into training and test set. The model is built on the training set and the training error is calculated. Then the model is applied to the test data. Thereby, the new registrations for the test time period are predicted and compared with the real values and based on this the test error is calculated.

to 19% led to an expert estimated sales increase of approximately 100.000 automobiles in 2006. Hence, this number was subtracted in 2006 and added in 2007. Furthermore, the exogenous parameters were normalized by the z-Transformation.

The normalized data are passed on to an external data pre-processing procedure. The method used is the Wrapper Approach with an exhaustive search. Since we use a T-fold cross-validation (leave-one-out) to select the best feature set, it should be noted that we implicitly assumed independency of the parameters. As regression method for the feature evaluation a Linear Regression is applied in the case of linear trend estimation and a Support Vector Machine in the case of non-linear trend estimation.

The elimination of the special effects in monthly data is not applicable because the monthly influences can be disregarded.

Step 3: Seasonal Component: The estimation of the seasonal component is done by the Phase Average method. In this contribution, the seasonal component is estimated before the trend component. The reason is that the Support Vector Machine absorbs a part of the seasonal variation, leading to faulty results. In order to remove the influence of trends on the data, the trend is estimated univariately by using the method presented in section 3.1. The univariate trend component is subtracted and the seasonal component can be estimated on the revised time series.

For the analysis of yearly data, the seasonal component cannot be computed because it measures the seasonal variation within one year.

Step 4: Time Lag and trend component: To analyse monthly and quarterly data, the time lag of each exogenous parameter is calculated by a correlation analysis. The exogenous parameters are smoothed using the estimated time lag. In this case, it is of advantage to limit the time lag because the influence of the exogenous parameters on the main time series is temporary bounded. The limit chosen was one year for monthly and quarterly data. Hence, the time lag is 0, if it is greater than the limit.

Afterwards, the trend component of the training set is estimated multivariately by Linear Regression (linear case) or by a Support Vector Machine (non-linear case). To optimize some specific parameters for the Support Vector Machine, a *Grid Search* algorithm is applied. Thereby, a bootstrapping with ten replications is performed for the evaluation.

Step 5: Calendar component: The calendar component of the training set is estimated univariately by the method presented in section 3.1.

The calendar component is computed only for monthly data because the variation of the working days in the case of yearly and quarterly data can be disregarded.

Step 6: ARMA-Model: In the stationary case the error component is estimated by a second-order ARMA process. Otherwise it set to zero (cf. 3.1).

Step 7: Training Error: The absolute training error is derived from the difference between the original values of the main time series and the values estimated by the model. These values are given by the sum of the values of the trend, seasonal, and calendar component. In the case of a stationary error component, the values estimated by the ARMA model are added. The mean ratio between the absolute training errors and the original values gives the *Mean Absolute Percentage Error (MAPE)*.

Let x_i, $i=1,...,T$, be the original time series after the elimination of special effects and z_i, $i=1,...,T$, the estimated values. Then, the error functions considered are represented by the following formulas:

Mean Absolute Error

$$E_{MAE} = \frac{1}{T} \sum_{i=1}^{T} |x_i - z_i|$$

Mean Absolute Percentage Error

$$E_{MAPE} = \frac{1}{T} \sum_{i=1}^{T} \frac{|x_i - z_i|}{x_i}$$

Step 8: Forecast: The predictions for the test time period are obtained by summing up the corresponding seasonal component, the trend component based on the exogenous parameters of the new time period and the respective multivariate regression method, and the calendar component. In the case of a stationary error component, the values predicted by the ARMA process are added, too.

Step 9: Test error: The differences of the predictions and the original values of the test set lead to the test errors. Its computation conforms exactly to the computation of the training errors.

5 Results

The results are obtained from the execution of the workflow presented in chapter 4.

In a first step, all ten exogenous parameters are used in the model. Secondly, a feature selection is performed by the Wrapper Approach and the same workflow is executed with only the selected parameters. The training period consists either of 14 or 15 years leading to a test set of two years or one year, respectively. In each case, an MLR as well as an SVM is used for the multivariate trend estimation. This leads to eight different evaluation workflows for each type of data collection.

In order to be able to assess the error rates resulting from the evaluation workflows, upper bounds for the training and test errors are required as reference values for the evaluation. Therefore, a Principal Component Analysis (PCA) is applied to the original exogenous parameters, and the same evaluation workflow is performed with the PCA-transformed parameters. The PCA results are an indicator of what can be expected from the multivariate model. However, in this contribution, the transformed parameters cannot be used, because they are not explicable and the influences of the original parameters cannot be reproduced anymore.

The results of the yearly, monthly, and quarterly data which generate the smallest errors using the PCA-transformed exogenous parameters are shown in Table 2. A training period of 14 years is used. In all cases the SVM gives much better results than the MLR [22]. To optimize the parameters of the SVM, the Grid search algorithm is applied.

Table 2. Errors of the best models using the PCA transformed exogenous parameters and training periods of 14 years

Non-linear trend estimation (SVM using PCA-transformed exogenous parameters)

	Parameters	Mean Percentage Training Error	Mean Percentage Test Error
Yearly Data	C=8, γ=0.25, ε=0.1	0.31%	2.04%
Monthly Data	C=1, γ=0.125, ε=0.1	3.29%	11.23%
Quarterly Data	C=8, γ=0.25, ε=0.1	0.54%	4.86%

5.1 Yearly Model

The Feature Selection for the linear trend approximation showed that the Gross Domestic Product, Unemployment Rate, Price Index, Private Consumption, and Industrial Investment Demand were the only parameters which significantly influenced the main time series. For the non-linear trend approximation, the respective parameters were the same, except for the Price Index being exchanged with the Latent Replacement Demand.

Table 3 shows the results of the various models. By comparison with the PCA analysis for yearly data (see Table 2), one can clearly see that the quality of the linear trend models is inferior. Considering the fact that we face high test errors although we start from low training errors, one can assume that the training set was too small for this specific problem and the model was overfitted. Especially, data points which are not in close proximity to the training set are hard to predict correctly.

Table 3. Training and test errors in the model using yearly data

Linear trend estimation (MLR)

	Mean Training Error				Mean Test Error			
	All Parameters		Reduced Parameters		All Parameters		Reduced Parameters	
	14 Years	15 Years	14 Years	15 Years	14 Years	15 Years	14 Years	15 Years
Absolute	33 146	35 165	96 802	91 144	207 169	418 335	128 609	236 653
Percentage	0.98%	1.05%	2.85%	2.69%	6.31%	12.87%	3.94%	7.28%

Non-linear trend estimation (SVM with C=16, γ=0.25, ε=0.1)

	Mean Training Error				Mean Test Error			
	All Parameters		Reduced Parameters		All Parameters		Reduced Parameters	
	14 Years	15 Years	14 Years	15 Years	14 Years	15 Years	14 Years	15 Years
Absolute	3 497	3 762	4 270	3 857	48 823	37 952	51 531	65 136
Percentage	0.10%	0.11%	0.13%	0.12%	1.47%	1.16%	1.54%	2.00%

In contrast, the results for the non-linear models shown in Table 3 have a better quality compared with the PCA analysis. That originates from the saturation effect generated by the regression in conjunction with the Support Vector Machine. It leads to the fact that data points far off can still be reasonably predicted. Another advantage (and consequence) of this approach is the fact that a parameter reduction does not severely lower the quality of the predictions down to a threshold value of five parameters. Models with such a low number of parameters offer the chance to easily explain the predictions, which appeals to us.

A general problem, however, is the very limited amount of information that leads to a prediction of, again, limited use (only annual predictions, no details for short-term planning). Therefore, an obvious next step is to test the model with the best statistics available, i.e. with monthly data.

5.2 Monthly Model

In this case, the Feature Selection resulted in the following: For the linear trend model, only the parameters Model Policy and Latent Replacement Demand were relevant while in the non-linear model, new car registrations were significantly influenced by the Gross Domestic Product, Disposal Personal Income, Interest Rate Model Policy, Latent Replacement Demand, Private Consumption, and Industrial Investment Demand, i.e. a superset of parameters of the linear case.

The results given in Table 4 are again first compared to the PCA analysis, cf. Table 2. As for the yearly data, the non-linear models are superior to both the results for the PCA analysis and for the linear model. Most accurate predictions can be achieved for the non-linear model with all parameters. However, the deviations are deemed too high and are therefore unacceptable for accurate predictions in practice. One reason for this originates from the fact that most parameters are not collected and given monthly, but need to be estimated from their quarterly values. Additionally, the time lag of the parameters can only be roughly estimated and is assumed to be a constant value for reasons of feasibility.

Table 4. Training and Test errors in the model used monthly data

Linear trend estimation (MLR)

	Mean Training Error				Mean Test Error			
	All Parameters		Reduced Parameters		All Parameters		Reduced Parameters	
	14 Years	15 Years	14 Years	15 Years	14 Years	15 Years	14 Years	15 Years
Absolute	10 920	11 832	11 082	11 516	38 243	29 767	35 560	33 948
Percentage	3.92%	4.22%	3.95%	4.11%	13.84%	12.72%	12.86%	14.24%

Non-linear trend estimation (SVM with C=4, γ=0.125, ε=0.1)

	Mean Training Error				Mean Test Error			
	All Parameters		Reduced Parameters		All Parameters		Reduced Parameters	
	14 Years	15 Years	14 Years	15 Years	14 Years	15 Years	14 Years	15 Years
Absolute	8 986	9 563	9 473	9 975	21 816	26 748	21 992	28 966
Percentage	3.18%	3.35%	3.35%	3.50%	8.42%	11.45%	8.54%	12.25%

The shortcomings of the model are most clearly visible in the case of the linear model where most of the parameters have been dropped out. Consequently, models of quarterly data basis have been investigated. They promise to be a suitable compromise between sufficient amount and stability of the provided data.

5.3 Quarterly Model

In this case, the Feature Selection for the linear trend approximation singled out the following parameters as being significant for the model: Interest Rate, Price Index, Latent Replacement Demand, and Industrial Investment Demand. Surprisingly, all parameters were found to be relevant in the non-linear case.

Table 5. Training and Test errors in the model used quarterly data

Linear trend estimation (MLR)

	Mean Training Error				Mean Test Error			
	All Parameters		Reduced Parameters		All Parameters		Reduced Parameters	
	14 Years	15 Years	14 Years	15 Years	14 Years	15 Years	14 Years	15 Years
Absolute	28 815	28 869	36 278	35 838	68 418	50 607	36 426	32 099
Percentage	3.42%	3.44%	4.32%	4.28%	8.31%	6.21%	4.45%	3.92%

Non-linear trend estimation (SVM with C=8, γ=0.125, ε=0.1)

	Mean Training Error				Mean Test Error			
	All Parameters		Reduced Parameters		All Parameters		Reduced Parameters	
	14 Years	15 Years	14 Years	15 Years	14 Years	15 Years	14 Years	15 Years
Absolute	3 507	3 503	---	---	22 316	17 035	---	---
Percentage	0.42%	0.42%	---	---	2.70%	2.13%	---	---

The results for the training and test errors for the models based on quarterly data are given in Table 5. Again, the linear model is inferior compared to the PCA (cf. Table 2) and compared to the non-linear model. The difference between training and test errors for the linear model with all parameters is still severe. Furthermore, the total error of the linear model with reduced parameters might look small. However, a closer look reveals that this originates only from error cancellation [22]. Altogether this indicates that the training set is again too small to successfully apply this model for practical use.

The results for the non-linear model, in turn, are very satisfying. They are better than the results from the PCA analysis and provide the best absolute test errors of all investigated models. This also indicates that all parameters are meaningful contributions for this kind of macro-economic problem. In a previous work, we have shown that a reduction to the six most relevant parameters would more than double the test error [22].

5.4 Summary

It can be clearly stated that the SVM provides a superior prediction (less test errors) compared to the MLR. This illustrates that the mutual influence of the parameters is essential to achieve accurate forecasts. In order to identify the overall best models, the errors of the training and test sets are accumulated to annual values. The results for the best model based on yearly, monthly, and quarterly data are visualized in Figure 5.

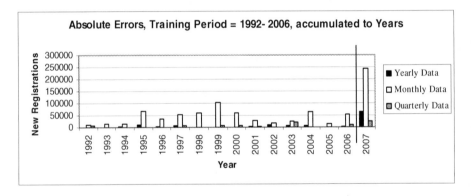

Fig. 5. Graphical illustration of the absolute errors of the best models for a 15 years training period, cumulated to years: On yearly and monthly data the non-linear model with reduced parameters, on quarterly data the non-linear model with all parameters

During the training period, the best model for the monthly data is significantly worse compared to both other models. The same holds for the test period. Here, the best quarterly model is significantly superior to the best yearly model, with roughly half of the test error. It can be observed that the quarterly model does not only deliver a minor test error, but at the same time provides higher information content than the yearly model. Both models generate very low errors during the training period, showing again that the set of parameters is well adapted to our problem. The only drawback of the best quarterly model is the fact that all exogenous parameters are necessary, making the model less explicable.

6 Discussion and Conclusion

Based on the results of Chapter 5, the three questions mentioned at the beginning of this contribution can now be answered.

1. Is it possible to create a model which is easy to interpret and which at the same time provides reliable forecasts?

To answer this question, a more discriminate approach must be taken. Considering only the used additive model, the answer is "yes", because the additive model has given better results than the Principal Component Analysis.

By looking at the different methods used in our model, answering the question becomes more difficult. Simple and easily explicable univariate estimations are used for the seasonal, calendar and error component but a more difficult multivariate method for the greatest and most important component, the trend. Thereby, the results given by the more easily explicable Multiple Linear Regression are less favorable than the results given by the less explicable Support Vector Machine. But in general, the chosen model is relatively simple and gives satisfying results in consideration of the quality of the forecast.

2. Which exogenous parameters influence the sales market of the German automobile industry?

Here, it has to be differentiated between the yearly, monthly and quarterly data. In the yearly model, only a few exogenous parameters are needed to get satisfying results. But it is not possible to generalize the results, because of the very small data set. Also, in the monthly model also less exogenous parameters can be used. But most of the exogenous parameters are not published monthly, so that the exact values of these parameters are not given, leading to inadequate results. However, in the quarterly model, where the highest number of exogenous parameters is explicitly given, a reduction of the exogenous parameters in our tested model is not possible without decreasing the quality of the results.

3. Which collection of data points, yearly, monthly or quarterly data, is the most suitable one?

Yearly, monthly, and quarterly data are regarded. The problems in the yearly model are the very small data set and the small information content of the forecast. The problems presented by the monthly model include training and test errors which are much higher than in the yearly and quarterly models. Probable causes for the weakness of the monthly model are the inexact nature of the monthly data since most of the exogenous parameters are not collected monthly. The problems of the yearly model as well as the problems of the monthly model can be solved by using the quarterly model. Therefore, the quarterly model is the superior method, even though no reduction of exogenous parameters is possible in this model.

To conclude, it should be pointed out that forecasts are always plagued by uncertainty. There can be occurrences (special effects) in the future, which are not predictable or whose effects can not be assessed. The current financial crisis, which led to lower sales in the year 2008, is an example for such an occurrence. Because of this fact, forecasts can only be considered as an auxiliary means for corporate management and have to be interpreted with care [23].

References

1. Lewandowski, R.: Prognose- und Informationssysteme und ihre Anwendungen. de Gruyter, Berlin (1974)
2. Lewandowski, R.: Prognose- und Informationssysteme und ihre Anwendungen Band II. de Gruyter, Berlin (1980)
3. Dudenhöffer, F.: Prognosemethoden für den PKW-Markt: Das Beispiel Dieselfahrzeuge. In: WISU-Wirtschaftsstudium, pp. 1092–1100 (2002)
4. Dudenhöffer, F., Borscheid, D.: Automobilmarkt-Prognosen: Modelle und Methoden. In: Automotive Management. Strategie und Marketing in der Automobilwirtschaft, pp. 192–202 (2004)
5. Bäck, T., Hammel, U., Lewandowski, R., Mandischer, M., Naujoks, B., Rolf, S., Schütz, M., Schwefel, H.-P., Sprave, J., Theis, S.: Evolutionary Algorithms: Applications at the Informatik Center Dortmund. In: Genetic Algorithms in Engineering and Computer Science, pp. 175–204 (1997)
6. Statistisches Landesamt des Freistaates Sachsen, http://www.statistik.sachsen.de/21/14_01/14_01_definitionen.pdf (last accessed Feburary 2009)
7. Hüttner, M.: Markt- und Absatzprognosen. Kohlhammer, Stuttgart (1982)
8. Stier, W.: Methoden der Zeitreihenanalyse. Springer, Heidelberg (2001)
9. Stier, W.: Verfahren zur Analyse saisonaler Schwankungen in ökonomischen Zeitreihen. Springer, Heidelberg (1980)
10. Box, G.E.P., Jenkins, G.M.: Time Series Analysis forecasting and control. Holden-Day, San Francisco (1976)
11. Leiner, B.: Einführung in die Zeitreihenanalyse. R. Oldenbourg Verlag, München - Wien (1982)
12. Kessler, W.: Multivariate Datenanalyse. Wiley-VHC (2007)
13. Vapnik, V.: The Nature of Statistical Learning Theory. Springer, Heidelberg (1995)
14. Schölkopf, B., Smola, A.: Learning with Kernels. MIT Press, Cambridge (2002)
15. Christianini, N., Shawe-Taylor, J.: An Introduction to Support Vector Machines and other kernel-based methods. Cambridge University Press, Cambridge (2000)
16. Chen, K., Wang, C.: Support vector regression with genetic algorithms in forecasting tourism demand. In: Tourism Management, pp. 1–13 (2006)
17. Trafalis, T.B., Ince, H.: Support Vector Machine for Regression and Applications to Financial Forecasting. In: International joint conference on neutral networks, vol. 6, pp. 348–353 (2000)
18. Yale School of Public Health, http://publichealth.yale.edu/faculty/labs/guan/Papers%20under%20review/KPSS.pdf (last accessed Feburary 2009)
19. Dunteman, G.H.: Principal Component Analysis. Sage Publications, Thousand Oaks (1989)
20. Kohavi, R., John, G.H.: Wrappers for feature subset selection. Artificial Intelligence Journal, Special Issue on Relevance, 273–324 (1997)
21. Witten, I.H., Frank, E.: Data Mining. Morgan Kaufmann Publishers, San Francisco (2005)
22. Brühl, B.: Absatzprognosen für die Automobilindustrie in der Bundesrepublik Deutschland. Diploma Thesis, University of Cologne (2008)
23. Taleb, N.N.: The Fourth Quadrant: A Map of the Limits of Statistics. Edge 257 (2008), http://www.edge.org (last accessed April 2009)

Screening Paper Runnability in a Web-Offset Pressroom by Data Mining

A. Alzghoul[1], A. Verikas[1,2], M. Hållander[1], M. Bacauskiene[2], and A. Gelzinis[2]

[1] Intelligent Systems Laboratory, Halmstad University,
Box 823, S-30118 Halmstad, Sweden
[2] Kaunas University of Technology,
Studentu 50, LT-51368, Kaunas, Lithuania

Abstract. This paper is concerned with data mining techniques for identifying the main parameters of the printing press, the printing process and paper affecting the occurrence of paper web breaks in a pressroom. Two approaches are explored. The first one treats the problem as a task of data classification into *"break"* and *"non break"* classes. The procedures of classifier design and selection of relevant input variables are integrated into one process based on genetic search. The search process results in a set of input variables providing the lowest average loss incurred in taking decisions. The second approach, also based on genetic search, combines procedures of input variable selection and data mapping into a low dimensional space. The tests have shown that the web tension parameters are amongst the most important ones. It was also found that, provided the basic off-line paper parameters are in an acceptable range, the paper related parameters recorded online contain more information for predicting the occurrence of web breaks than the off-line ones. Using the selected set of parameters, on average, 93.7% of the test set data were classified correctly. The average classification accuracy of the break cases was equal to 76.7%.

Keywords: Classifier, GA, Mapping, Variable selection, Web break.

1 Introduction

To stand high competition in the market related to the printing industry, companies are striving to get the best possible return from their equipment [1,2]. In the printing industry, runnability is considered as being one of the most important factors affecting printing process productivity. High runnability is defined as a printing process without any faults, interruptions, and stops. Web[1] breaks, web instability, register errors and wrinkling are examples of runnability problems affecting productivity and the quality of the products [3]. Runnability problems cause huge financial losses. Paper web breaks are considered as the most significant runnability problem. Web breaks entail huge financial losses caused by the disruption of the production process due to the necessity of removing the

---

[1] Observe that "web" has nothing to do with "www" in this paper.

P. Perner (Ed.): ICDM 2009, LNAI 5633, pp. 161–175, 2009.

ruined paper and restarting the machine, possible machinery damage, and, in some cases, extra penalties due to not delivering on time [4]. A large number of factors, such as human faults, paper defects, too large variations of the paper thickness, formation, moisture, speed of printing, web tension may trigger a web break. Due to the large number of parameters and the rare occurring of the paper web breaks, the problem is very difficult to solve. Various approaches have been taken aiming to explain the reasons behind paper web breaks [3,5]. Most of the attempts focus on finding the parameters showing high correlation with the web break occurring.

This work aims at finding the most important parameters affecting web breaks by using soft computing techniques. Two approaches are explored. The first one treats the problem as a task of data classification into *"break"* and *"non break"* classes. The procedures of classifier design and selection of relevant input variables are integrated into one process based on genetic search. The search process results in a set of variables providing the lowest average loss incurred in taking decisions. The second approach, also based on genetic search, combines procedures of input variable selection and data mapping into a low dimensional space. The genetic search process results into a variable set providing the best mapping according to some quality function. The quality function used is such that the disparity between the breaks and non-breaks data is emphasized. We rely on genetic search in this work, since genetic algorithms have proved themselves powerful tools for exploring high-dimensional spaces of variables [6].

2   Related Work

Various approaches have been taken aiming to explain the reasons behind paper web breaks [3,5,7]. Most of the attempts focus on finding the parameters showing high correlation with the web break occurring. There are many parameters, which may be responsible for the runnability problems and some of them are difficult or even impossible to measure. Therefore, some researchers use modeling and simulation to analyze the paper structure, printing process, and the paper-press interaction analytically [5]. Other researchers based their analysis solely on data, collected during the production process. For example, Parola et al. [3] gathered a large amount of data from a printing press and a paper mill and used data mining techniques to analyze the printing press runnability. The goal of this work was to identify the factors responsible for the runnability problems in a pressroom. Miyaishi and Shimada [7], developed an artificial neural network (a multilayer perceptron) to tackle the web breaks problem on a commercial newsprint paper machine. The network was trained to predict the occurrence of a web break. The magnitude of the network weights in the input layer has been used to reduce the number of the input variables by keeping those related to the largest input weights.

There are various factors, which may trigger a paper web break. *Printing press related* factors, *unknown* factors, and *paper related* factors are the main groups of factors. Usually, press related factors compose the largest group of factors trigging paper web breaks. Thus, the optimization of the press operator actions, could be one of the most important measures to reduce the number of paper web breaks. There is a huge variety of parameters, which may influence printing press and paper related factors trigging a paper web break. Previous studies have shown that the most important parameters influencing paper web break factors are: *web strength, slack area in the paper, web tension,* and *pale spots* [5]. However, these parameters are not so easy to control because the web tension varies from time to time, especially when a paper roll is changed. Moreover, the web strength is not constant across the whole web and sometimes the web contains holes or weaknesses.

Parameters affecting printing press related factors causing paper web breaks may be quite pressroom specific, depending on the printing press and paper interaction. However, some general conclusions may also be drawn from such pressroom specific studies. Pressrooms experiencing frequent paper web breaks are valuable information sources for such studies. This work is done using data from such a pressroom. The work aims at finding the most important parameters affecting web breaks by using soft computing techniques. Two main approaches are explored. The first one treats the problem as a task of data classification into *"break"* and *"non break"* classes. The procedures of classifier design and the selection of relevant input variables (features) are integrated into one process based on genetic search [8]. The search process results in a set of input variables providing the lowest average loss incurred in taking decisions. The second approach, also based on genetic search, combines procedures of input variable selection and data mapping into a low dimensional space. The curvilinear component analysis is employed for implementing the mapping [9]. The genetic search process results into a variable set providing the best mapping according to some quality function. The quality function used is such that the disparity between the breaks and non-breaks data is emphasized. The integration of classification and variable selection or mapping and variable selection processes allows finding the most important variables according to the quality functions used to assess the classification or mapping results.

3   The Pressroom and the Data

3.1   The Pressroom

Offset printing is the most common printing technology. Offset printing is an indirect lithographic printing technology, where ink is transferred from an ink tray onto a printing plate and then onto a printing surface via a flexible blanket. Four primary colours: cyan, magenta, yellow, and black are usually used to create colour printed pictures in offset printing [10]. There is a separate printing plate four each colour.

Regarding paper feed into a printing press, offset printing can be categorized into *Sheet-fed offset* and *Web-fed offset*. There are two types of web-fed offset, the so called: *Coldset web*, where the ink is drying due to absorption and evaporation, and *Heatset web*, where hot air dryers are used to dry the ink. In this work, a four colour (cyan, magenta, yellow, and black) heatset web-fed offset printing press has been used to collect the data.

3.2   The Data

There are two types of variables: *paper related* and *printing press related* variables. All the paper related variables are recorded at a paper mill, online as well as off-line. There are 110 paper-related variables in total. The first 84 paper variables are computed from online measurements recorded during the paper production process by the commercial measuring system called Measurex. There are three types of measurements available from the Measurex system, namely *moisture content*, *thickness* and *dry weight*. The measuring head equipped with the three sensors traverses across the web and makes snapshots of moisture content, thickness and dry weight. The measurements are collected into three matrices. The matrices are then considered as images and 28 co-occurrence matrices-based Haralick's coefficients [11,12], characterizing image texture, are computed for each of the images. Fourteen coefficients are obtained from one co-occurrence matrix computed in the vertical (MD) direction and the other fourteen from one co-occurrence matrix computed in the horizontal (CD) direction. The ij element of the co-occurrence matrix reflects the probability of finding the grey levels i and j in the image in a given distance and a given direction. Homogeneity, abrupt transition, the amount of local variation, uniformity, variance are the main properties expressed by Haralick's coefficients. The other paper related variables are obtained from the winding machine, paper defects detection system or recorded at the paper lab.

Web tension, ink registration, web position, moisture content, and *printing speed* are the groups of press-related variables. There are 31 press-related variables in total. All the printing press related data have been collected online. The printing press related and paper related data were linked by reading a bar code available on each paper roll. There is only one dependent variable—a binary $(-1/1)$ web break variable (no break/break for a given paper roll). Thus, the ith paper roll is characterized by the ith data point given by $(\mathbf{x}_i, y_i)$, where the vector $\mathbf{x}_i$ has 141 components and the scalar y_i is a binary variable.

There were 759 data points available for the experiments. Amongst those, 112 web break cases. Not all the 141 variables were available for all the 759 paper rolls investigated: some entries in the data matrix $\mathbf{X}$ of size 759×141 were missing. Therefore, single groups and several groups of variables have been used in different experiments.

4   Methods

Since the work is aiming at finding a small set of the most important variables affecting the occurrence of web breaks, the task can be treated as a variable selection problem. We start the work by exploring linear relations between the independent and the dependent variables. The existence of strong linear relations would indicate a possibility of finding a simple linear model. Correlation analysis and the multivariate linear regression are the two techniques used to examine the linear relations in this work. We use the Pearson correlation coefficient to assess the strength of the linear relationships. To estimate the parameters of the linear model, the least squares technique is applied. Next, two genetic algorithms (GA) based approaches are applied to solve the task. In the first one, the task is considered as a two-class (break/non-break) classification problem–classification-based genetic search. In the second approach, relations between various process parameters and the occurrence of web breaks are explored through mapping the multi-dimensional process data into a low dimensional space–mapping-based genetic search.

4.1   Exploring Linear Relations

To determine the statistical significance of the Pearson correlation coefficient r, the $p-value$ has been used in this work. The $p-value$ expresses the probability of obtaining the computed correlation coefficient value by chance.

Given a vector $\mathbf{x} = (x_1, x_2, ..., x_n)^T$ of independent variables, the multivariate linear regression model for predicting the dependent variable y can be written as

$$y = \sum_{j=1}^{n} x_j \beta_j = \mathbf{x}^T \boldsymbol{\beta} \tag{1}$$

The optimal values of the parameters β_j are given by $\boldsymbol{\beta} = (\mathbf{X}^T \mathbf{X})^{-1} \mathbf{X}^T \mathbf{y}$, where $\mathbf{X}$ is a $N \times n$ matrix of input data and $\mathbf{y}$ is the N-vector output. We use the $z-score$, to test the statistical significance of β_j,

$$z_j = \frac{\beta_j}{\sigma \sqrt{d_j}} \tag{2}$$

where d_j is the j diagonal element of the matrix $(\mathbf{X}^T \mathbf{X})^{-1}$ and σ is the standard deviation of the noise. If we assume that σ is known, then z_j has a standard normal distribution and a $1 - 2\alpha$ confidence interval for β_j is

$$(\beta_j - z^{(1-\alpha)} \sigma \sqrt{d_j}, \beta_j + z^{(1-\alpha)} \sigma \sqrt{d_j}) \tag{3}$$

where $z^{(1-\alpha)}$ is the $1 - \alpha$ percentile of the normal distribution, for example $z^{(1-0.025)} = 1.96$. Thus, the approximate 95% confidence interval is given by $\beta_j \pm 2\sigma \sqrt{d_j}$.

4.2   Genetic Algorithm

Information representation in a chromosome, generation of initial population, evaluation of population members, selection, crossover, mutation, and reproduction are the issues to consider when designing a genetic search algorithm.

A **chromosome** contains all the information needed to build a support vector machine (SVM) classifier. We divide the chromosome into three parts. One part encodes the regularization constant C, one the kernel width parameter σ, and the third one encodes the inclusion/noninclusion of features. To generate the **initial population**, the features are masked randomly and values of the parameters C and σ are chosen randomly from the interval $[C_0 - \Delta C, C_0 + \Delta C]$ and $[\sigma_0 - \Delta\sigma, \sigma_0 + \Delta\sigma]$, respectively, where C_0 and σ_0 are the very approximate parameter values obtained from the experiment. The **fitness function** used to evaluate a chromosome is given by the difference between the maximum possible and the average loss incurred when using the model encoded in the chromosome to classify the validation set data.

The **selection process** of a new population is governed by the fitness values. A chromosome exhibiting a higher fitness value has a higher chance to be included in the new population. The selection probability of the ith chromosome p_i is given by

$$p_i = \frac{\Lambda_m - \overline{\Lambda}_i}{M\Lambda_m - \sum_{j=1}^{M} \overline{\Lambda}_i} \tag{4}$$

where Λ_m is the maximum possible loss, $\overline{\Lambda}_i$ is the average loss obtained using the classifier encoded in the ith chromosome and M is the population size. We assume that there is no loss incurred by the correct decisions. There is a loss of $\Lambda_{nb} = 1$ if a non-break case is classified as break and a loss of $\Lambda_{bn} > 1$ if a break case is classified as a non break one.

The **crossover operation** for two selected chromosomes is executed with the probability of crossover p_c. Crossover is performed separately in each part of a chromosome. The crossover point is randomly selected in the "feature mask" part and two parameter parts.

The **mutation operation** adopted is such that each gene is selected for mutation with the probability p_m. The mutation operation is executed independently in each chromosome part. If the gene selected for mutation is in the feature part of the chromosome, the value of the bit representing the feature in the feature mask (0 or 1) is reversed. To execute mutation in the parameter part of the chromosome, the value of the offspring parameter determined by the selected gene is mutated by $\pm\Delta\gamma$, where γ stands for C or σ, as the case may be. The mutation sign is determined by the fitness values of the two chromosomes, namely the sign resulting into a higher fitness value is chosen. The way of determining the mutation amplitude $\Delta\gamma$ is somewhat similar to that used in [13] and is given by

$$\Delta\gamma = w\beta(\max(|\gamma - \gamma_{p1}|, |\gamma - \gamma_{p2}|)) \tag{5}$$

where γ is the actual parameter value of the offspring, $p1$ and $p2$ stand for parents, $\beta \in [0,1]$ is a random number, and w is the weight decaying with the iteration number:

$$w = k\alpha^t \tag{6}$$

where t is the iteration number, $\alpha = 0.95$ and k is a constant. The constant k defines the initial mutation amplitude. The value of $k = 0.4$ worked well in our tests.

In the **reproduction process**, the newly generated offspring replaces the chromosome with the smallest fitness value in the current population, if a generated random number from the interval [0,1] is smaller than the reproduction probability p_r or if the fitness value of the offspring is larger than that of the chromosome with the smallest fitness value.

4.3   Classification-Based Genetic Search

The problem is treated as a task of data classification into the *break* and *non break* classes. During the genetic search a classifier is designed using N labelled samples. Since both hyper-parameters of the classifier and the input variables are encoded in a chromosome, the procedures of classifier design and selection of relevant input variables are integrated into one process, based on genetic search. The genetic search results into a set of best chromosomes (according to the fitness function given by Eq.(4)) containing information on the most important variables and the hyper-parameters values.

SVM [14] is one of the most successful and popular classifiers. The advantages of SVM are the following: the ability to find the global minimum of the objective function, no assumptions made about the data, the complexity of SVM depends on the number of support vectors, but not on the dimensionality of the transformed space. SVM is well-suited for integration into the genetic search process. Therefore, we have chosen SVM as a base classifier in this work.

4.4   Mapping-Based Genetic Search

In this approach, relations between various process variables and the occurrence of web breaks are explored through mapping the multi-dimensional process data into a two-dimensional space. We expect that the web break and non break cases will be mapped into more or less distinctive areas of the space. The emergence of such distinctive areas of break and non break cases is promoted in the genetic search-based mapping process through the optimization of a specifically designed fitness function. The search aims at finding an optimal feature subset according to the fitness function given by the difference between the maximum possible and the average loss incurred when using the model encoded in the chromosome to classify the validation set data represented by two components of the two-dimensional space, Eq.(4). An SVM is used to make the classification.

Multi-Dimensional Scaling (MDS) [15], Sammon mapping (SM) [16], Generative Topographic Mapping (GTM) [17], Self-Organizing Maps (SOM) [18], Curvilinear Component Analysis (CCA) [9] are the most popular nonlinear mapping techniques. Previous studies have shown that CCA is superior to other nonlinear techniques, such as MDS, SOM, performing similar tasks. CCA is a relatively fast and immune to noise mapping technique, if compared to nonlinear MDS, for example [9]. Therefore, CCA was chosen to perform the mapping.

As one can notice, an SVM is used to assess the obtained solution in both the classification- and mapping-based approaches. In the mapping-based approach, the SVM is always trained with 2-D mapped data, while in the classification based approach, the dimensionality of data used to train the SVM is equal to the number of selected salient input variables. One can expect lower classification accuracy from the mapping-based approach. However, the mapping-based technique is rather transparent and easy to understand for the user.

4.5 Support Vector Machine

The 1-norm SVM [19] was used in this work. Assuming that $\boldsymbol{\Phi}(\mathbf{x})$ is the non-linear mapping of the data point $\mathbf{x}$ into the new space, the 1-norm soft margin SVM can be constructed by solving the following minimization problem [19]:

$$\min_{\mathbf{w},b,\gamma,\boldsymbol{\xi}} -\gamma + C\sum_{i=1}^{N} \xi_i \tag{7}$$

subject to

$$y_i(\langle \mathbf{w}, \boldsymbol{\Phi}(\mathbf{x}_i)\rangle + b) \geq \gamma - \xi_i, \quad \xi_i \geq 0, \quad \parallel \mathbf{w} \parallel^2 = 1, \quad i = 1, ..., N \tag{8}$$

where $\mathbf{w}$ is the weight vector, $y_i = \pm 1$ is the desired output (± 1), N is the number of training data points, $\langle \rangle$ stands for the inner product, γ is the margin, ξ_i are the slack variables, b is the threshold, and C is the regularization constant controlling the trade-off between the margin and the slack variables. The discriminant function for a new data point $\mathbf{x}$ is given by:

$$f(\mathbf{x}) = \mathcal{H}\left[\sum_{i=1}^{N} \alpha_i^* y_j k(\mathbf{x}, \mathbf{x}_i) + b^*\right], \tag{9}$$

where $k(\mathbf{x}, \mathbf{x}_i)$ stands for the kernel and the Heaviside function $\mathcal{H}[y(\mathbf{x})] = -1$, if $y(\mathbf{x}) \leq 0$ and $\mathcal{H}[y(\mathbf{x})] = 1$ otherwise. In this work, the Gaussian kernel, $\kappa(\mathbf{x}_i, \mathbf{x}_j) = \exp\{-||\mathbf{x}_i - \mathbf{x}_j||^2/\sigma\}$, has been used. The optimal values α_i^*, b^* of the parameters α_i and b are found during training.

4.6 Curvilinear Component Analysis

CCA aims to map the data in such a way that the local topology is preserved. The mapping is implemented by minimizing a *cost function* based on matching the inter-point distances in the input and output spaces [9].

Let the Euclidean distances between a pair of data points (i, j) be denoted as $\chi_{ij} = d(\mathbf{x}_i, \mathbf{x}_j)$ and $\vartheta_{ij} = d(\mathbf{y}_i, \mathbf{y}_j)$ in the input and the output space, respectively. Then, the cost function minimized to obtain the mapping is given by [9]:

$$E = \frac{1}{2}\sum_{i}\sum_{j\neq i}(\chi_{ij} - \vartheta_{ij})^2 F(\vartheta_{ij}, \lambda_y) = \frac{1}{2}\sum_{i}\sum_{j\neq i} E_{ij} \tag{10}$$

where the weighting function $F(\vartheta_{ij}, \lambda_y)$, with the neighbourhood parameter λ_y, is used to favor local topology preservation. $F(\vartheta_{ij}, \lambda_y)$ should be a bounded and decreasing function, for example, a sigmoid, unit step function, decreasing exponential or Lorentz function [9]. In this work, a step function was used:

$$F(\vartheta_{ij}, \lambda_y) = \begin{cases} 1, \text{ if } \vartheta_{ij} \leq \lambda_y \\ 0, \text{ if } \vartheta_{ij} > \lambda_y \end{cases} \tag{11}$$

5   Experimental Tests

The value of loss $\Lambda_{bn} = 6$, has been used for a break case classified as a non break one. Six is the approximate number of times the non break cases exceed the break ones in the collected data set. In all the tests, to assess the average loss and the average classification accuracy, CA, (for example, when evaluating the fitness function value or comparing different models), we repeat the experiments 10 times using different random split of the data into training 70% and test 30% sets. The estimated CA is then given by the average CA of the test set data, calculated from these 10 runs. The following parameter values have been used in the genetic search: $p_c = 0.95$, $p_m = 0.01$, and $p_r = 0.05$. The values were found from experiments.

5.1   Exploring Linear Relations

Table 1 presents values of the correlation coefficient exhibiting the $p-$values lower than 0.05. The $p-$value expresses the probability of obtaining such correlation by chance. The first three variables express variations of dry weight calculated from online Measurex measurements. There are seven other paper related variables. The last three variables characterize web tension, where axis 1 and axis 2 mean that web tension is measured before and after the printing nip, respectively. The correlation coefficient values for all the variables presented in Table 1 are relatively low, meaning that the linear relations between the independent variables and the occurrence of web breaks are not strong. One can notice that the correlation coefficient values computed for the web tension related variables are slightly higher than for the other ones. Using variables presented in Table 1 an attempt was made to create an SVM classifier, based on the genetic search procedures, for categorizing the data into the *break* and *non break* classes. The obtained classification accuracy was far below that reported in the next section.

When using the linear regression model, nine variables significant at the 95% confidence level ($z-$score> 1.96) were found. Table 2 presents the selected variables along with the $z-$score values. The SVM classifier created using the selected variables provided a considerably higher performance than those based on the variables selected via the correlation analysis. However, the classification accuracy obtained was significantly lower than the highest accuracy obtained using the GA based approaches discussed in the following sections.

5.2   Genetic Search Alternatives

The genetic sear based techniques were applied to the following four groups of variables

Table 1. The coefficient of correlation between the independent variables and the occurrence of web breaks for the p−values lower than 0.05

Variable	r
Measurex data	
Local variation of dry weight in MD	0.23
Uniformity of dry weight in MD	0.30
Homogeneity of dry weight in CD	0.26
Other paper related data	
Tensile index in MD	−0.24
Tear strength in CD	−0.25
Surface roughness	−0.27
Anisotropy	−0.26
Paper elongation in MD	−0.27
Pinhole area	0.28
Moisture content	−0.28
Variance of winding hardness	0.26
Printing press data	
Web tension variance, axis 1	0.36
Web tension variance, axis 2	0.31
Variance of web tension moment	0.41

Table 2. The variables included into the linear model along with the z−score values

Variable	z−score
Measurex data	
Moisture content uniformity in CD	−2.29
Thickness uniformity in CD	3.21
Dry weight homogeneity in MD	−2.57
Other paper related data	
Tensile index in CD	3.21
Tear strength in CD	3.26
Surface roughness	−4.68
Printing press data	
Slope of main registry adjustments	−2.41
Variance of web tension moment	7.52
Max of sliding web tension mean	2.58

i. Paper related data, excluding Measurex. Winding machine data are often missing in similar studies.
ii. Measurex data. To our knowledge, Measurex data have not been used to screen web breaks in earlier studies.

iii. All paper related data. To our knowledge, such a variety of variable types has not been used in earlier studies.
iv. All paper and printing press related data. A unique attempt to exploit such a comprehensive set of parameters.

Next, we summarize the results obtained for these four groups of variables.

5.3   Paper Related Data, Excluding Measurex

As can be seen from Table 1, there are eight variables from this group showing the statistically significant correlation with the occurrence of web breaks on the 95% significance level. However, *tensile index in MD, surface roughness, anisotropy*, and *paper elongation in MD* is a group of very correlated parameters. The classification-based genetic search has chosen the following six parameters: *ash content, surface roughness, winding hardness, winding speed, edge/non edge roll*, and *distance from the tambor center*. Thus, only one parameter from the group of very correlated parameters was included into the final set. Observe that paper is produced in big rolls called tambors, which are then cut into smaller rolls. Using the selected parameters, 85.3% of the test set data were classified correctly. However, the classification accuracy for the break cases was only 40.0%. The classification accuracy obtained using the mapping-based genetic search was lower.

5.4   Measurex Data

Three groups of measurements, *moisture content, thickness*, and *dry weight*, constitute the Measurex data. The classification-based genetic search resulted into five parameters characterizing: *abrupt change of dry weight, abrupt change of thickness, abrupt change of moisture, non-homogeneity of dry weight*, and *non-homogeneity of thickness*. Using the selected parameters, 85.2% of the test set data were classified correctly. The classification accuracy for the break cases was 56.7%. The results indicate that, provided the basic off-line paper parameters are in an acceptable range, *Measurex* data may provide more information for predicting the occurrence of web breaks than all other paper related parameters. Again, the classification accuracy obtained using the mapping-based genetic search was lower.

5.5   All Paper Related Data

In the next set of experiments, the genetic search was applied to all the paper related variables. *Dry weight homogeneity, thickness homogeneity, abrupt dry weight change*, and *abrupt moisture change* were the four parameters chosen from the Measurex group. Only *ash content* and *number of paper defects* were the parameters chosen from the group of other paper related parameters. Using the selected set of parameters, 90.6% of the test set data were classified correctly. The classification accuracy for the break cases was equal to 71.7%. Fig. 1 illustrates the paper related data mapped onto the first two CCA components calculated

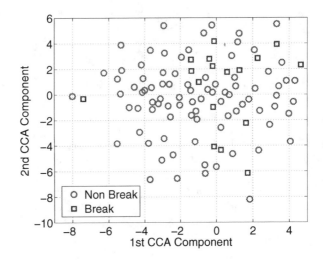

Fig. 1. The paper related data mapped onto the first two CCA components calculated using the four selected Measurex and two other paper related parameters

using the six selected paper parameters. As can be seen from Fig. 1, a break free region can be found in the plot. Observe that some web breaks occur due to various unforeseen events, which have little to do with the paper, the printing press or the operator actions. For instance, due to ink drop on the paper web. The leftmost break case in Fig. 1 is an example of such a web break. Observe also that the data in the 6D space may be much less overlapping compared to the view seen in the two-dimensional plot.

5.6   The Whole Set of Variables

When applying the classification-based genetic search to the whole set of parameters, eight variables were selected. Three variables selected from the Measurex group reflect *uniformity of dry weight, abrupt change of dry weight,* and *abrupt change of thickness.* Two variables selected from the group of other paper related parameters are *winding machine speed* and *distance from the tambor center.* The group of printing press related parameters is represented by: *web tension moment variance axis 2, maximum value of the sliding web tension mean,* and *number of ink registry adjustments.* Using the selected set of parameters, 93.7% of the test set data were classified correctly. The classification accuracy for the break cases was equal to 76.7%. Fig. 2 illustrates the data mapped onto the first two CCA components calculated using the selected parameters. The CCA plot shown in Fig. 2 clearly reveals that the break cases are concentrated on the upper part of the plot. Table 3 summarizes the average test data classification accuracy (CA) obtained for the classification-based genetic search technique using the different groups of variables.

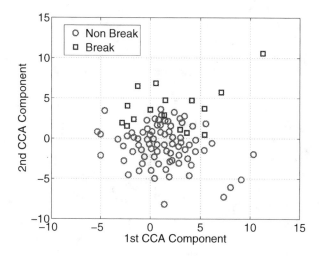

Fig. 2. The data mapped onto the first two CCA components calculated using the variables selected from the whole set of parameters

Table 3. The average test data classification accuracy obtained for the classification-based genetic search technique using the different groups of variables

Group of variables	Overall CA,%	Break CA,%
Paper related without Measurex	85.3	40.0
Measurex	85.2	56.7
All paper related	90.6	71.7
All paper and press related	93.7	76.7

The mapping-based search technique relies on classification of 2-dimensional mapped data, while the classification-based technique usually selects more than two variables to design a classifier. Therefore, the classification accuracy obtained for the mapping-based genetic search technique was considerably lower than the accuracy achieved by the classification-based genetic search technique. However, the mapping-based technique proved to be a good data visualization tool, since mappings produced by the technique exhibit data ordering property [20,21] and are not constrained by the discrete grid, as in the case of SOM.

The analysis performed has shown that the web tension parameters are amongst the most important ones. It was found that there is a statistically significant correlation between parameters characterizing the winding machine data and the web tension parameters. For example, a statistically significant correlation was observed between the *hardness variance* and *web tension axis 2 variance*, between the *winding speed variance* and *web tension mean axis 2*. The binary *edge/non edge paper roll* variable has not been selected amongst the

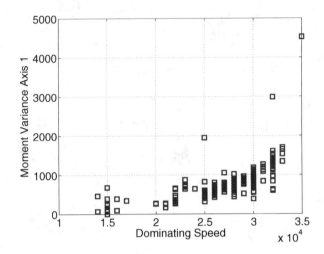

Fig. 3. The tension moment variance (axis 1) as a function of the dominating speed

most important ones. However, there is an indication that edge paper rolls create larger *web tension on axis 2* and require more ink registry adjustments. There is a clear relation between the *dominating printing speed* and the *web tension moment (axis 1) variance*. Fig. 3 illustrates the relation.

6   Conclusions

Several parameters characterizing paper and the printing process have statistically significant correlation with the occurrence of web breaks. However, the correlation is not strong. The analysis of nonlinear relations between various parameters and the occurrence of web breaks has shown that the web tension parameters *web tension moment variance axis 2* and *maximum value of the sliding web tension mean* are amongst the most important ones. There is a statistically significant correlation between parameters characterizing the winding machine data and the web tension parameters. Surprisingly enough, the binary *edge/non edge paper roll* variable has not been selected amongst the most important ones. However, there is an indication that edge paper rolls create larger web tension on axis 2 and require more ink registry adjustments. There is a clear relation between the *dominating printing speed* and the *web tension moment (axis 1) variance*.

Parameters calculated from the *Measurex* data have been found to be rather informative. There is an indication that, provided the basic off-line paper parameters are in an acceptable rage, the Measurex data contain more information for predicting the occurrence of web breaks than the other paper related data. Using the selected set of parameters, 93.7% of the test set data were classified correctly. The classification accuracy for the break cases was equal to 76.7%.

References

1. Perner, P.: Knowledge-based image inspection system for automatic defect recognition, classification and process diagnosis. Machine Vision and Applications 7, 135–147 (1994)
2. Klapproth, U., Perner, P., Böhme, B.: Einsatz eines expertensystems zur diagnose von druckfehlern im offsetdruck. Messen steuern regeln 34, 115–120 (1991)
3. Parola, M., Kaljunen, T., Beletski, N., Paukku, J.: Analysing printing press runnability by data mining. In: Prceedings of the TAGA Conference, Montreal, Canada, pp. 435–451 (2003)
4. Roisum, R.: Runnability of paper. Part 1: Predicting runnability. TAPPI Journal 73, 97–101 (1990)
5. Hristopulos, D.T., Uesaka, T.: A model of machine-direction tension variations in paper webs with runnability applications. Journal of Pulp and Paper Science 28, 389–394 (2002)
6. Bacauskiene, M., Verikas, A., Gelzinis, A., Valincius, D.: A feature selection technique for generation of classification committees and its application to categorization of laryngeal images. Pattern Recognition 42, 645–654 (2009)
7. Miyaishi, T., Shimada, H.: Using neural networks to diagnose web breaks on a newsprint paper machine. TAPPI Journal 81, 163–170 (1998)
8. Holland, J.H.: Adaptation in Natural and Artificial Systems: An Introductory Analysis with Applications to Biology, Control, and Artificial Intelligence. University of Michigan Press, Ann Arbor (1975)
9. Demartines, P., Herault, J.: Curvilinear component analysis: A self-organizing neural network for nonlinear mapping of data sets. IEEE Trans Neural Networks 8, 148–154 (1997)
10. Verikas, A., Malmqvist, K., Malmqvist, L., Bergman, L.: A new method for colour measurements in graphic arts. Color Research & Application 24, 185–196 (1999)
11. Haralick, R.M., Shanmugam, K., Dinstein, I.: Textural features for image classification. IEEE Trans System, Man and Cybernetics 3, 610–621 (1973)
12. Gelzinis, A., Verikas, A., Bacauskiene, M.: Increasing the discrimination power of the co-occurrence matrix-based features. Pattern Recognition 40, 2367–2372 (2007)
13. Leung, K.F., Leung, F.H.F., Lam, H.K., Ling, S.H.: Application of a modified neural fuzzy network and an improved genetic algorithm to speech recognition. Neural Computing & Applications 16, 419–431 (2007)
14. Vapnik, V.N.: Statistical Learning Theory. Wiley, New York (1998)
15. Borg, I., Groenen, P.J.F.: Modern Multidimensional Scaling: Theory and Applications. Springer, New York (1997)
16. Sammon, J.W.: A nonlinear mapping for data structure analysis. IEEE Tans. Computers 18, 401–409 (1969)
17. Bishop, C.M., Svensen, M., Williams, C.K.I.: GTM: The generative topographic mapping. Neural Computation 10, 215–234 (1998)
18. Kohonen, T.: The self-organizing map. Proceedings of the IEEE 78, 1461–1480 (1990)
19. Shawe-Taylor, J., Cristianini, N.: Kernel Methods for Pattern Analysis. Cambridge University Press, Cambridge (2004)
20. Verikas, A., Malmqvist, K., Bacauskiene, M., Bergman, L.: Monitoring the deinking process through neural network-based colour image analysis. Neural Computing & Applications 9, 142–151 (2000)
21. Bauer, H.U., Pawlezik, K.R.: Quantifying the neighbourhood preservation of selforganizing feature maps. IEEE Trans Neural Networks 3, 570–579 (1992)

Evaluation of Distraction in a Driver-Vehicle-Environment Framework: An Application of Different Data-Mining Techniques

Fabio Tango and Marco Botta

University of Torino, Department of Computer Science
{tango,botta}@di.unito.it

Abstract. Distraction during driving task is one of the most serious problems affecting traffic safety, being one of the main causes of accidents. Therefore, a method to diagnose and evaluate Distraction appears to be of paramount importance to study and implement efficient counter-measures. This research aims at illustrating our approach in diagnosis of Distraction status, comparing some of the widely used data-mining techniques; in particular, Fuzzy Logic (with Adaptive-Network-based Fuzzy Inference System) and Artificial Neural Networks. The results are compared to select which method gives the best performances.

Keywords: Fuzzy Logic, Adaptive-Network-based Fuzzy Inference System, Neural Networks, Machine Learning, Distraction, Traffic Safety.

1 Introduction

Since its introduction over a century ago, automobiles have enabled individual mobility for an ever growing part of human population. In particular, passenger-cars provide multi-purpose and flexible transportation, playing an important economic, social and cultural role in society. Unfortunately, motorised traffic has several drawbacks as well, including problems of *accessibility / traffic jam*, *sustainability* and *safety*. Specifically, traffic safety is one of the most urgent and big problem that European society has to deal with, due to the dimension of the phenomenon, to the high number of people involved (dead or severely injured) and to the consequent social costs.

It is well known that the great majority of road accidents (> 80%) are caused by human error [1] or, more generally speaking, by human (wrong) behaviour. More recent data have identified inattention (including distraction, "looked but did not see" and falling asleep at the wheel) as the primary cause of accidents, accounting for at least 25% of the crashes [2]. Moreover, in a jointly report of the National Highway Traffic Safety Administration and of the Virginia Tech Transportation Institute, driver's inattention has been found as the leading factor for most crashes and near crashes: this report states that 80% of collisions and 65% of near-collisions involved driver's inattention, from drowsiness to cell-phone use, within three seconds before the event. Any activity that distracts the drivers or competes for their attention to the primary task has the potential to degrade user's performance and so can have serious consequences for road safety. The American Automobile Association Foundation for

P. Perner (Ed.): ICDM 2009, LNAI 5633, pp. 176–190, 2009.

Traffic Safety defines driver distraction as occurring *"when a driver is delayed in the recognition of information needed to safely accomplish the driving task because some event, activity, object or person within or outside the vehicle compelled or tended to induce the driver's shifting attention away from the driving task"* ([3], p. 21).

Section 2 presents different methods for Distraction (DIS) evaluation, focusing on the one followed by our research. Section 3 gives a brief overview about the experimental set-up and the tests carried out to collect data for DIS model development. Its formulation is the focus of Section 4, where the details of both the implemented algorithms are given. Data processing and related results are pointed out in Section 5. The paper ends with Section 6 and 7, which illustrate the conclusions (including a critical analysis) and the next steps for future researches.

2 Different Methods for Distraction Evaluation

Several studies have examined the relative levels of distraction afforded by various in-vehicle systems or activities. All in all, these studies suggest that the more complex sources of distraction (e.g., navigation systems) afford greater levels of driver distraction than do more 'simple' sources (e.g., eating or smoking) which require fewer steps to be completed. A number of researchers have found that using in-vehicle systems can increase drivers' reaction times to a periodically braking lead vehicle by 30 percent ([4] for the effects of cellular phone on driving and [5] for some studies about the effects by operating a route guidance system while driving). In all these studies, in order to assess the driver's performance, one of the most used indicator was the Reaction Time (RT).

In the current state of the art, the use of indicators about the primary driving task, (i.e. standard deviation of the steering angle and lateral position) is very common to evaluate the DIS parameter. We have therefore included these measurements for the development of our model; however, due to the variety of conditions and scenarios (different road segments, different types of roads, and so on), the data of steering angle and lateral position are very "noisy" and with a large variability; thereby, it would be definitely difficult to understand when they were due to the secondary task activities and when to the natural driving manoeuvres carried out by the drivers in their task. Thus, we have decided to use the reaction time as target to develop the Distraction model, but evaluated on the secondary task and not directly on the primary driving task. From the literature, it is usually regarded as a measure of the performance achieved in the secondary task and thus, it denotes the level of cognitive workload required by the fulfilment of the secondary task. Our hypothesis is that when the reaction time (RT) is low, this means good performances on the secondary task and therefore the driver is able to manage safely both the primary and the secondary tasks; in other words, the distraction induced by the secondary task on the primary one is minimal. On the other hand, if the RT has high values, this means that the driver is splitting the cognitive and visual resources between the primary and the secondary task; this event does not lead to a high driving performance and – above all – it means that the secondary task is inducing a source of distraction with respect to the primary task (because the driver is managing the sharing of attention between primary and secondary task with a very low level of performance). In our experiments, the secondary task was represented by the activity required to the subjects and based on the HASTE methodology [6]. Such a procedure is in line with the so-called *compensatory theory*, for which when the user has to deal with multiple tasks of different levels of difficulty, he/she tries to balance the resources (searching for the trade-off) in order to achieve a general good level of global performances (both on primary and on secondary task) [7].

3   The Experimental Design and Set-Up

The experiments have been carried out with "real subjects" using the static driving simulator in the *Human Machine Interaction Laboratory of University of "Modena e Reggio Emilia"* (Italy). They were designed to reproduce different driving conditions, as requested by the Distraction model development [8]. The level of difficulty of secondary task interaction was defined according to [9].

3.1   Test Methodology

The simulator used is our experiment is the OKTAL Driving Simulator (www.oktal.fr) which has the following main features:

- Real car cabin equipped including steering wheel, ordinary pedals and a dashboard emulating the vehicle instruments panel (i.e. tachometer, RPM);
- Projector device and frontal projector screen for the display of the driving scenario;
- SCANER2 and EVARISTE (www.scaner2.com) Simulation Software platforms responsible for the simulation of the vehicle model dynamics controlled by the driver, the traffic and weather conditions, scenario projection, data recording of hundreds of vehicle, drivers' interaction data as well as for the design and development of virtual driving scenarios

Concerning the test, twelve participants between 19 and 40 years of age were involved in the experiment. Drivers were supposed to drive a car on at least 4 days per week, and to drive at least 10 000 km per year

Participants were asked to:

- Drive along a simulated track, respecting road rules;
- Perform a secondary detection task when required to.

Concerning the second bullet-point, an LCD monitor was placed nearby the driver's seat, randomly presenting visual stimuli at two different levels of cluttering, associated to different degrees of task difficulty; a touch screen allowed the user to call whether a target stimulus was presented on the left, centre or right of the screen. The secondary task was provided at 2 levels of difficulty:

- Easy;
- Difficult.

The difficult level is defined by the number of symbols (i.e. circles) displayed on the screen; among them, a symbol was different (based on [6]). The driver was asked to touch the screen side the different symbol was placed. In particular, this "look and find" task was divided into several steps:

- to locate in which side the different symbol is placed;
- to verbalize the answer (saying "Right" or "Left" of the display);
- to touch the monitor in the verbalized side.

Each driver was asked to complete each passage described in the above mentioned list, otherwise the answer should have been considered not correct. Moreover, participants were instructed *to perform both tasks as rapidly and accurately as possible, still keeping safety as the highest priority.*

The experimental track was 24000 m (= 24 km) long and articulated into nine segments, each of them corresponding to different driving conditions. These ones were defined by different combinations of environmental and road traffic situations, as well as two degrees of secondary task difficulty. Hence, this track was accordingly divided into 9 sections, each 2750 m long; in particular:

- 2000m of driving condition;
- 750m of gradual transition from one condition to the following

Subjects were told to give priority to the primary task (so, avoid accidents and depart from the lane / road): hence, with reference to what stated in the previous section, a high value of RT is due to the fact that they were not able to share performances and so secondary task was a source of distraction. Moreover, analyzing the data of steering angle and lateral position (during straight segment) it is possible to observe that the highest values are in correspondence to high values of RT. Again, this leads to the conclusion that in this specific moment the drivers were not able to share attention between the two tasks.

3.2 Data Collection

Because of our choice to model the Distraction using a Machine Learning approach, a reference indicator (e.g. a target output) was needed for training the system. As mentioned above, this is represented by drivers' Reaction Time (RT) for Distraction, that is, the time spent by the driver to react to the secondary task stimulus, which was computed according to [9] and it measures the time (milliseconds) between the activation of the secondary task and the response of the driver. Assuming that the driver is distracted from the primary task when he/she is asked to complete the secondary one, a high value of Reaction Time means the driver is dedicating his/her time to complete a complex secondary task, while a low value means he/she is completing an easy task. The assumption is: the more difficult the secondary task is, the more the driver could be impaired in the primary task control. At the end, a time-series profile is obtained for the training data-sets.

The variables needed for the computation of this parameter were recorded in order to be used for the development of Distraction, as developed in the framework of the Driver-Vehicle-Environment (www.aide-eu.org); in particular, the inputs and the output to the model are listed below, with the related unit of measures:

- Steering Angle (SA) $\Rightarrow$ [SA] = deg
- Lateral Position (LP) $\Rightarrow$ [LP] = m
- Lateral Acceleration (LA) $\Rightarrow$ [LA] = deg/s^2
- Speed of the host-vehicle (S) $\Rightarrow$ [S] = m/s

- Deceleration Jerk (DJ) $\Rightarrow$ [DJ] = m/s^3
- Level of difficulty of secondary task (IVIS) $\Rightarrow$ [IVIS] = number
- Reaction Time on secondary task (RT) $\Rightarrow$ [RT] = s

Here, the Deceleration Jerk (DJ) is defined as the number of abrupt onset of the brakes in a time/spatial window when a deceleration change higher than 10 m/s^3 occurs.

For each parameter in the list, the mean and the standard deviation have been computed, as method to group the data. In particular, Standard Deviation of Steering Angle (and the same for the Mean of Steering Angle) is a measure of the steering variation, as accumulated on a segment of the track; since the secondary task was active on the straight segment, SDSA reflects this increment of the steering variation on the primary task (induced by the secondary task). We have avoided to use the punctual measures (speed, steering angle, etc.) because of two main reasons. Firstly, we would not have been a too big amount of data. Secondly, each minimum variation of steering and speed (or reaction time) would be used as valid data, while they represent rather a noise and a disturbance to the measure. Using the averages or the standard deviation, such a noise can be minimized Following the ordinary procedure for neural networks, each data set has been split in three different subsets:

- *Training data* (around 60% of the whole dataset) $\Rightarrow$ These are presented to the network during training and the network is adjusted according to its error.
- *Checking data* (around 15% of the whole dataset) $\Rightarrow$ These are used to measure network generalization and to halt training when generalization stops improving.
- *Testing data* (around 25% of the whole dataset) $\Rightarrow$ These have no effect on training and so provide an independent measure of network performance during and after training.

The datasets used for our models have been split accordingly.

Finally, we discuss here the sampling method of the data. This type of simulator can output data at 20 Hz (1 data-point each 0.05s); in order to build our training sets, data have been collected and sampled each 100m of traveled scenario. Considering an average speed of 60 km/h (= 16.67 m/s), this means that each data-point has been collected every around 6s. In the simulator, data were recorded during the driving experiments at a frequency rate of 20 Hz.

4   The Formulation of Distraction Model

In order to develop the Distraction models, we have followed the flow of procedure and the methodology sketched in the figure:

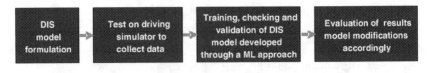

Fig. 1. Visual representation of the methodology followed in our research

Starting for the list of data presented in the previous Section, an analysis of the correlation matrix have been carried out and in this way we have obtained the set of variables to be used for the model development; they are:

- Standard Deviation of Steering Angle (SDSA)
- Standard Deviation of Lateral Position (SDLP)
- Standard Deviation of Lateral Acceleration (SDLA)
- Deceleration Jerk (DJ)
- Level of difficulty of secondary task (IVIS)

It is worth to note here that the input variables include both the lateral behaviour (steering angle and lane position) and the longitudinal one (vehicle velocity and acceleration/deceleration). After all this process, the following combinations of input variables have been considered:

- *Combination C1* = IVIS, SDSA
- *Combination C2* = IVIS, SDSA, DJ
- *Combination C3* = IVIS, SDSA, SDLP, SDLA

Two models have been developed and then the related performances compared. They have been built by means of some of the most commonly used techniques in data-mining; there are some reasons for this choice, listed as following:

- From a more "philosophical" point of view, one of the most ambitious goals of automatic learning systems is to mimic the learning capability of humans; this concept seems to fit absolutely well all the needs and constraints typical of user's (and specifically driver's) model (humans capability of driving is widely based on experience and possibility to learn from experience)
- From a more technical viewpoint, data that are collected about vehicle dynamics and external environment, characterising the driver, are definitely non-linear. From literature, several studies have demonstrated that in such situations machine learning approaches can outperform the traditional analytical methods. Moreover, also human's driver mental and physical behaviour is non-deterministic and highly non-linear, hence there are seldom "basic" available principles, governing such behaviours.

The first model is based on the **Adaptive-Network-based Fuzzy Inference System** (ANFIS) and second on the **Artificial Neural Networks** (ANN). Since training multiple times generate different results, due to the different initial conditions and sampling, we have carried out several runs for each training session (10 times).

4.1 The ANFIS Algorithms

Sometimes, a logic based on two truth values, True and False, can be inadequate when describing human reasoning, so **Fuzzy Logic** (FL) uses all values inside the interval [0 1] (where $0 \equiv$ False and $1 \equiv$ True). In this context, FL is a fascinating area of research: in "mathematical terms", it is a convenient way to map an input space to an output space, where "convenient" is referred to the matter of mapping inputs to the appropriate outputs. In particular, since a Fuzzy Logic System (FLS) is able to handle simultaneously numerical data and linguistic knowledge, it is a non-linear mapping of an input data (feature) vector into a scalar output (i.e. it maps numbers into numbers). We could say that *in FL, the truth of any statement becomes a matter of degree*. A key-point is represented by the **Membership Functions** (MF), which are curves that define how each point in the input space is mapped to a membership value in the interval-time [0, 1]. There are several types of MF and one of the most important point is to choose the most appropriate one. Another basic concept in FL is the **if-then** rules, which are rule-based systems allowing the translation of a human solution. The output of each rule is a fuzzy set, but in general we want the output for an entire collection of rules to be a single number. Since a fuzzy inference system maps an input vector to a crisp output value, we need this **defuzzification** process. Many defuzzification techniques have been proposed in the literature. Perhaps the most popular defuzzification method is the *centroid calculation*, which returns the centre of area under the curve. The centroid defuzzification method finds the "balance" point of the solution fuzzy region by calculating the weighted mean of the output fuzzy region. There are also other methods for defuzzification process, which include the *maximum*, the *means of maxima*, *height* and *modified height* method, and so on.

As aforementioned, in traditional Fuzzy systems, the membership functions are fixed and somewhat arbitrarily chosen. Moreover, fuzzy inference is applied to modelling systems whose rule structure is essentially predetermined by the user's interpretation of the characteristics of the variables in the model. Hence, one of the key-points in Fuzzy Theory is the choice and tuning of membership functions, which are done very often arbitrarily and manually. One possibility is to use the architecture and learning procedure called ANFIS, which is **a fuzzy inference system** (FIS) in the **framework of adaptive networks** [11], [12]. In other words, ANFIS can be regarded as a basis for constructing a set of fuzzy *"if-then" rules* with appropriate membership functions, in order to generate the stipulated pairs of inputs/outputs and to take into account the human knowledge, by using a hybrid learning procedure. We have used the ANFIS model with two types of membership functions: *G-Bell* (Bell-shape function) and *Gaussian*. For the defuzzification method, we have used *centroid calculation*, described before; this is the most widely used, because the defuzzified values tend to move smoothly around the output fuzzy region; however, the disadvantage consists in the fact that it is not easy to implement computationally. Ordinarily, there are two type of FIS (at least, they are the most common used): *Sugeno-type* and *Madani-type*; in our model, we have selected the first one.

4.2 The ANN Algorithms

An **Artificial Neural Networks** (ANN) system is an information processing paradigm inspired by the human brain and it consists in a large number of highly interconnected processing elements, working together to cope with specific problems [10]. One of the most important type of Neural Networks (NN) – and the one used inside our work – is the **Feed-forward Neural Networks** architecture, which comprehends two types of networks: the *Multi-layer Feed-forward Perceptron* (MLP) and the *Radial Basis Function* (RBF). In this paper, we focused on the first one. A *feed-forward network* has a layered structure. Each layer consists of units which receive their input from units from a layer directly below and send their output to units in a layer directly above the unit. One fundamental aspect in NN is the definition of the appropriate network topology. The basic characteristics of the neural network to use has been so defined. The Hidden Layer was equal to 1; the Transfer Function for Hidden Layer was Tangent-Sigmoid; the Transfer Function for Output Layer was Linear; the Optimisation method for training was Levenberg-Marquardt; the Training parameter values (most important) were: goal = 0; Min Gradient = 1e-10; $\mu = 0.001$.

5 Data Processing and Main Results

In order to "measure" the performances of our models, two main indexes have been considered:

- MSE (RMSE) = **Mean Squared Error** (**Root Mean Squared Error**) is *the (root of the) average squared difference between outputs and targets.* Lower values are better. Zero means no error (asymptotic value).
- NDEI = **Non Dimensional Error Index** is defined as *the ratio between MSE parameter and the standard deviation of the particular variable considered.* NDEI is an indicator of the dispersion between the error and the standard deviation.

After collecting data from simulator tests, in the post-processing phase we have clustered them using three type of data-windows:

- Win = 0 ⇒ namely, the data are left as they are
- Win = 2 ⇒ two rows are grouped and then the window is shifted
- Win = 3 ⇒ three rows are grouped and then the window is shifted

With reference to sub-section 4.1. we have used and compared the three combinations of data C1, C2 and C3 for different values of "*Win*" (this is true for both the algorithms developed).

5.1 Results for ANFIS Model

The first model of Distraction to be analyzed is the ANFIS one, with grid-partition and sub-clustering of data; the following table shows the situation:

Table 1. Results of ANFIS model for Distraction, for different combinations of data

Combination of data	Value of Win	Error Testing	Range of Influence
C1	0	0.39802	0.2
	2	0.62045	0.8
	3	0.49811	2.1
C2	0	0.61728	0.1
	2	0.54152	1.9
	3	0.57068	2.5
C3	0	0.64468	1.85
	2	0.70826	2.1
	3	0.72302	1.95

In the first column the combination of variables is presented. The second column indicates the value of data window used The third column represent the *testing error*, which is the difference between the testing data output value and the output of the fuzzy inference system corresponding to the same testing data input value. It gives the indication about how good is the ANFIS model to generalize with respect to unseen data.

For rows with *Win = 0*, the membership functions defined for the fuzzy model were *G-Bell types* (Bell-shape function), while for all the other rows, they were *Gaussian*.

With reference to Table 1, the best performance is obtained by combination C1 with Win = 0 (row 1), for which the testing error is equal to 0.39802. Some deeper investigations have been made for the best combinations of *SDSA* and *IVIS* variables, considering *Win = 0* and *Win = 2*; in particular, the index NDEI has been computed with the following results:

Table 2. NDEI values for combination C1

Combination of Data	Value of Win	Range of Influence	NDEI
	2	0.2	1.0345
		0.8	1.0253
C1		2.0	1.1165
	0	0.2	0.4806
		1.8	0.4815

Where *Range of Influence* is the radius of clustering and it is a parameter to be set for the ANFIS training mode; *Win* is the window value used to group the data.

The results showed in Table 2 can be summarised as follows:

- Combination C1 (IVIS and SDSA) as inputs to the model; Reaction Time, regarded as an indicator of distraction for output target for training the model
- MSE = 0.16 and NDEI = 0.4806
- ANFIS model developed with a sub-clustering technique with the following settings:
    - Range of influence = 0.2;
    - Win = 0 (no mobile window used)

The "big issue" when training an automatic learning system is to obtain a system able to **generalise well**. With generalisation, it is meant the ability of a (learning) machine to recognise and classify / predict correctly even data sample $P(X;Y)$ not originally present in the dataset D (used for training and checking). Therefore, it is needed to make a trade-off between the *capability* and the *approximation smoothness* of the learning system. In particular, the former means the ability of a function to learn correctly any training dataset (and it is related to *VC dimension*); the latter – smoothness of approximation – is the ability of the system to learn the general "trend" of the data and not to follow the data points too precisely; in other words, it is the problem of overfitting, for which the learning function learns too well the training data set, simply memorising the data, without any capacity to approximate the general trend of the function and so to predict new data. The generalisation capability of our model is illustrated in the following figure:

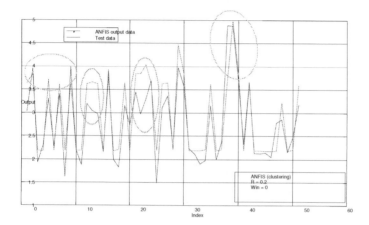

Fig. 2. Output data (dot line) vs. test data (straight line)

From Figure 2 and in accordance with the values of NDEI found, there is a general very good agreement between the test data and the output data generated by the AN-FIS model (as highlighted by the green dot circles in the figures), whereas red straight-line circles show when the accordance is less good.

5.3   Results for ANN Model

We have considered the Multi-Layer Perceptron (MLP). For each combination presented in sub-section 4.1, we have computed the performances of the network, after training, checking and testing, for different window values of data (similarly to the ANFIS case, we used $Win = 0, 2, 3$) and different number of Hidden Neurons (HN) in the HL The main results are shown in the following table:

Table 3. MSE values for different Win values, for combination C1 and for different HN

Number of HN	MSE Value	Win Value	Combination of data
30	3.39353e-2	0	C1
15	3.35976e-2	2	
20	3.39008e-2	2	
15	\3.35866e-2\	3	

We have considered only the performance values of combination C1, since the others gave out worst results. The best conditions are provided by raw 4. Now, we examine NDEI index, shown in the following table:

Table 4. NDEI values for different numbers of Hidden Neurons and Win values

Number of HN	Win Value	NDEI Value
30	0	0.74341
15	2	0.63796
20	2	0.69198
15	3	0.68908

The best configuration is represented by raw 2 of Table 4, considering the combination of SDSA and IVIS as inputs, with a mobile window of 2.

Starting from these results we have considered the generalisation capability of the network model:

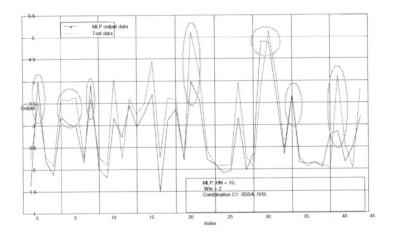

Fig. 3. Output data (dot line) vs. test data (straight line)

With reference to Figure 3, the green dot circles illustrate a very good agreement of output data of the model, with respect the test data, whereas the red straight-line circles show the situation where this agreement is less good; all in all, it is possible to say that the network capability of generalisation are satisfactory.

To sum up, the topology of the network is the one typical of a feed-forward back-propagation Neural Network, with *Lavenberg-Marquardt* technique used for training method. The number of HN is equal to 15. The maximum number of epochs for training have been set equal to 1000. Two transfer functions are used: *Hyperbolic tangent sigmoid* transfer function (TANSIG) for HL and *linear* transfer function (PURELIN) for OL.

5.3 Comparison of the Main Results Achieved

In order to compare the best results obtained for each technique, we consider the NDEI index; in fact, although MSE (or RMSE) has low values (that is, close to 0) nonetheless this fact does not necessarily mean that the regression of the model gives out good results. Generalisation is one of the most important feature of a ML model: the capacity to avoid overfitting problems is fundamental. The NDEI index, since it takes into consideration the following fact: it is possible that a model makes few errors, but these ones can be big; while another model can make smaller errors, but the general trend of data is (more) poorly recognised. For our goals, of course we prefer the first solution, for which the NDEI parameter is lower.

In the following list, we show the results

- NDEI for ANFIS = 0.48060
- NDEI for MLP = 0.63796

The best performances have been obtained considering $Win = 0$ for ANFIS, whereas, for MLP, it has been achieved, using $Win = 2$. For all, the combination C1 has been adopted (that is, SDSA and IVIS are the variables used as inputs to the model). Among these two models, the one with the lowest NDEI value is the ANFIS. Therefore, our research assessed that a good model of distraction is the one obtained with ANFIS technique, presenting these characteristics:

- Inputs = SDSA, IVIS – clustering of data
- Radius of clustering: R = 0.2
- Win = 0
- Target = reaction time on secondary task

Finally, about the computational time required for training ANFIS and MLP models, they are very similar and, with the MATLAB toolboxes used in this research, they are of the order of 5s.

6 Conclusions

The results achieved for Distraction modelling prove that our initial choice to consider Fuzzy Inference Systems (FIS) as starting point for the model was the "winning" one; in fact, FIS appears to be quite suitable to "mimic" human behaviour and status; furthermore, the combination with a machine learning technique, such as Neural Networks for the ANFIS algorithm implementation, allowed us to deal with data which are highly non-linear. We have chosen a method for Distraction estimation based on the reaction time to a secondary task (SURT method), which simulates the use of an IVIS application in a real car (for example, the insertion of the destination in a navigation system).

6.1 Discussion

Literature has plenty of works about the detection of driver's distraction, but for what concerns the use of ML techniques in this field, there are two or three researches which are closer to our activity; they are the work of Liang and colleagues [13], the one of Kutila [14] and finally the one of SAVE-IT project [15]. All these works have the goal to detect driver's cognitive distraction, in order to adapt in-vehicle systems accordingly.

In particular, Liang considered Support Vector Machines (SVM) to develop a real-time model for detecting cognitive distraction and he used drivers' eye movements and driving performance data, which were collected in a simulator experiment, where subjects had to complete a secondary task simulating an IVIS task. These data were used to train and test both SVM and logistic regression models, in order to compare them: the results show that the SVM models were able to detect driver distraction with an average accuracy of 81.1%, outperforming more traditional logistic regression models.

In the work of Kutila, the aim was to investigate the feasibility of different techniques and methods, for monitoring the driver's momentary distraction state and level of vigilance during a driving task. The study provides a multidisciplinary review by adopting new aspects and innovative approaches to state-of-the-art monitoring applications for adapting them as an in-vehicle environment. The hypothesis is that detecting the level of distraction and/or fatigue of a driver can be performed by means of a set of image processing methods, enabling eye-based measurements to be fused with other safety-monitoring indicators such as lane-keeping performance or steering activity. The performance tests show that even an 80% classification capability was achieved. Moreover, Kutila selected a SVM type classifier as the best choice for this classification problems, but pointing out that the drawback of SVM is its sensitivity to outliers in data, especially in the training or validation stages. Perhaps the Neural Networks would allow a more advanced processing methodology, like better robustness to "abnormal behaviour" during a cognitive workload.

Eventually, Lee and colleagues, in SAVE-IT project, have found that SVMs provide a viable means of detecting cognitive distraction in real-time and outperformed the more traditional approach of logistic regression. They included both eye and driving measures as inputs to a distraction detection algorithm. These data were collected in a driving simulator with relatively homogenous traffic and roadside objects. They also explored other data mining/machine learning approaches, like Bayesian Networks (BNs) and Hidden Markov Models (HMMs). Considering the latter to predict driver's behaviour by utilising eye movement, HMM gave a limited success. For what concerning the former, they first compare two types of BN; static (SBN) and dynamic (DBN); DBN models were more sensitive than SBN models, but there were no significant differences in accuracy and response bias. Then, comparing SVM and SBN technique, the SVM models detected driver cognitive distraction more accurately than the SBN models. This suggested that the SVM learning technique has advantages over the BN technique.

A deeper discussion is out of the scope of this paper, due to room problem. Anyway, all the works using ML to detect driver's distraction use basically SVM algorithms and in some cases Bayesian Networks . Everyone uses driving simulator for the experiments, where the collected data are about the eye-movements (and related parameters), driver's action and vehicle dynamics.

First of all, in our research we did not deal with a classification problem – like all others related activities – but with a regression problem (a time series profile – reaction time – to be reconstructed). We used a methodology of distraction evaluation that has the big advantage to make possible its estimation without using specific sensors or devices (like camera looking at head movements or for eye tracking), which are often expensive and "invasive" towards drivers (and quite complex to manage). In addition, we have explored also other ML algorithms, specifically Neural Networks and Fuzzy Logic. Our results show that NN (ANFIS in particular) can perform very well, so they can be an interesting new approach to investigate. Moreover – at the best of our knowledge – the application of Fuzzy Inference Systems to distraction detection problem is quire new in literature.

6.2 Limitations

There are also some limitations in our approach. We collected some variables in the experiments at driving simulator, regarded as good candidates to diagnose driver's distraction. In other words, we try to diagnose certain parameters of the human mind (DIS in this case) by looking at symptoms. However, some other variables can be included, in particular the ones more related to environmental conditions (such as visibility, density of traffic, etc.). Even if they have not been included since it is quite difficult to get this information for a real-time system, nevertheless, these factors shall be taken into consideration in some way in the future. The risk is, otherwise, to implement systems which are trained on "average conditions", but due to this characteristic, at the end they do not really fit the variety of scenarios and situations that a driver can meet.

Another limit consists in the fact that we represented the IVIS with the levels of difficulty induced by a secondary task (see Ch. 3) as one of the indicator / input for Distraction model. Actually, not only the number of active IVIS, but also the kind of interaction, as well as the concrete design of the Human Machine Interfaces (HMI), are very relevant for inducing distraction on the primary task and thus this could be an aspect that needs to be further investigated.

7 Future Researches

Possible future researches are oriented towards three directions. The first takes into consideration the extension of the experimental conditions (more specific situations, broader number of traffic scenarios, etc.) in which the model is trained, including the influence of HMI in inducing driver's distraction on primary task.

The second line of research considers some alternatives to the techniques used to model the distraction; in fact, up to now, we have deemed ANFIS (or MLP) as the most appropriated one for our goals. A very interesting (and promising) opportunity is to consider the use of SVM for Regression problem (SVR, in short) that is, for function estimation. In order to deal with this problem, in our work we have considered neural network methods (including ANFIS method), but good results can be achieved only if many parameters are optimally tuned by hand (thus depending largely on the skill and expertise of the experimenter) or anyway by a specific experimental activity for this purpose (which can be long and require a lot of efforts). Therefore, one alternative is to use SVR, which may achieve similar performance of neural networks in many cases, even if they are not a "silver bullet". However, as they have only few critical parameters (e.g. regularization and kernel width), we think that state-of-the-art results can be achieved with relatively little effort and from this basis, it is possible to start with improvements. Many different types algorithms and SW tools are available; at the moment, we are evaluating the best choice (some preliminary SVR models of Distraction are running and we are checking the results).

Finally, the third step of our work will be the implementation of the Distraction model in a specific system on-board vehicle, in order to diagnose the possible status of the driver in "real-time" and thus to take the appropriate counter-measures, in order to eliminate the problems or – at least – to minimise the consequences.

References

1. Treat, J., et al.: Tri-level Study of the Causes of Traffic Accidents: Final Report, vol. 1. Technical Report Federal Highway Administration, US DOT (1979)
2. Beirness, D.J., et al.: The Road Safety Monitor: Driver Distraction. Traffic Injury Research Foundation, Ontario, Canada (2002)
3. Treat, J.R.: A study of the Pre-crash Factors involved in Traffic Accidents. The HSRI Review 10(1), 1–35 (1980)
4. Burnset, P.C.: How dangerous is Driving with a Mobile Phone? Benchmarking the Impairment to Alcohol. TRL Report TRL547. TRL Limited, Berkshire, United Kingdom (2002)
5. Srinivasan, R., Jovanis, P.P.: Effect of In-vehicle Route Guidance Systems on Driver Workload and Choice of Vehicle Speed: Findings from a Driving Simulator Experiment. In: Ergonomics and safety in intelligent driver interfaces. Lawrence Erlbaum Associates, Publishers, Mahwah (1997)
6. Roskam, A.J., et al.: Deliverable D1: Development of experimental protocol. HASTE European Project (2002), http://www.its.leeds.ac.uk/projects/haste
7. Hockey, G.R.J.: Compensatory Control in the Regulation of human Performance under Stress and high Workload: a cognitive energetical Framework. Biological Psychology 45, 73–93 (1997)
8. Tango, F., et al.: Field Tests and Machine Learning Approaches for refining Algorithms and Correlations of Driver's Model Parameters. Applied Ergonomics Journal, Special Issue: Cacciabue P.C. et al (eds). Springer, Heidelberg (in press)
9. Mattes, S.: The Lane Change Task as a Tool for Driver Distraction Evaluation. In: IHRA-ITS Workshop on Driving Simulator Scenarios – Dearborn, Michigan (October 2003)
10. Haykin, S.: Neural Networks: a comprehensive Foundation. Prentice-Hall, Englewood Cliffs (1999)
11. Jang, J.-S.R.: ANFIS: Adaptive-Network-based Fuzzy Inference System. IEEE Transactions on Systems, Man and Cybernetics 23 (May 2003)
12. Jang, J.-S.R.: Neuro-Fuzzy Modeling: architectures, analyses and applications. PhD Dissertation, University of Barkeley – CA (1992)
13. Liang, Y., et al.: Real-time Detection of Driver Cognitive Distraction using Support Vector Machines. IEEE Transaction on Intelligent Transportation Systems 8(2) (June 2007)
14. Reyes, M., Lee, J.D.: The influence of IVIS distractions on tactical and control levels of driving performance. In: Proceedings of the Human Factors and Ergonomics Society 48th Annual Meeting (CD), pp. 2369–2373. Human Factors and Ergonomics Society, Santa Monica (2004)

SO_MAD: SensOr Mining for Anomaly Detection in Railway Data

Julien Rabatel[1,2], Sandra Bringay[1,3], and Pascal Poncelet[1]

[1] LIRMM, Université Montpellier 2, CNRS
161 rue Ada, 34392 Montpellier Cedex 5, France
[2] Fatronik France Tecnalia, Cap Omega, Rond-point Benjamin Franklin - CS 39521
34960 Montpellier, France
[3] Dpt MIAp, Université Montpellier 3, Route de Mende
34199 Montpellier Cedex 5, France
{rabatel,bringay,poncelet}@lirmm.fr

Abstract. Today, many industrial companies must face problems raised by maintenance. In particular, the anomaly detection problem is probably one of the most challenging. In this paper we focus on the railway maintenance task and propose to automatically detect anomalies in order to predict in advance potential failures. We first address the problem of characterizing normal behavior. In order to extract interesting patterns, we have developed a method to take into account the contextual criteria associated to railway data (itinerary, weather conditions, etc.). We then measure the compliance of new data, according to extracted knowledge, and provide information about the seriousness and possible causes of a detected anomaly.

Keywords: behavior characterization, anomaly detection, sequential patterns.

1 Introduction

Today, many industrial companies must face problems raised by maintenance. Among them, the anomaly detection problem is probably one of the most challenging. In this paper we focus on the railway maintenance problem and propose to automatically detect anomalies in order to predict in advance potential failures. Usually, data is available through sensors and provides us with important information such as temperatures, accelerations, velocity, etc. Nevertheless, data collected by sensors are difficult to exploit for several reasons. First, a very large amount of data usually available at a rapid rate must be managed. Second, they contain a very large amount of data to provide a relevant description of the observed behaviors. Furthermore, they contain many errors: sensor data are very noisy and sensors themselves can become defective. Finally, when considering data transmission, very often lots of information are missing.

Recently, the problem of extracting knowledge from sensor data have been addressed by the data mining community. Different approaches focusing either on

P. Perner (Ed.): ICDM 2009, LNAI 5633, pp. 191–205, 2009.
© Springer-Verlag Berlin Heidelberg 2009

the data representation (e.g., sensors clustering [1], discretization [2]) or knowledge extraction (e.g., association rules [2], [3], [4], [5], sequential patterns [6], [7], [8]) were proposed. Nevertheless, they usually do not consider that contextual information could improve the quality of the extracted knowledge. The development of new algorithms and softwares is required to go beyond the limitations. We propose a new method involving data mining techniques to help the detection of breakdowns in the context of railway maintenance. First, we extract from sensor data useful information about the behavior of trains and then we characterize normal behavior. Second, we use the previous characterization to determine if a new behavior of a train is normal or not. We are thus able to automatically trigger some alarms when predicting that a problem may occur.

Normal behavior strongly depends on the context. For example, a very low ambient temperature will affect a train behavior. Similarly, each itinerary with its own characteristics (slopes, turns, etc..) influences a journey. Consequently it is essential, in order to characterize the behavior of trains as well as to detect anomalies, to consider the surrounding context. We have combined these elements with data mining techniques. Moreover, our goal is not only to design a system for detecting anomalies in train behavior, but also to provide information on the seriousness and possible causes of a deviation.

This paper is organized as follows. Section 2 describes the data representation in the context of train maintenance. Section 3 shows the characterization of normal behaviors by discovering sequential patterns. Experiments conducted with a real dataset are described in Section 5. Finally, we conclude in Section 6.

2   Data Representation

In this section, we address the problem of representing data. From raw data collected by sensors, we design a representation suitable for data mining tasks.

2.1   Sensor Data for Train Maintenance

The data resulting from sensors for train maintenance is complex for the two following reasons: *(i)* very often errors and noisy values pervade the experimental data; *(ii)* multisource information must be handled at the same time. For instance, in train maintenance following data must be considered.

Sensors. Each sensor describes one property of the global behavior of a train which can correspond to different information (e.g., temperature, velocity, acceleration).

Measurements. They stand for numerical values recorded by the sensors and could be very noisy for different reasons such as failures, data transfer, etc.

Readings. They are defined as the set of values measured by all the sensors at a given date. The information carried out by a reading could be considered as

Table 1. Extract from raw data resulting from sensors

TIME	Sensor 1	Sensor 2	Sensor 3	...
2008/03/27 06: 36: 39	0	16	16	...
2008/03/27 06: 41: 39	82.5	16	16	...
2008/03/27 06: 46: 38	135.6	19	21	...
2008/03/27 06: 51: 38	105	22	25	...

the state of the global behavior observed at the given moment. Due to the data transfer, some errors may occur and then readings can become incomplete or even missing.

We consider that the handled data are such as those described in Table 1, where a **reading** for a given date (first column) is described by **sensor measurements** (cells of other columns).

2.2   Granularity in Railway Data

Data collected from a train constitutes a list of readings describing its behavior over time. As such a representation is not appropriate to extract useful knowledge, we decompose the list of readings at different levels of granularity and then we consider the three following concepts journeys, episodes and episode fragments which are defined as follows.

Journey. The definition of a journey is linked to the railway context. For a train, a journey stands for the list of readings collected during the time interval between the departure and the arrival. Usually, a journey is several hours long and has some interruptions when the train stops in railway stations. We consider the decomposition into journeys as the coarsest granularity of railway data.

Let $minDuration$ a minimum duration threshold, $maxStop$ a maximum stop duration, and J be a list of readings $(r_m, ..., r_i, ...r_n)$, where r_i is the reading collected at time i. J is a journey if:

1. $(n - m) > minDuration$,
2. $\nexists (r_u, ..., r_v, ...r_w) \subseteq J \mid \begin{cases} (w - u) > maxStop, \\ and\ \forall\ v \in [u, w],\ velocity(v) = 0. \end{cases}$

Episode. The main issue for characterizing train behavior is to compare elements which are similar. However, as trains can have different routes the notion of journey is not sufficient (for instance, between two different journeys, we could have different number of stops as well as a different delay between two railway stations). That is the reason why we segment the journeys into episodes to get a finer level of granularity. To obtain the episodes, we rely on the stops of a train (easily recognizable considering the train velocity).

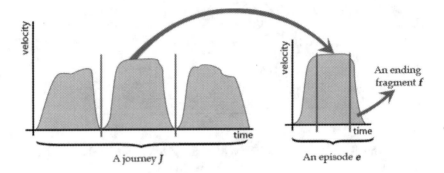

Fig. 1. Segmentation of a journey into episodes

An episode is defined as a list of readings $(r_m, ... r_i, ..., r_n)$ such as:

- $velocity(m) = 0$ and $velocity(n) = 0$[1],
- if $m < i < n$, $velocity(i) \neq 0$.

Figure 1 describes a segmentation of a journey into episodes by considering the velocity changes. This level of granularity is considered as the most relevant because it provides us with a set of homogeneous data. However, we can segment episodes in order to obtain a more detailed representation and a finer granularity level.

Episode Fragment. The level of granularity corresponding to the fragments is based on the fact that the behavior of a train during an episode can easily be divided into three chronological steps. First, the train is stationary (i.e., *velocity* 0) then an acceleration begins. We call this step the *starting* step. More formally, let $E = (r_m, ..., r_n)$ be an episode. The *starting fragment* $E_{starting} = (r_m, ... r_k)$ of this episode is a list of readings such as:

$$\forall i, j \in [m, k], i < j \Leftrightarrow velocity(i) < velocity(j).$$

At the end of an episode, the train begins a deceleration ending with a stop. This is the *ending step*. More formally, let $E = (r_m, ..., r_n)$ be an episode. The *ending fragment* $E_{ending} = (r_k, ... r_n)$ of this episode is a list of readings such as:

$$\forall i, j \in [k, n], i < j \Leftrightarrow velocity(i) > velocity(j).$$

The *traveling fragment* is defined as the sublist of a given episode between the *starting fragment* and the *ending fragment*. During this fragment, there are accelerations or decelerations, but no stop. More formally, let E be an episode, $E_{starting}$ its starting fragment, and E_{ending} its ending fragment. Then, the *traveling fragment* of E, denoted as $E_{traveling}$, is a list of readings defined as :

$$E_{traveling} = E - E_{starting} - E_{ending}.$$

---

[1] Here, the velocity of the train at time t is denoted as $velocity(t)$.

Figure 1 shows the segmentation of an episode into three fragments: the starting fragment, the traveling fragment and the ending fragment.

From now we thus consider that all the sensor data are stored in a database, containing all information about the different granularity levels. For example, all the sensor readings composing the fragment shown in Figure 1 are indexed and we know that a particular fragment f is an ending fragment included in an episode e, belonging to the journey J. J is associated with the itinerary I and the index of e in I is 2 (i.e., the second portion of this route).

3 Normal Behavior Characterization

In this section, we focus on the data mining step in the knowledge discovery process and more precisely on the extraction of patterns characterizing normal behavior.

3.1 How to Extract Normal Behavior?

The objective of the behavior characterization is, from a database of sensor measurements, to provide a list of patterns depicting normal behavior. We want to answer the following question: *which patterns often appear in the data?* Such a problem, also known as pattern mining, has been extensively addressed by the data mining community in the last decade.

Among all the data mining methods, we can cite the sequential pattern mining problem. The sequential patterns were introduced in [9] and can be considered as an extension of the concept of association rule [10] by handling timestamps associated to items. The research for sequential patterns is to extract sets of items commonly associated over time. In the " *basket market* " concern, a sequential pattern can be for example: " *40 % of the customers buy a television, then buy later on a DVD player* ". In the following we give an overview of the sequential pattern mining problem.

Given a set of distinct attributes, an *item*, denoted as i, is an attribute. An itemset, denoted as I, is an unordered collection of items $(i_1 i_2...i_m)$. A sequence, denoted as s, is an ordered list of itemsets $\langle I_1 I_2...I_k \rangle$. A sequence database, denoted as DB, is generally a large set of sequences. Given two sequences $s = \langle I_1 I_2...I_m \rangle$ and $s' = \langle I'_1 I'_2...I'_n \rangle$, if there exist integers $1 \leq i1 < i2 < ... < im \leq n$ such that $I_1 \subseteq I'_{i_1}$, $I_2 \subseteq I'_{i_2}$, ..., $I_m \subseteq I'_{i_m}$, then the sequence s is a *subsequence* of the sequence s', denoted as $s \sqsubseteq s'$, and s' supports s.

The *support* of a sequence is defined as the fraction of total sequences in DB that support this sequence. If a sequence s is not a subsequence of any other sequences, then we say that s is *maximal.*

A sequence is said to be frequent if its support is greater than or equal to a threshold minimum support (*minSupp*) specified by the user.

The sequential pattern mining problem is, for a given threshold *minSupp* and a sequence database DB, to find all maximal frequent sequences.

Sequential Patterns and Sensor Data. The discovery of sequential patterns in sensor data in the context of train maintenance requires choosing a data format adapted to the concepts of sequence, itemsets and items defined earlier.

So, from now we consider a sequence as a list of readings, an itemset as a reading, and an item as the state of a sensor. The order of the itemsets in a sequence is given by the timestamps associated to each reading.

Items are Si_{vt}, where Si is a sensor and vt is the value measured by the sensor at time t. For example, data described in Table 1 are translated into the following sequence:

$$\langle (S1_0 S2_{16} S3_{16})(S1_{82.5} S2_{16} S3_{16})(S1_{135.6} S2_{19} S3_{21})(S1_{105} S2_{22} S3_{25}) \rangle.$$

In addition, we use generalized sequences and time constraints ([11], [12]). More precisely, a time constraint called *maxGap* is set in order to limit the time between two consecutive itemsets in a frequent sequence. For instance, if *maxGap* is set to 15 minutes, the sequence $\langle (S1_{low})(S2_{low}, S3_{high}) \rangle$ means that the state described by the second itemset occurs at most 15 minutes after the first one.

A sequence corresponds to a list of sensor readings. So, a sequence database can be created, where a sequence is a journey, an episode or an episode fragment, depending on the chosen level of granularity (see Section 2).

3.2   Contextualized Characterization

Structural and Environmental Criteria
With the data mining techniques described in the previous section we are able to extract patterns, describing a set of episodes, which currently occur together. However, they are not sufficient to accurately characterize train behavior. Indeed, the behavior of a train during a trip depends on contextual[2] criteria. Among these criteria, we can distinguish the two following categories:

- **Structural criteria**, providing information on the journey structure of the studied episode (route[3], episode index in the route).
- **Environmental criteria**, providing information on the contextual environment (weather conditions, travel characteristics, etc.).

Example 1. *Table 2 presents a set of episodes, identified by the id column. In this example, the sensor data were segmented by selecting the level of granularity corresponding to episodes (see Section 2). Each episode is associated with environmental criteria (the duration of the episode and the exterior temperature) and structural criteria (the global route, and the index of the episode in this route).*

---

[2] The notion of context stands for the information describing the circumstances in which a train is traveling. This information is different from behavioral data that describe the state of a train.

[3] Here, a route is different from a journey. The route of a journey is the itinerary followed by the train during this journey. Therefore, several journeys may be associated with a single route (e.g., Paris-Montpellier).

Table 2. Episodes and contextual information

id	Environmental Dimensions		Structural Dimensions	
	Duration	Exterior Temperature	Route	Index
e_1	low	high	J1	E1
e_2	low	low	J1	E2
e_3	high	high	J2	E1
e_4	low	low	J1	E1
e_5	high	low	J1	E2

For example, the duration of the episode e_2 is short, and this episode was done with a low exterior temperature. In addition, e_2 is part of the itinerary denoted by J1, and is the second portion of J1.

Let us consider a more formal description of the context of an episode. Each episode is described in a set of n dimensions, denoted by D. There are two subsets D_E and D_S of D, such as:

- D_E is the set of environmental dimensions. In Table 2, there are two environmental dimensions: *Duration* and *Exterior Temperature*.
- D_S is the set of structural dimensions, i.e., the *Route* dimension whose value is the route of the episode, and the dimension *Index* for the index episode in the overall route.

Data Mining and classes

Now we present how the influence of these criteria on railway behavior are handled in order to extract knowledge. The general principle is the following: *(i)* we divide the data into classes according to criteria listed above and *(ii)* we extract frequent sequences in these classes in order to get contextualized patterns.

Based on the general principle of contextualization, let us see how classes are constructed. Let c be a class, defined in a subset of D denoted by D_C. A class c is denoted by $[c_{D_1}, ..., c_{D_i}, ..., c_{D_k}]$, where c_{D_i} is the value of c for the dimension D_i, $D_i \in D_C$. We use a *joker value*, denoted by $*$, which can substitute any value on each dimension in D_C. In other words, $\forall A \in D, \forall a \in Dim(A), \{a\} \subset *$.

Thus, an episode e belongs to a class c if the restriction of e on D_C[4] is included in c:

$$\forall A \in D_C, e_A \subseteq c_A.$$

Environmental criteria are linked with hierarchical relationships that can be represented by a lattice, while the structural criteria are linked by relations of composition represented as a tree.

---

[4] The restriction of e in D_C is the description of e, limited to the dimensions of D_C.

Environmental Lattice. The set of environmental classes can be represented in a multidimensional space containing all the combinations of different environmental criteria as well as their possible values. Environmental classes are defined in the set of environmental dimensions denoted by D_E.

Class	Duration	Ext. Temperature
[*,*]	*	*
[low,*]	low	*
[high,*]	high	*
[low,high]	low	high
[*,high]	*	high
...	...	...

Fig. 2. Some environmental classes

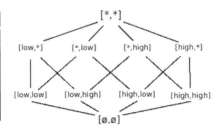

Fig. 3. Environmental Lattice

Example 2. *Figure 2 shows some of the environmental classes corresponding to the dataset presented in Table 2.*

A class c is denoted by $[c_{ExtT}, c_{Dur}]$, where ExtT stands for the dimension Exterior Temperature and Dur for the dimension Duration. For example, the class denoted by $[low, *]$ is equivalent to the context where the temperature is low (i.e., $c_{ExtT} = low$), for any duration (i.e., $c_{Dur} = *$).

Using the dataset of Table 2, we can see that the set of episodes belonging to the class $[low, *]$ is $\{e_1, e_2, e_4\}$. Similarly, the set of episodes belonging to the class $[low, high]$ is $\{e_1\}$.

Environmental classes and their relationships can be represented as a lattice. Nevertheless, we first have to define a generalization/specialization order on the set of environmental classes.

Definition 1. *Let c, c' be two classes. $c \geq c' \Leftrightarrow \forall A \in D, v_A \subset u_A$. If $c \geq c'$, then c is said to be more general than c' and c' is said to be more specific than c.*

In order to construct classes, we provide a *sum operator* (denoted by $+$) and a *product operator* (denoted by $\bullet$).

The sum of two classes gives us the most specific class generalizing them. The **sum operator** is defined as follows.

Definition 2. *Let c, c' be two classes.*

$$t = c + c' \Leftrightarrow \forall A \in D, t_A = \begin{cases} c_A \ if \ c_A = c'_A, \\ * \ elsewhere. \end{cases}$$

The product of two classes gives the most general class specializing them. The **product operator** is defined as follows.

Definition 3. *Let c, c' be two classes. Class z is defined as follows: $\forall A \in D, z_A = c_A \cap c'_A$. Then,*

$$t = c \bullet c' \Leftrightarrow \begin{cases} t = z \ if A \in D \mid z_A = \emptyset \\ < \emptyset, ..., \emptyset > \ elsewhere. \end{cases}$$

We can now define a lattice, by using the generalization/specialization order between classes and the operators defined above. The ordered set $\langle CS, \geq \rangle$ is a lattice denoted as CL, in which Meet ($\bigwedge$) and Join ($\bigvee$) elements are given by:

1. $\forall A \subset CL, \bigwedge A = +_{a \in A} a$
2. $\forall A \subset CL, \bigvee A = \bullet_{a \in A} a$

Figure 3 illustrates the lattice of environmental classes of the dataset provided in Table 2.

Structural Hierarchy. Structural hierarchy is used to take into account information that could be lost by manipulating episodes. Indeed, it is important to consider the total journey including an episode, and the position of this episode in the journey. Some classes are presented in Figure 4.

id	Route	Index	Fragment
[*]	*	*	*
[I1]	I1	*	*
[I2]	I2	*	*
[I1,E1]	I1	E1	*
[I1,E1,begin]	I1	E1	begin
[I1,E1,middle]	I1	E1	middle
...	...	...	...

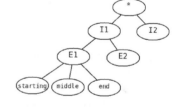

Fig. 4. Some structural classes   **Fig. 5.** Structural Hierarchy

Example 3. *Using the dataset of Table 2, we can see that the set of episodes belonging to the class $[I1]$ is $\{e_1, e_2, e_4, e_5\}$. Similarly, the set of episodes belonging to the class $[I1, E1]$ is $\{e_1, e_4\}$.*

Therefore, we create the hierarchy described in Figure 5, such as the higher the depth of a node is, the more specific is the symbolized class.

Structural classes, defined in D_S, are represented as a tree. The branches of the tree are symbolizing a relationship "is included". A tree is particularly appropriated here because it represents the different granularity levels defined earlier, i.e., from the most general to the finer level: journeys, episodes and fragments.

Let C be the set of all structural classes, and H^C the set of classes relationships. $H^C \subseteq C \times C$, and $(c_1, c_2) \in H^C$ means that c_2 is a subclass of c_1.

A lattice can not represent a structural hierarchy as elements are included in only one element of higher granularity: a fragment belongs to a single episode and an episode belongs to a single route.

The root of this hierarchy, denoted as $*$, is the most general class, i.e., it contains all the episodes stored in the database.

Nodes of depth 1 correspond to the various routes made by trains (i.e., $I1$ or $I2$ in Figure 5). The next level takes into account the position of the episode in the path represented by the father node. The class $[I1, E1]$ contains all the first episodes of the itinerary $I1$, i.e., episodes whose index is 1 in $I1$.

So far, the classes we have defined contain episodes. However, we have seen that it is possible to use a finer granularity in the data (see Section 2). To obtain more detailed knowledge, we consider classes of fragments in the leaves of the hierarchy. Thus, class $[I1, E1, starting]$ contains *starting fragments* of the episodes contained in $[I1, E1]$.

With the environmental lattice and structural tree presented in this section, we can index knowledge from normal behavior according to their context. Thus, when new data are tested to detect potential problems, we can precisely evaluate the similarity of each part of these new data with comparable elements stored in our normal behavior database.

3.3   Behavior Specificity

In the previous section, we showed how to extract frequent behaviors in classes. To go further, we want to know the specific behavior of each class (i.e., of each particular context). We therefore distinguish the specific patterns for a class (i.e., not found in class brothers in a hierarchy), and the general patterns appearing in several brother classes. General patterns possibly may be specific to a higher level of a class hierarchy.

Using this data representation, we can define the following concepts. Let DB be a sequence database, s a sequence, c a class in the structural hierarchy or the environmental lattice described in Section 3.2.

Notation 1. *The number of sequences (i.e., episodes or fragments according to the chosen level of granularity) contained in c is denoted as $nbSeq(c)$.*

Notation 2. *The number of sequences contained in c supporting s is denoted as $suppSeq_c(s)$.*

Definition 4. *The support of s in c is denoted as $supp_c(s)$ and is defined by:*

$$supp_c(s) = \frac{suppSeq_c(s)}{nbSeq(c)}.$$

Let *minSupp* be a minimum support threshold, s is said to be frequent in c if $supp_c(s) \geq minSupp$.

Definition 5. *s is a **specific pattern** in c if:*

- *s is frequent in c,*
- *s is non-frequent in the classes c' (where c' is a brother of c in the classes hierarchy).*

The notion of specificity provides additional information to the experts. A pattern specific to a class describes a behavior that is linked to a specific context. In particular, this information can be used in order to detect anomalies. For example, if we test the episodes being in $[J_1, E_1, starting]$ and that we meet a lot of behaviors specific to another class, then we can think that these behaviors are not normal in this context and we can investigate the causes of this anomaly more efficiently.

The hierarchical management of knowledge provides advantages in our industrial context. First, extracted sequential patterns are very informative for the experts. This characteristic is necessary and important because we do not want only to develop a system for detecting anomalies, but also to provide new knowledge to experts in the railway field. Moreover, the use of contextual classes and hierarchies allows us to generate very precise knowledge. Indeed, for each extracted frequent behavior (i.e., each sequential pattern), the expert has the ability to know precisely what are the contextual criteria that promote the appearance of such behavior. In addition, with such precision, we can build a system to detect anomalies that is described in the next section.

4   Anomaly Detection

In this section, we present how anomaly detection is performed. We consider that we are provided with both one database containing normal behavior on which knowledge have been extracted (see Section 3) and data corresponding to one journey.

The main idea is organized as follows. First, we define a measure to evaluate the compliance of a new journey in a given contextual class. In case of any detected anomaly, we make use of the class hierarchy to provide more detailed information about the problem.

4.1   Detection of Abnormal Behaviors

This involves processing a score to quantify if the sequence corresponding to new data to be tested is consistent with its associated class. We consider that the consistency of an episode with a class c depends on the number of patterns of c included in the episode. The most the episode contains patterns of c, the most this episode is consistent with c. We thus introduce a similarity score called the *Conformity Score* which is defined as follows.

Definition 6. *Let s be a sequence to be evaluated in a class c. We denoted by P the set of patterns of c, and P_{incl} the set of patterns of c being included in s. So, the Conformity Score of s in c, denoted by $score_c(s)$, is such as:*

$$score_c(s) = \frac{|P_{incl}|}{|P|}.$$

4.2   Anomaly Diagnosis

The detection is performed as follows.

First, we measure the conformity score of a new episode e in the most precise level of each of the hierarchies (i.e., structural and environmental). If the score is high enough in the two classes then the behavior of the train during e is considered as normal. Otherwise, it is possible to distinguish more clearly what is the cause of the anomaly. For example, the anomaly may have structural or environmental reasons.

To obtain more information about the detected anomaly, it is possible to go up in the hierarchy and test the conformity score of e in "parent" classes of problematic classes. For example, if the episode e has a poor score in the class $[low, high]$, then we evaluate its score in the classes $c_1 = [low, *]$ and $c_2 = [*, high]$. If $score_{c_1}(e)$ is inadequate and $score_{c_2}(e)$ is sufficient, then the provided information is that e is in compliance with the normal behaviors related to the exterior temperature, but not with those related to travel duration.

By defining a minimum conformity score $minCo$ we can also determine whether the emission of an alarm is necessary or not. If $score_c(e) < minCo$, then the episode e is considered as problematic, and an alarm is emitted. However, we can send an alarm in relation to the seriousness of the detected problem. For example, a score of 0.1 probably corresponds to a more important issue than a score of 0.45.

We have seen in the previous section that the knowledge obtained is useful for the experts' understanding of the train behavior. Moreover, the contextualization of the normal behavior characterization is relevant to detect anomalies with precision, but also to provide information about the detected anomalies.

5   Experiments

In order to evaluate our proposal, several experiments were conducted on a real dataset. They correspond to the railway data collected on 12 trains where each train has 249 sensors. A reading is collected every five minutes. 232 temperature sensors and 16 acceleration sensors are distributed on the different components (e.g., wheels, motors, etc..) and a sensor measures the overall speed of the train.

5.1   Experimental Protocol

The experimental protocol follows the organization of the proposals:

1. **Characterization of Normal Behavior.** (see Section 3) We have studied the impact of contextualization on the characterization of normal behavior, and the benefit of the search for specific patterns in various contexts.
2. **Anomaly Detection.** (see Section 4) We have evaluated the *conformity score* by applying it on normal and abnormal data.

5.2   Normal Behavior Characterization

The discovery of sequential patterns has been performed with the PSP algorithm, described in [13]. We have used a C++ implementation, which can manage time constraints.

Table 3. Number of frequent sequences and specific patterns, according to the environmental class

Class	Frequent Sequences	Specific Patterns
[*, *]	387	387
[low, *]	876	634
[high, *]	616	411
[low, high]	6430	5859

Table 3 shows, for an extract from the hierarchy presented in Section 3.2, the number of frequent sequences found in each class, and the corresponding number of specific sequences, extracted with a minimum support set to *0.3*. We can note that filtering specific patterns reduces the amount of stored results. Moreover, the fact that each class, including most restrictive ones (i.e., the leaves of the hierarchy), contain specific patterns shows both the importance of the context in railway behavior, and the usefulness of our approach.

We can note that for the most general class, denoted as [*, *], the number of specific patterns and the number of frequent sequences are equal. Indeed, this class does not have a brother in the environmental hierarchy. Moreover, we notice in these results that the more a class is specific, the more it contains frequent sequences and specific patterns. In fact, we extract frequent behaviors which heavily depend on surrounding context. Therefore, the more a class is general, the less behaviors are frequent, as they vary much more from one journey to another.

5.3   Anomaly Detection

We have noted in the previous section that we can extract very precise knowledge about normal train behavior. This knowledge is now used in an anomaly detection process, through the methods described in Section 4.

To validate our approach, we use normal behavior from real data and we ensure that they do not generate anomalies. To this end, we have segmented our real data set into two subsets: *(i)* data on which we perform the characterization of normal behavior; *(ii)* data to test.

Figure 6 shows the average score of episodes, depending on the class in the environmental hierarchy. In each class, we have tested 15 randomly selected episodes. Calculated scores are high, in particular in the most specific classes.

We have also measured the conformity of a random selection of 15 episodes of the class [*high*, *] with the normal behavior of the class [*low*, *high*]. The

Class	Average Score
[*,*]	0.64
[low,*]	0.85
[high,*]	0.74
[low,high]	0.92

Class	Score
[low,high]	0.05
[low,*]	0.17
[*,high]	0.41
[*,*]	0.3

Fig. 6. Average score of episodes

Fig. 7. Conformity Score of a degraded episode

average score decreases to 0.32, confirming that the behavior of a train when the outside temperature is high does not correspond to what is expected when the temperature is low.

Then, in order to detect anomalies in really abnormal behavior, we simulate defects. Indeed, the available real dataset does not contain enough abnormal behaviors to perform valid experiments. Simulated data are created by degrading data in the test dataset. For example, we apply our methods on an episode created by increasing all wheel temperatures by $15^\circ C$. Figure 7 presents the conformity score of this episode in its environmental classes, from general to specific. The true score of this episode before being corrupted, in its more specific environmental class (e.g., $[low, high]$) is 0.94. However, once the episode is degraded, the conformity score becomes *0.12*. By studying the previous level of the environmental lattice, we can note a difference between the score of the episode in class $[low, *]$ (i.e., low exterior temperature) and in class $[*, high]$ (i.e., long episode duration). The score is higher in the class $[*, high]$. This can be explained by the fact that such heating of wheels sometimes occurs when a train was traveling during a long time, and is not totally unusual in this context. However, this is not the case in the context of the class $[low, *]$.

6   Conclusion

In this paper, we have proposed a new strategy for detecting anomalies from sensor data in the context of railway maintenance. We have addressed the problem of characterizing train behavior with sequential pattern mining. First, we extract patterns to describe normal behavior. Then, we use the previous characterization to determine if a new behavior is normal or not. We are thus able to automatically trigger some alarms.

Our contribution is twofold: *(i)* as the behavior of trains depends on environmental conditions, we have proposed a method of characterization, focusing on contextual knowledge. *(ii)* to detect anomalies, we have developed a conformity measure of new data. When an anomaly is detected we are able to provide information on the seriousness and possible causes of a deviation. In order to validate our approach, we have applied it to real railway data.

This preliminary work opens up interesting prospects. How to make extracted knowledge accessible for the experts, by designing a graphical user interface?

How to provide more precise information about detected anomalies? How to apply these methods in a real-time context?

References

1. Rodrigues, P.P., Gama, J.: Online prediction of streaming sensor data. In: Gama, J., Roure, J., Auguilar-Ruiz, J.S. (eds.) Proceedings of the 3rd International Workshop on Knowledge Discovery from Data Streams (IWKDDS 2006), in conjuntion with the 23rd International Conference on Machine Learning (2006)
2. Yairi, T., Kato, Y., Hori, K.: Fault detection by mining association rules from house-keeping data. In: Proceedings of the 6th International Symposium on Artificial Intelligence, Robotics and Automation in Space (2001)
3. Halatchev, M., Gruenwald, L.: Estimating missing values in related sensor data streams. In: Haritsa, J.R., Vijayaraman, T.M. (eds.) Proceedings of the 11th International Conference on Management of Data (COMAD 2005). Computer Society of India (2005)
4. Chong, S.K., Krishnaswamy, S., Loke, S.W., Gaben, M.M.: Using association rules for energy conservation in wireless sensor networks. In: SAC 2008: Proceedings of the 2008 ACM symposium on Applied computing. ACM, New York (2008)
5. Ma, X., Yang, D., Tang, S., Luo, Q., Zhang, D., Li, S.: Online mining in sensor networks. In: Jin, H., Gao, G.R., Xu, Z., Chen, H. (eds.) NPC 2004. LNCS, vol. 3222, pp. 544–550. Springer, Heidelberg (2004)
6. Guralnik, V., Haigh, K.Z.: Learning models of human behaviour with sequential patterns. In: Proceedings of the AAAI 2002 workshop "Automation as Caregiver" (2002)
7. Cook, D.J., Youngblood, M., Heierman III, E.O., Gopalratnam, K., Rao, S., Litvin, A., Khawaja, F.: Mavhome: An agent-based smart home. In: PERCOM 2003: Proceedings of the First IEEE International Conference on Pervasive Computing and Communications. IEEE Computer Society Press, Los Alamitos (2003)
8. Wu, P.H., Peng, W.C., Chen, M.S.: Mining sequential alarm patterns in a telecommunication database. In: Jonker, W. (ed.) VLDB-WS 2001 and DBTel 2001. LNCS, vol. 2209, p. 37. Springer, Heidelberg (2001)
9. Agrawal, R., Srikant, R.: Mining sequential patterns. In: Yu, P.S., Chen, A.S.P. (eds.) Eleventh International Conference on Data Engineering. IEEE Computer Society Press, Los Alamitos (1995)
10. Agrawal, R., Imieliński, T., Swami, A.: Mining association rules between sets of items in large databases. SIGMOD Rec. 22(2) (1993)
11. Srikant, R., Agrawal, R.: Mining sequential patterns: Generalizations and performance improvements. In: Apers, P.M.G., Bouzeghoub, M., Gardarin, G. (eds.) EDBT 1996, vol. 1057. Springer, Heidelberg (1996)
12. Masseglia, F., Poncelet, P., Teisseire, M.: Efficient mining of sequential patterns with time constraints: Reducing the combinations. Expert Systems with Applications 36(2, Part 2) (2009)
13. Masseglia, F., Cathala, F., Poncelet, P.: The psp approach for mining sequential patterns. In: Żytkow, J.M., Quafafou, M. (eds.) PKDD 1998. LNCS, vol. 1510. Springer, Heidelberg (1998)

Online Mass Flow Prediction in CFB Boilers

Andriy Ivannikov[1,2], Mykola Pechenizkiy[1], Jorn Bakker[1], Timo Leino[3],
Mikko Jegoroff[3], Tommi Kärkkäinen[2], and Sami Äyrämö[2]

[1] Department of Computer Science, Eindhoven University of Technology,
P.O. Box 513, NL-5600 MB, Eindhoven, The Netherlands
{m.pechenizkiy,j.bakker}@tue.nl
[2] Department of Mathematical Information Technology, University of Jyväskylä,
P.O. Box 35, FIN-40014, Jyväskylä, Finland
aivanni@cc.jyu.fi, {tka,samiayr}@mit.jyu.fi
[3] VTT, Technical Research Centre of Finland,
P.O. Box 1603, FIN-40101 , Jyväskylä, Finland
{Timo.Leino,Mikko.Jegoroff}@vtt.fi

Abstract. Fuel feeding and inhomogeneity of fuel typically cause process fluctuations in the circulating fluidized bed (CFB) process. If control systems fail to compensate for the fluctuations, the whole plant will suffer from fluctuations that are reinforced by the closed-loop controls. This phenomenon causes a reduction of efficiency and lifetime of process components. Therefore, domain experts are interested in developing tools and techniques for getting better understanding of underlying processes and their mutual dependencies in CFB boilers. In this paper we consider an application of data mining technology to the analysis of time series data from a pilot CFB reactor. Namely, we present a rather simple and intuitive approach for online mass flow prediction in CFB boilers. This approach is based on learning and switching regression models. Additionally, noise canceling, and windowing mechanisms are used for improving the robustness of online prediction. We validate our approach with a set of simulation experiments with real data collected from the pilot CFB boiler.

1   Introduction

Continuous and growing increase of fluctuations in electricity consumption brings new challenges for the control systems of boilers. Conventional power generation will face high demands to ensure the security of energy supply because of increasing share of renewable energy sources like wind and solar power in power production. This can lead to frequent load changes which call for novel control concepts in order to minimize emissions and to sustain high efficiency during load changes.

From the combustion point of view the main challenges for the existing boilers are caused by a wider fuel selection (increasing share of low quality fuels), increasing share of bio fuels, and co-combustion. In steady operation, combustion is affected by the disturbances in the feed-rate of the fuel and by the incomplete

P. Perner (Ed.): ICDM 2009, LNAI 5633, pp. 206–219, 2009.

mixing of the fuel in the bed, which may cause changes in the burning rate, oxygen level, and increase of CO emissions. This is especially important, when considering the new biomass based fuels, which have increasingly been used to replace coal. These new biofuels are often rather inhomogeneous, which can cause instabilities in the feeding. These fuels are usually also very reactive. Biomass fuels have much higher reactivity compared to coals and the knowledge of the factors affecting the combustion dynamics is important for optimum control. The knowledge of the dynamics of combustion is also important for optimizing load changes [9].

Different needs for facilitating intelligent analysis of time series data from CFB sensors measurements, which would lead to better understanding of underlying processes in the CFB reactor, were considered in [6]. In general, a data-mining approach can be used to develop a model for optimizing the efficiency of an CFB boiler. E.g. the selection of process variables to optimize combustion efficiency has been studied in [10]. Another typical problem, which we address in this work, is online reliable prediction of system parameters which can not be measured reliably in a direct way.

In this paper we focus on the problem of online mass flow prediction. This is an important problem since knowing mass flow is necessary for the control optimization, but measuring it directly in the CFB operational settings is difficult. Currently, the mass flow is calculated offline using relatively long time period averages. We propose a rather simple and intuitive approach for online mass flow prediction, which is based on learning and switching regression models. We validate our ideas with a set of simulation experiments with real data collected from the pilot CFB boiler.

The rest of the paper is organized as follows. Section 2 present a wider context of sensor data mining for developing better understanding and control of CFB reactors. In Section 3 we discuss the problem of obtaining mass flow signal from the CFB pilot boiler. In Section 4 we outline our approach for online mass flow prediction. The experimental results are discussed in Sections 5. We briefly conclude with a summary and discussion of the further work in Section 6.

2   A Data Mining Approach for CFB Understanding and Control

The supercritical CFB combustion utilizes more cleanly, efficiently, and sustainable way coal, biofuels, and multifuels, but need advanced automation and control systems because of their physical peculiarities (relatively small steam volume and absence of a steam drum). Also the fact that fuel, air, and water mass flows are directly proportional to the power output of the boiler sets tight demands for the control system especially in CFB operation where huge amount of solid material exist in the furnace.

When the CFB boilers are becoming larger, not only the mechanical designs but also the understanding of the process and the process conditions affecting heat transfer, flow dynamics, carbon burnout, hydraulic flows etc. have been important factors. Regarding the furnace performance, the larger size increases the horizontal dimensions in the CFB furnace causing concerns on ineffective mixing of combustion air, fuel, and sorbent. Consequently, new approaches and tools are needed in developing and optimizing the CFB technology considering emissions, combustion process, and furnace scale-up.

Fluidization phenomenon is the heart of CFB combustion and for that reason pressure fluctuations in fluidized beds have been widely studied during last decades. Other measurements have not been studied so widely. Underlying the challenging objectives laid down for the CFB boiler development it is important to extract as much as possible information on prevailing process conditions to apply optimization of boiler performance. Instead of individual measurements combination of information from different measurements and their interactions will provide a possibility to deepen the understanding of the process.

A very simplified view on how a CFB boiler operates is presented in the upper part of Fig. 1. Fuel (mixture of fuels), air, and limestone are the controlled inputs to the furnace. Fuel is utilized to heat production; air is added for enhancing the combustion process and limestone is aimed at reducing the sulfur

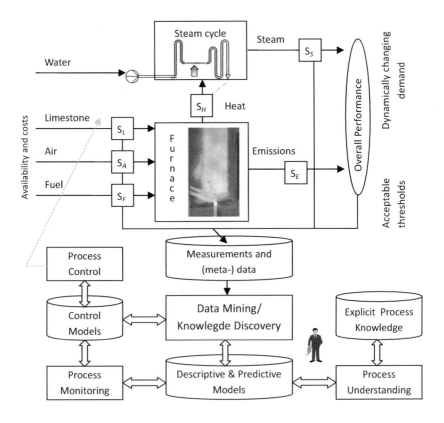

Fig. 1. A simplified view of a CFB boiler operation with the data mining approach

dioxides (SO2). The produced heat converts water into steam that can be utilized for different purposes. The measurements from sensors SF, SA, SL, SH, SS and SE that correspond to different input and output parameters are collected in database repository together with other meta-data describing process conditions for both offline and online analysis. Conducting experiments with pilot CFB reactor and collecting their results into database creates the necessary prerequisites for utilization of the vast amount of DM techniques aimed to identifying valid, novel, potentially useful, and ultimately understandable patterns in data that can be further utilized to facilitate process monitoring, process understanding, and process control.

3   The Problem of Obtaining Mass Flow Signal

The combustion and emission performance of different types of solid fuels and their mixtures are studied at VTT's 50 kW CFB pilot boiler (Fig. 2). The height of the riser of the boiler is 8 m and the inner diameter 167 mm. The reactor is equipped with several separately controlled electrically heated and water/air cooled zones in order to control the process conditions, for example, oxygen level, temperature and load almost independently. Several ports for gas and solid material sampling are located in the freeboard area.

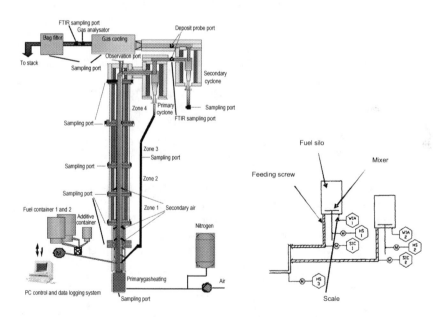

Fig. 2. A schematic diagram of the laboratory scale CFB-reactor (left) and its fuel feeding system (right)

The fuel can be fed into the reactor through two separate fuel feeding lines (Fig. 2). In each line there is a fuel screw feeder on the bottom of the silo and also a mixer which prevents arching of the fuel in the silo. The fuel silos are mounted on the top of scales which enables the determination of mass flow rates for solid fuels as a weight loss against time. Some of the phenomena that make the scale signal fluctuate with constant screw feeder rotational speed are:

- the quality of the fuel changes (e.g. moisture content, particle size);
- the fuel grades in the silo;
- some of the particles jam in between the screw rotor and stator causing a peak in the mass signal;
- fuel addition causes a step change in the signal

The amount of the fuel in the silo has an effect on the mass flow. This causes that mass flow decreases in between fillings as the level of fuel in the tank decreases

Due to the fluctuation in the scale signal no reliable online data can be obtained from the mass flow of fuel to the boiler. The measurement system cannot be easily improved and therefore the mass flow is calculated offline using longer time period averages. The aim of the study was to remove the problematic fluctuation and fuel filling peaks from the data using signal processing methods in order to obtain online mass flow signal which could be used among others for mass flow control of fuel at the CFB-pilot.

4   Our Approach: Context-Sensitive Learning

In case of CFB boiler (like in many other dynamic environments) the data flows continuously and the target concept (that is the mass flow in this study) could change over time due to the different operational processes (like fuel loading) or changes in these processes themselves. Therefore, learning algorithms must recognize abrupt (in case of mass flow) change in the target concept and adjust a model accordingly.

For this study, it is assumed that the transitions from fuel feeding processes to fuel consuming processes are known. The focus will be on a learning algorithm that is able to track the mass flow given the high frequency nature of the signal and the changing fuel types for each of the processes.

In this section we consider several practical aspect, including the use of domain knowledge and experimental settings metadata, and noise toleration in online mass flow prediction.

4.1   Integration of Domain Knowledge and Experimental Settings Metadata

The overall measurements during the experiment are presented in Fig. 3.

The data was recorded with 1 Hz sampling rate. In each test the type of fuel and/or the rpm of the feeding screw were varied. The rotation of the mixing

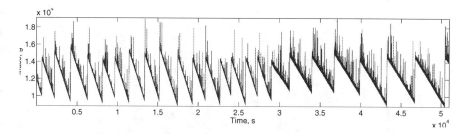

Fig. 3. Measurements of the fuel mass in the tank

screw was kept constant at 5.5 rpm. The three major sources of noise in the measurements are mixing and feeding screws and the occasional jamming of the fuel particle between the stator and rotor in the screw. The rotation of the screws causes vibrations to the system that are reflected by the scales as high frequency fluctuations around the true mass value. In Fig. 4 the evolution of the frequency content of the measurements is shown by means of the short-time Fourier transform [5], from which the influence of the screws is evident. Namely, the rotating parts induce oscillations to the measurements of the same frequency as the rotation frequency. The frequency content due to the screws is identified from the figure as contrasting vertical curves.

The jamming of the fuel particle causes an abnormally large upward peak to the measurements that can be seen from Fig. 3. The speed of the mass change in the tank at a given time depends not only on the rpm of the feeding screw and the speed of the replenishment of the fuel in the tank, but also on the amount of the fuel in the tank. The more fuel is in the tank the more fuel gets in the screw, since the weight of the fuel at the higher levels of the tank compresses (increases the density) the fuel in the lower levels and in the screw. The size and grade of fuel also have an effect on the compression rate of the fuel. Therefore, we assume that the mass flow signal has nonzero second derivative. Thus the nature of the phenomena measured by the scales can be modeled using the following equation:

$$y(t - t_0) = \frac{a \cdot (t - t_0)^2}{2} + v_0 \cdot (t - t_0) +$$
$$m_0 + A \cdot \sin(\omega_{feed} \cdot (t - t_0) + \alpha_{feed}) +$$
$$B \cdot \sin(\omega_{mix} \cdot (t - t_0) + \alpha_{mix}) + e(t - t_0), \tag{1}$$

where $y(t - t_0)$ denotes the time series of the output of the scales at time $t - t_0$, a is acceleration of the mass change, v_0 stands for the speed of the mass change at time t_0, m_0 is the initial mass at time t_0; A and B, ω_{feed} and ω_{mix}, α_{feed} and α_{mix} are amplitude, frequency and phase of the fluctuations caused by feeding and mixing screws, respectively; $e(t - t_0)$ denotes the random peaked high amplitude noise caused by the jamming of the fuel particle at time $t - t_0$.

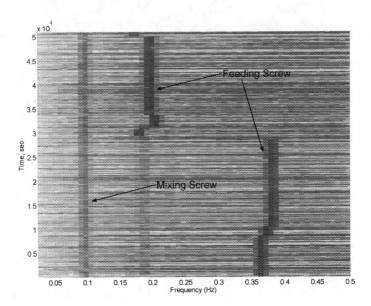

Fig. 4. Spectrogram of the fuel mass measurements computed using Short-Time Fourier Transform. The color denotes the power spectral density (PSD).

Formally, the problem of denoising/approximating the true mass flow signal consists in extracting the component related to the mass flow from the measured time series:

$$m(t - t_0) = \frac{a \cdot (t - t_0)^2}{2} + v_0 \cdot (t - t_0) + m_0, \qquad (2)$$

where $m(t - t_0)$ denotes the value of the mass at time $t - t_0$. One solution to this problem is to use stochastic gradient descent [1] to fit the model (1) without $e(t - t_0)$ term to the measured data with the high amplitude peaks skipped from the learning process. This is closely related to fitting the model (2) to the same data in the mean-least-squares sense as the noise fluctuations are symmetric relatively to the true mass signal. Alternatively a linear regression approach with respect to the second order polynomial can offer a better local stability and faster convergence. As the accurate mass flow measurements are required on-line by a control system the choice of the linear regression method seems more reasonable.

The linear regression can be performed by using the Vandermonde matrix [3], whose elements are powers of independent variable x. In our case the independent variable is time $x_i = t_{i-1} - t_0$, $i = 1, \ldots, T$, where T denotes the number of the time samples. If the linear regression is done for a polynomial of order n $(p^n(x) = p_n x^n + p_{n-1} x^{n-1} + \ldots + p_1 x + p_0)$ the Vandermonde matrix is computed from the observed time series of the independent variable as follows:

$$v_{i,j} = x_i^{n-j+1}, \; i = 1, \ldots, T, \; j = 1, \ldots, n+1, \qquad (3)$$

where i and j run over all time samples and powers, respectively. Provided the Vandermonde matrix the problem of polynomial interpolation is solved by solving the system of linear equations $\mathbf{V}\mathbf{p} \cong \mathbf{y}$ with respect to $\mathbf{p}$ in the least square sense:

$$\widehat{\mathbf{p}} = \operatorname{argmin}_{\mathbf{p}} \sum_{i=1}^{T}(\sum_{j=1}^{n+1} V_{i,j}p_{n-j+1} - y_i)^2 \qquad (4)$$

Here, $\mathbf{p} = [p_n \; p_{n-1} \; \ldots \; p_1 \; p_0]^T$ denotes the vector of the coefficients of the polynomial, and $\mathbf{y} = [y(x_1) \; y(x_2) \; \ldots y(x_T)]^T = [y_1 \; y_2 \; \ldots y_T]^T$ is the time series of the dependent variable that is indication of the scales. Provided that the $n+1$ columns of the matrix $\mathbf{V}$ are linearly independent, this minimization problem has a unique solution given by solving the normal equation [4]:

$$(\mathbf{V}^T\mathbf{V})\widehat{\mathbf{p}} = \mathbf{V}^T\mathbf{y}. \qquad (5)$$

4.2 Application

In this section we consider practical aspects of application of the least square approximation to our case study. For the modeling of the mass flow we have chosen second order polynomial. The estimated coefficients of the polynomial have the following meanings: $\widehat{p}_0 = \widehat{m}_0$, $\widehat{p}_1 = \widehat{v}_0$, $\widehat{p}_2 = \widehat{a}$. We distinguish the two types of the periods in the experiment: the consumption (fuel is only consumed) and the consumption with fuel replenishment. When one period of the CFB operating changes to another (i.e. a new portion of fuel is being added) the process of mass flow approximation and prediction starts over again, as the model of the mass flow changes. Thus, the most problematic unstable regions are the transitions intervals, when the parameters of the model change their values. First, we describe the on-line approach and then move to a semi-on-line one, which offers a more accurate estimates of the mass flow parameters in the beginning of the period. When a period of the session starts the samples of measurements start to accumulate in the buffer. The data in the buffer are used to approximate the mass flow signal, i.e., to fix the parameters of the approximation/interpolation model. Only the samples that do not contain high amplitude peak due to jamming are placed to the buffer. As the amplitude of these peaks is abnormally high, they can easily be detected on-line as exceeding a certain threshold that should be higher for the replenishment period. The time when a period changes to another is precisely known as it is manipulated by the system operator. At these times the buffer is emptied and starts to accumulate new data. The first measurement within the experiment is taken as the approximation of the mass flow signal at that time $\widehat{m}_1 = y_1$ and as the first point that is placed to the buffer. In contrast, the first approximation of the mass within a following period is taken as the last approximation of the mass from the previous period that is obvious. In addition the last approximation of the mass from the previous period $\widehat{m}_{i_c}$ is placed to the buffer as the first point for the new period, where i_c denotes the number of the sample when the change of the periods occurs. When a new sample arrives the parameters of the model are estimated based on the points that are in the buffer independently of whether the current point was placed to the buffer or not. The current approximation of the mass signal is computed based on the current model. Depending on the number of data points in the buffer different approximation models apply:

1. If the number of points in the buffer amounts to one, then the current approximation of the mass is taken as the approximation of the mass at previous iteration $\widehat{m}_i = \widehat{m}_{i-1}$.
2. If the number of points in the buffer is $1 < T \leq 4$, then $\widehat{p}_2$ is set to zero and the parameters $\widehat{p}_0$ and $\widehat{p}_1$ of the first order polynomial are estimated from the available data. The approximation of the mass at the current time sample x_i is taken as $\widehat{m}_i = \widehat{p}^1(x_i)$, where $\widehat{p}^1(x)$ is the current approximation of mass flow model by the first order polynomial.
3. If the number of points in the buffer is larger than four then the second order polynomial model is fitted to the data, and the current approximation of the mass is computed as $\widehat{m}_i = \widehat{p}^2(x_i)$, where $\widehat{p}^2(x)$ is the current approximation of mass flow model by the second order polynomial.

In practice, the operational settings often allow a delay between the data arrival and the evaluation of the signal of interest at this time sample. This means that the estimate of the signal at a given time sample is obtained based on the data that are accumulated also during the future times. This allows the more accurate estimates of the signal to be computed. Note that in our case this will have an effect of increased accuracy mainly for the beginning of the period, when the amount of the data in the buffer is small yet. The extension of the previous considerations to a case with a certain delay time is straightforward: the estimation of the mass signal is delayed in respect to the current time and the filling of the buffer by the delay τ expressed in time sample units. A minor concern is related to the last time points within a period, for which the specified delay can not be applied as the end of the period is reached. As we already mentioned, when the amount of accumulated data is large the influence of the time delay becomes insignificant in respect to the estimate of the signal. Thus, the last approximations of the mass within the period can be obtained without a delay. Alternatively, if the delay τ is used, for the last τ approximations of the mass within the period the following rule can be used $\widehat{m}^\tau_{i_c-\tau+k} = \widehat{m}^{\tau-k}_{i_c-\tau+k}$, $k = 1, \ldots, \tau$. Here, $\widehat{m}^\tau_i$ denotes the estimate of the mass at time sample i using the delay τ. The latter means that the last approximations of the mass signal for delay τ are taken as approximations of the signal from the approaches with smaller delay. This option can be useful for the replenishment period, which is usually short, as even at the end of this period the amount of data in the buffer is small. In this case it is important to use as large delay as possible, to obtain smoother and more accurate signal approximation.

5   Experimental Results

As the main points of interest are the points, where the change of the period occurs, for the analysis we took an interval of the experiment containing both types of the transition points (see Fig. 5).

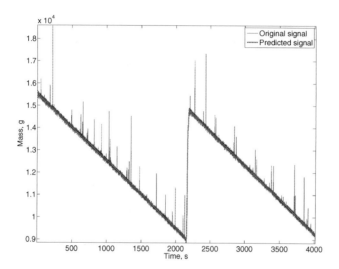

Fig. 5. Online mass flow prediction with the switching regression models and zero delay time

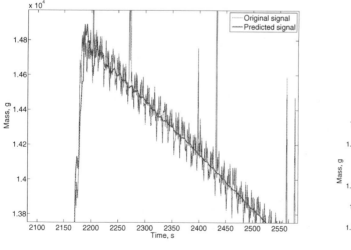
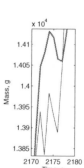

Fig. 6. Online mass flow prediction with the switching regression models and zero delay time. Zooming to transition point.

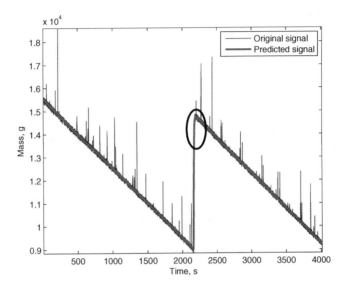

Fig. 7. Online mass flow prediction with the switching regression models and 20 samples delay time

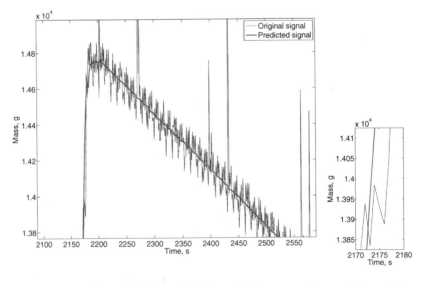

Fig. 8. Online mass flow prediction with the switching regression models and 20 samples delay time. Zooming to transition point.

The computations of mass denoising were done for the delay time varying from 0 to 20 samples/seconds with step 1. In Fig. 5 and 6 the resulting approximation of mass for zero delay is shown in different scales.

It takes about $100 - 150$ seconds from the beginning of each consumption period for the model of the mass signal stabilizes. Since the replenishment period lasts during shorter times $(3 - 40$ seconds), the model may exhibit instability even at the end of the period. However, overall the approximation of the mass during the replenishment period is satisfactory, because changes of the mass are rather steep and prominent against the background of the noise. For comparison, the effect of increased delay time is shown in Fig. 7 and 8, where the delay time 20 seconds was used.

With this delay time the model is relatively well stabilized already in the beginning of each period, which offers clear accuracy advantage. To indicate the rate of convergence of the model with respect to the delay time, we computed the mean-square error between the mass signal approximations for each pair of the consecutive delays (see Fig. 9):

$$MSE(\tau) = \frac{1}{T} \sum_{i=1}^{T} (\widehat{m}_i^{\tau} - \widehat{m}_i^{\tau-1})^2. \tag{6}$$

It can be clearly seen from the figure that for the small delays $(1 - 4$ samples) the approximation of the signal improves dramatically, and for the larger delays the improvement slows down.

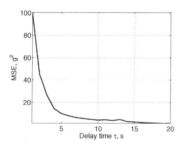

Fig. 9. Mean-square error between the mass signal approximations for each pair of the successive delays

6   Conclusions and Further Work

Prediction of mass flow in CFB boilers in online settings is an important and challenging problem having some connections to the problem of learning under the presence of sudden concept drifts.

In this paper we presented a rather simple regression learning approach with filtering of outliers and the dynamic switching of predictors depending on the current state of the fuel feeding/reloading process that was assumed to be known.

Our experimental study demonstrated that this approach performs reasonably well. A delay of less than 5 seconds allows to predict the mass flow accurately enough to be used as a reliable indicator for the CFB control system.

The directions of our further work in mining CFB sensor data include (1) studying the effect of fuel feeding speed, and the effect of using different mixtures of fuels, and (2) external validation of our online mass flow prediction approach being implemented as part of the control system of the pilot CFB boiler in operational settings with different conditions. We anticipate that local reliability estimation of mass flow prediction [7] may be helpful for the domain experts. We also plan to adopt and develop further the dynamic integration of regression models [8] which may help to improve the reliability of predictions.

Currently the implementation assumes that the transitions between fuel feeding and fuel consumption phases are known. However, to adapt the outlier removal and prediction procedures to more general settings, when the knowledge about the state changes is not available, the method for automatic on-line identification of the state changes would be of benefit.

Hence, automating the predictions of the mass flow by developing an algorithm that is able to detect these state transitions is one of the next steps. In data mining and machine learning these changes are generally known as concept drift, that is the changes in the (hidden) context inducing more or less radical changes in the target concept [11]. The challenge is to keep track of the drift and adjust the model accordingly. This might be possible by using statistics of the performance of the model [2] or by keeping a set of models and select the best. And all of this needs to be done by an online method.

Acknowledgements

This research is partly supported by TEKES (Finnish Funding Agency for Technology and Innovations) DYNERGIA project and NWO (The Netherlands Organization for Scientific Research) HaCDAIS project.

References

1. Duda, R.O., Hart, P.E., Stork, D.G.: Pattern Classification, 2nd edn. Wiley Interscience, Hoboken (2001)
2. Gama, J., Castillo, G.: Learning with local drift detection. In: Li, X., Zaïane, O.R., Li, Z.-h. (eds.) ADMA 2006. LNCS, vol. 4093, pp. 42–55. Springer, Heidelberg (2006)
3. Horn, R.A., Johnson, C.R.: Topics in matrix analysis. Cambridge University Press, Cambridge (1991)
4. Lawson, C.L., Hanson, R.J.: Solving Least Squares Problems. Prentice-Hall, Englewood Cliffs (1974)
5. Nawab, S., Quatieri, T.: Short time fourier transform. In: Lim, J., Oppenheim, A. (eds.) Advanced topics in signal processing, pp. 289–337. Prentice-Hall, Englewood Cliffs (1988)
6. Pechenizkiy, M., Tourunen, A., Kärkkäinen, T., Ivannikov, A., Nevalainen, H.: Towards better understanding of circulating fluidized bed boilers: a data mining approach. In: Proceedings ECML/PKDD Workshop on Practical Data Mining, pp. 80–83 (2006)

7. Rodrigues, P.P., Gama, J., Bosnic, Z.: Online reliability estimates for individual predictions in data streams. In: ICDM Workshops, pp. 36–45. IEEE Computer Society Press, Los Alamitos (2008)
8. Rooney, N., Patterson, D.W., Anand, S.S., Tsymbal, A.: Dynamic integration of regression models. In: Roli, F., Kittler, J., Windeatt, T. (eds.) MCS 2004. LNCS, vol. 3077, pp. 164–173. Springer, Heidelberg (2004)
9. Saastamoinen, J.: Modelling of dynamics of combustion of biomass in fluidized beds. Thermal Science 8(2), 107–126 (2004)
10. Song, Z., Kusiak, A.: Constraint-based control of boiler efficiency: A data-mining approach. IEEE Trans. Industrial Informatics 3(1), 73–83 (2007)
11. Widmer, G., Kubat, M.: Learning in the presence of concept drift and hidden contexts. Mach. Learn. 23(1), 69–101 (1996)

Integrating Data Mining and Agent Based Modeling and Simulation

Omar Baqueiro[1], Yanbo J. Wang[2], Peter McBurney[3], and Frans Coenen[3]

[1] Department of Structural Development of Farms and Rural Areas,
Institute Of Agricultural Development in Central and Eastern Europe (IAMO),
Theodor-Lieser-Str.2, Halle (Saale), D-06120, Germany
baqueiro@iamo.de
[2] Information Management Center, China Minsheng Banking Corp. Ltd.,
Room 606, Building No. 8, 1 Zhongguancun Nandajie, Beijing, 100873, China
wangyanbo@cmbc.com.cn
[3] Department of Computer Science, University of Liverpool,
Ashton Building, Ashton Street, Liverpool, L69 3BX, UK
{Mcburney,Coenen}@liverpool.ac.uk

Abstract. In this paper, we introduce an integration study which combines Data Mining (DM) and Agent Based Modeling and Simulation (ABMS). This study, as a new paradigm for DM/ABMS, is concerned with two approaches: (i) applying DM techniques in ABMS investigation, and inversely (ii) utilizing ABMS results in DM research. Detailed description of each approach is presented in this paper. A conclusion and the future work of this (integration) study are given at the end.

Keywords: Agents, Agent Based Modeling and Simulation, Data Mining, KDD (Knowledge Discovery in Databases) Process.

1 Introduction

Two promising fields of current studies in computer science are Data Mining (DM) and Agent Based Modeling and Simulation (ABMS). DM represents the process of identifying hidden and interesting knowledge in large amounts of data. It is *"a multidisciplinary field, drawing work from areas including database technology, machine learning, statistics, pattern recognition, information retrieval, neural networks, knowledge-based system, artificial intelligence, high-performance computing, and data visualization"* [22]. In the past decade, DM techniques have been widely applied in bioinformatics [45], e-commerce [38], financial studies [26], geography [32], marketing and sales studies [9][41], etc.

ABMS is *"a new modeling paradigm and is one of the most exciting practical developments in modeling since the invention of relational databases"* [34]. It has *"connections to many other fields including complexity science, systems science, systems dynamics, computer science, management science, social sciences in general, and traditional modeling and simulation"* [30]. In the past decade, ABMS has been used to

P. Perner (Ed.): ICDM 2009, LNAI 5633, pp. 220–231, 2009.

model real-life systems in a diversity of domains such as biology [14], manufacturing [42], computing [13] and economics [8] among others. This variety of applications demonstrates the acceptance of ABMS as a useful system modeling and simulation approach to gain knowledge regarding complex systems in such domains. The work presented in [6] cites three reasons for the importance of ABMS for social sciences: first, that other approaches have been proved not suitable for the modeling of these (social sciences related) systems; second, that the agent based approach is a natural representation of many social systems; and third, that the emergence property in agent based models is not easily achieved with other approaches.

In this paper, we introduce the integration study of DM and ABMS (at a *conceptual* level). It is concerned with two approaches/directions:

1. **Applying DM in ABMS** aiming to provide solutions to the *open problem* (further described in section 2.2) in ABMS investigation, based on DM techniques; and

2. **Applying ABMS in DM** with the objective to surpass the *data limitation* (further described in section 2.4) of DM research, based on the simulation results of ABMS.

It is also one objective of this work to make a call for a closer interaction of DM community and ABMS community. We believe that such interaction will benefit both fields. In the case of ABMS it can be used to provide more generalized methods for the validation of agent-based models.

The rest of this paper is organized as follows. In section 2 we describe both ABMS and DM in detail as the background relevant to this study. Section 3 presents the approach of "Applying DM in ABMS". In section 4, we propose the idea of "Applying ABMS in DM". Finally our conclusions and the open issues for further research are given in section 5.

2 Background

2.1 Agent Based Modeling and Simulation

Computer based modeling and simulation of (real-life) complex systems has been one of the driving forces in the development of computer systems. A general definition of a simulation is the imitation of the operation of a process or a real world system through time [7]. A computational model is the representation of a real-life system through a computer program, expressed by a set of algorithms and mathematical formulas implemented as code in a programming language.

In contrast with pure mathematical models, the objective of computational models is not usually to obtain analytical solutions to specific questions. Instead, computational models allow the design of experiments to test the developed models under different scenarios (with different parameter configurations). These experiments are carried out with the objective of testing the behavior of the modeled systems under a certain set of assumptions [7]. This experimentation allows the designer to obtain insight of certain aspects of a complex system which would not be possible to detect using mathematical analysis, or for problems for which there is no tractable mathematical representation.

In ABMS, a system is modeled as a set of autonomous entities, namely *agents*. Each of these agents is positioned in an environment (either virtual or real) from which the agent obtains information by the use of sensors and makes decisions based on its perceived state of the environment and its objectives. These decisions are then reflected as actions performed to modify the state of the environment (i.e. direct actions to the environment, communication with other agents, further reasoning).

An agent can have different behaviors according to the system it populates [50]. Agents also have three basic properties: (a) *reactivity* – the ability to respond to events in the environment; (b) *pro-activity* – the ability to demonstrate some behavior determined by its particular objectives, taking the initiative to satisfy its necessities; and (c) *sociability* – the ability to interact with other agents and humans to fulfill its objectives [49]. These properties give agent based systems a great versatility in comparison with typical object based systems by providing a new type of abstraction for the representation of problem domains.

2.2 The Open Problem in ABMS

Within the agent-based modeling community there have been continuous efforts to create standard processes for the development of agent-based models, such as the standardization of model description (described in [21] and [36]) via the ODD protocol, or the use of software engineering techniques (like UML diagrams [28]). However, one of the main issues inhibiting researches from fields outside computer science to accept ABMS as a tool for the modeling and simulation of real-life systems is the lack of standard verification and validation methodologies. In fact, verification and validation of multi-agent simulations is a concept which has been investigated only in conjunction with the development of specific models. It has only been in recent times that researchers have engaged in independent development of techniques for verification and validation (see [31], [48] and [51]).

Verification and validation are two independent actions that need to be performed in order to achieve the accreditation of a simulation [5]. Verification aims to test whether the implementation of the model is an accurate representation of the abstract model. Hence, the *accuracy* of transforming a created model into the computer program is tested in the verification phase. Model validation is used to check that the implemented model can achieve the proposed objectives of the simulation experiments. That is, to ensure that the built model is an accurate representation of the modeled phenomena to simulate.

In contrast with pure mathematical models which use analytical equations, there is still no consensus among the scientific community on the appropriate methods for verifying and validating an agent based simulation [31]. However, part of the reason for the lack of formalisms which validate agent based simulations is the inherent complexity that these systems try to represent. Some verification methods (discussed in [31]) are source code analysis, automatic theoretic verification and finite state verification. However, there is still some debate in the agent based modeling community on whether formal proofs of systems are useful [20]. Similarly, there is some debate on whether the verification of complex models with many parameters is possible [43].

2.3   Data Mining and Knowledge Discovery in Databases

Data Mining (DM) is an active research field in computer science, which is attracting more and more attention from a wide range of different groups of people. It aims to extract various types of hidden, interesting, previously unknown and potentially useful knowledge (i.e. rules, patterns, regularities, customs, trends, etc.) from sets of data, where the size of a collected dataset can be measured in gigabytes. In DM common types of mined knowledge include: association rules [1], classification rules [37], prediction rules [23], clustering rules [33], sequential patterns [46], emerging patterns [18], etc.

DM is considered to be the core stage/task of KDD (Knowledge Discovery in Databases), where KDD refers to the overall process of *knowledge discovery*. Piatetsky-Shapiro [35] clearly differentiates both terminologies of KDD and DM – *"sometimes 'knowledge discovery process' is used for describing the overall process, including all the data preparation and postprocessing while 'data mining' is used to refer to the step of applying the algorithm to the clean data"*. The KDD process has been well studied and analyzed, e.g. [12] [19] [22] [44] [47]. It was pointed out in [2] that *"the problem of knowledge extraction from large databases involves many stages and a unique scheme has not yet been agreed upon"*.

One possible outline of the KDD process can be presented as follows.

1. **Problem Specification:** In the first stage of the KDD process, a domain-oriented understanding of the target mining task/application is identified, which clarifies the goal of the application.
2. **Resourcing:** The second stage of the KDD process aims to create a suitable set of data on which the target application can be performed. It may be possible to find several large databases available that appear to be task-relevant for the target application, and which were originally built for other purposes and are irrelevant to the target application.
3. **Data Cleaning:** The purpose of this stage is as Han and Kamber [22] explain *"to remove noise and inconsistent data"* from a given dataset. Furthermore, the missing and distorted data [16] and data outliers are also cleansed in this stage.
4. **Data Integration:** In this stage, the cleaned data sets from different resources are combined into an integrated data set with a unified data view.
5. **Pre-processing:** The data collected may be in an unstructured format, i.e. texts, images, videos, etc. In this stage the collected data is transformed into a structured/semi-structured (e.g. XML, SGML) representation that follows the data to be further operated upon in the KDD process. For simplicity, especially when the volume of the collected data is considered too large, this stage then selects the most significant data for the target application for further usage, and other data is discarded.
6. **Data Mining:** The purpose of this stage is to identify the most valuable information in the prepared data by utilizing *"data analysis and knowledge discovery techniques under acceptable computational efficiency limitations, and produces a particular enumeration of patterns over the data"* [52].

7. **Interpretation and Evaluation of Results:** The validity of each pattern discovered is interpreted and measured. From this the overall quality of the mining performance can be evaluated. In this stage, the discovered valuable knowledge is initially interpreted in a user-readable form (especially when the user is strongly involved in the evaluation), where the patterns, rule symbols, and/or variables are precisely and concisely expressed in human language. Suitable patterns (valuable knowledge) are then caught in this stage.

8. **Future Application:** The set of valuable knowledge mined, interpreted and measured from the stages 6 and 7 is then available to be applied for domain-oriented decision marking in the future.

It can be noted that the above stages are usually applied iteratively; with results of one stage providing feedback that allows improvement to earlier stages.

2.4 The Data Limitation of DM and KDD

With regard to the KDD process, it can be indicated that stages 2 to 5 together determine the requirements of data for DM. Although Anand *et al.* [4] indicate that *"the amount of data being stored in databases has been on the increase since the 1970's partly due to the advances made in database technology since the introduction of the relational model for data by E. F. Codd"*, there is still some debate to the availability of data (for different application-domains) being stored electronically. Hence one major problem of DM and KDD is the difficulty to obtain enough data resources and collect sufficient amount of data for a mining task.

For some classical DM research tasks (i.e. classification, association rule mining, text mining, etc.), sufficient amounts of *real* data can be found (i.e. the UCI Machine Learning Repository [10], the LUCS-KDD Discretised Normalised Data Library [15], Usenet Articles [27], respectively), even though much of this data may be imperfect (being noisy, inconsistent, missing or distorted or containing data outliers). De Veaux and Hand [16] argue that *"anyone who has analyzed real data knows that the majority of their time on a data analysis project will be spent 'cleaning' the data before doing any analysis"*, and *"common wisdom puts the extent of this at 60-95% of the total project effort"*. Klein [25] argues that *"there is strong evidence that data stored in organizational databases have a significant number of errors"* and *"between one and ten percent of data items in critical organizational databases are estimated to be inaccurate"*. Thus another problem of DM and KDD is that even there are enough data resources and sufficient amount of data, it is quite difficult to accurately detect and correct erroneous data in a given dataset.

Besides, the automated approaches for data integration and pre-processing may further damage the quality of the collected data. For example, during the data integration and transformation phases the significance of a data item for a particular mining task/application may be varied; and certainly the inaccurate selection of the most significant data may even cause none of any valuable pattern (knowledge) to be extracted.

3 Applying DM in ABMS

3.1 Previous Work

In [40], a method for the validation of ABMS using DM techniques is proposed. In that work, the authors present the results of the analysis of a simulation data obtained from the repeated execution of experiments while varying a single parameter. They analyze the obtained data using clustering techniques finding new patterns in the data.

As far as we know, the work by Remondino and Correndo [39], [40] is the only attempt to apply DM techniques to ABMS. In these works, the authors differentiate between *endogenous* and *exogenous* application of DM techniques. The endogenous use is concerned with providing the agents participating in the simulation with the DM techniques in order to improve their performance. For example in [3], the authors established *data mining agents* in the context of a multi-agent system, making use of DM techniques to perform distributed mining.

On the other hand, the exogenous application focuses on using DM techniques to analyze the data resulting from the simulation. Exogenous use of DM in ABMS is exemplified by the application described by Remondino and Correndo [40], where they perform a series of simulation experiments using an agent based model of a biological phenomenon. Afterwards, they proceeded to perform a DM analysis with the obtained data to detect if there was any novel pattern.

3.2 Proposed Ideas

Although we generally agree with the use of DM in ABMS discussed in [39] and [40], we believe that DM techniques can be further used to aid in the formalization of validation processes for ABMS. To achieve this, we claim that techniques such as the identification of classification rules, clustering rules, association rules and sequential patterns in the data obtained from the simulation experiments can be used. Specifically, we believe that using DM techniques would be possible to *abstract* experimental results by creating higher level description elements representing the behavior observed in the simulations, without having to focus in the specific obtained data values.

This would allow a more straightforward comparison between different experimental results, obtained from the same model or from independent models representing the same phenomena. Such comparative analysis would be possible by the comparison of the high level description elements (such as sequential patterns of agent behavior or identified clusters) obtained from the raw experimental data. Similarly, such an approach could be used to compare the results obtained from the experiments of the simulation and data from the real-life/world system that is being simulated. This considering that, even though the values obtained from the different experiments will not be exactly the same than those from the (real-life) system, it is sensible to expect that the data obtained from the simulation experiments and the data from the system will share some general descriptive patterns. Fig. 1 depicts the process followed when applying DM techniques for as part of the verification and validation of agent based models.

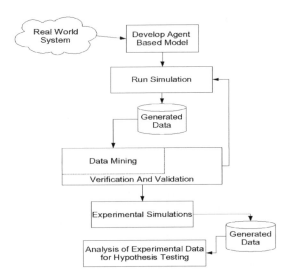

Fig. 1. Applying DM for the verification and validation of agent based models

4 Applying ABMS in DM

Beyond the topic of "Applying DM in ABMS", in this paper we inversely, for the first time, introduce the approach of "Applying ABMS in DM" (Fig. 2). The aim of this approach is to provide sufficient amount of data with good quality for various DM mechanisms and ensure the utilization of DM techniques in different domain-applications. Specifically, the simulation results of ABMS can be used as *quasi-real* data when there is the lack of *real* data for a domain-specific DM task.

The approach described in this section represents a new paradigm in data preparation of DM and KDD, which in turn create an improved KDD process. This new KDD process consists of five stages only, presented as follows.

1. **Problem Specification:** This stage contains the original problem specification contents (as previously described in section 2.3) plus the specification of an ABMS system for this application problem.

2. **Data Preparation:** Based on the determined ABMS system (from the stage 1), this stage then generates simulation results and forms these results into a corresponding dataset for the next (DM) stage. It can be pointed out that the data generated herein shows the following properties together surpassing the data limitation of DM and KDD (see section 2.4): (a) *quasi-real* – as mentioned above, ABMS is well established and a mature technology in the analysis of (real-life) complex system, its simulation results can be very similar to the real-life/world situations (real data); (b) *suitable-sized* – for different DM techniques/mechanisms, the amount of data may be sensitive, hence generating not too much but sufficient amount of data is considered; (c) *qualified* – the erroneous data can be controlled, so that there is neither noisy/inconsistent/missing/distorted data nor data outliner involved; and (d) *significant* – insignificant data will be avoided

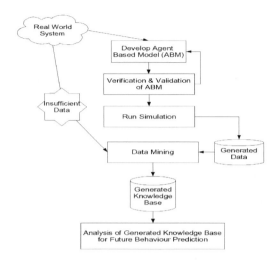

Fig. 2. Applying ABMS for the data preparation of data mining

during the simulation process by providing correct parameter configuration(s); thus the data significance for a particular application can be ensured.

3. **Data Mining:** keep this stage as previously described in section 2.3.

4. **Interpretation and Evaluation of Results:** Keep this stage as previously described in section 2.3.

5. **Future Application:** Keep this stage as previously described in section 2.3.

Further benefit of the proposed "Applying ABMS in DM" approach can be exemplified with the following case.

Assume that we are dealing with a classification problem – the automated categorization of an "unseen" data instance into pre-defined classes – in a financial situation, say that to classify whether an oil derivative is currently buyable, sellable or non-tradable. In this case, alternative classification mechanisms can be applied that include: decision tree induction [37], naive Bayes [17], k-nearest neighbor [24], neural networks [53], support vector machine [11], classification association rule mining [29], etc. In a DM context, it will be interesting to determine which (classification) mechanism is always well performed herein, so that it can be suggested in a default manner whenever a similar problem is handled. This offers the solution to avoid the problem of over-fitting – some mechanism is only good when applying it to a particular set of data, but not general applicable at all. With regard to an ABMS system available that simulates the real-life oil trading market, a set of simulation results can be generated by rationally and strategically varying the value of different parameters, where each simulation result is then formed into a dataset among which hidden rules can be mined and performance of each mechanism is then evaluated. Finally an overall averaging process can be applied to determine the best performed mechanism throughout various simulation scenarios (with different parameter configurations) in the agent based oil trading modeling and simulation.

5 Conclusions

In this paper, we proposed a general idea for the bidirectional integration of Data Mining (DM) and Agent Based Modeling and Simulation (ABMS) (i.e. "Applying DM in ABMS" and "Applying ABMS in DM"). A conceptual framework of this integration study was provided; and a broader view of the advantages that such integration can provide to both DM and ABMS was presented.

The integration of DM and ABMS provides promising mechanisms for the advancement in both fields. From the ABMS point of view, DM can provide necessary mechanisms for the validation of created models and to facilitate the comparison between data obtained from different simulations. From the DM point of view, ABMS can be used to generate necessary data when the real data obtained is not enough or does not have good quality. Moreover, ABMS can be used to obtain datasets from designed simulation scenarios (for which real-life data may not be available), in order to determine the default mining approach for a group of similar applications. This offers a solution to the over-fitting problem in DM.

Further research may identify some new approaches other than the proposed verification & validation and data preparation ideas under the headings of "Applying DM in ABMS" and "Applying ABMS in DM".

Acknowledgments. The authors would like to thank Kamal Ali Albashiri, Chuntao Jiang and Maya Wardeh of the Department of Computer Science at the University of Liverpool for their support with respect to the work described here. Omar Baqueiro acknowledges the support provided by the Mexican Council of Science and Technology (CONACYT) with the grant 187564.

References

1. Agrawal, R., Imielinski, T., Swami, A.: Mining Association Rules between Sets of Items in Large Databases. In: Proceedings of the 1993 ACM SIGMOD International Conference on Management of Data, Washington, DC, USA, pp. 207–216. ACM Press, New York (1993)
2. Ahmed, S.: Strategies for Partitioning Data in Association Rule Mining. Ph.D. Thesis, University of Liverpool, Liverpool, UK (2004)
3. Albashiri, K.A., Coenen, F., Leng, P.: Agent Based Frequent Set Meta Mining: Introducing EMADS. In: Artificial Intelligence in Theory and Practice II – Proceedings of IFIP 20th World Computer Congress, TC 12: IFIP AI 2008 Stream, Milano, Italy, pp. 23–32. Springer, Heidelberg (2008)
4. Anand, S.S., Bell, D.A., Hughes, J.G.: Evidence based Discovery of Knowledge in Databases. IEE Colloquium on Knowledge Discovery in Databases, Digest No: 1995/021(A), 9/1–9/5, London, UK (1995)
5. Balci, O.: Verification Validation and Accreditation of Simulation Models. In: Proceedings of the 29th Conference on Winter Simulation, Atlanta, GA, USA, pp. 135–141. ACM Press, New York (1997)
6. Bankes, S.: Agent-based Modeling: A Revolution? In: Proceedings of the National Academy of Sciences, pp. 7199–7200 (1999)
7. Banks, J.: Handbook of Simulation. John Wiley & Sons, Chichester (1998)

8. Beam, C., Segev, A.: Automated Negotiations: A Survey of the State of the Art. Wirtschaftsinformatik 39(3), 263–268 (1997)
9. Berry, M.J.A., Linoff, G.: Data Mining Techniques for Marketing, Sales, and Customer Support. John Wiley & Sons, Inc., USA (1997)
10. Blake, C.L., Merz, C.J.: UCI Repository of Machine Learning Databases. Department of Information and Computer Science, University of California, Irvine, CA, USA (1998), http://www.ics.uci.edu/~mlearn/MLRepository.html
11. Boser, B.E., Guyon, I.M., Vapnik, V.N.: A Training Algorithm for Optimal Margin Classifiers. In: Proceedings of the 5th ACM Annual Workshop on Computational Learning Theory, Pittsburgh, PA, USA, pp. 144–152. ACM Press, New York (1992)
12. Brachman, R.J., Anand, T.: The Process of Knowledge Discovery in Databases: A Human Centered Approach. In: Advance in Knowledge Discovery and Data Mining, pp. 37–57. AAAI/MIT Press (1996)
13. Buyya, R., Abramson, D., Giddy, J., Stockinger, H.: Economic Models for Resource Management and Scheduling in Grid Computing. Concurrency Computation Practice and Experience 14(14), 1507–1542 (2002)
14. Castella, J., Trung, T., Boissau, S.: Participatory Simulation of Land-use Changes in the Northern Mountains of Vietnam: The Combined Use of an Agent-based Model, a Role-playing Game, and a Geographic Information System. Ecology and Society 10(1), 27 (2005)
15. Coenen, F.: The LUCK-KDD Discrietised/Normalised ARM and CARM Data Library. Department of Computer Science, University of Liverpool, Liverpool, UK (2003), http://www.csc.liv.ac.uk/~frans/KDD/Software?LUCS-KDD-DN
16. De Veaux, R.D., Hand, D.J.: How to Lie with Bad Data? Statistical Science 20(3), 231–238 (2005)
17. Domingos, P., Pazzani, M.: On the Optimality of the Simple Bayesian Classifier under Zero-one Loss. Machine Learning 29(2/3), 103–130 (1997)
18. Dong, G., Li, J.: Efficient Mining of Emerging Patterns: Discovering Trends and Differences. In: Proceedings of the Fifth ACM SIGKDD International Conference on Knowledge Discovery and Data Mining, San Diago, CA, USA, pp. 43–52. ACM Press, New York (1999)
19. Fayyad, U., Piatetsky-Shapiro, G., Smyth, P.: Knowledge Discovery and Data Mining: Towards a Unifying Framework. In: Proceedings of the Second International Conference on Knowledge Discovery and Data Mining, Portland, OR, USA, pp. 82–95. AAAI Press, Menlo Park (1996)
20. Glass, R.L.: Practical Programmer: Inspections, Some Surprising Findings. Commun. ACM 42(4), 17–19 (1999)
21. Grimm, V., Berger, U., Bastiansen, F., Eliassen, S., Ginot, V., Giske, J., Goss-Custard, J., Grand, T., Heinz, S.K., Huse, G., Huth, A., Jepsen, J.U., Jorgensen, C., Mooij, W.M., Muller, B., Pe'er, G., Piou, C., Railsback, S.F., Robbins, A.M., Robbins, M.M., Rossmanith, E., Ruger, N., Strand, E., Souissi, S., Stillman, R.A., Vabo, R., Visser, U., Deangelis, D.L.: A Standard Protocol for Describing Individual-based and Agent-based Models. Ecological Modelling 198(1-2), 115–126 (2006)
22. Han, J., Kamber, M.: Data Mining: Concepts and Techniques, 2nd edn. Morgan Kaufmann Publishers, San Francisco (2006)
23. Han, J., Kamber, M.: Data Mining: Concepts and Techniques. Morgan Kaufmann Publishers, San Francisco (2001)
24. James, M.: Classification Algorithms. Wiley Interscience, New York (1985)

25. Klein, B.D.: Data Quality in the Practice of Consumer Product Management: Evidence from the Field. Data Quality Journal 4(1) (1998)
26. Kovalerchun, B., Vityaev, E.: Data Mining in Finance: Advances in Relational and Hybrid Methods. Kluwer Academic Publishers, Dordrecht (2000)
27. Lang, K.: NewsWeeder: Learning to Filter Netnews. In: Machine Learning – Proceedings of the Twelfth International Conference on Machine Learning, Tahoe City, CA, USA, pp. 331–339. Morgan Kaufmann Publishers, San Francisco (1995)
28. Lisec, A., Ferlan, M., Lobnik, F., Sumrada, R.: Modelling the Rural Land Transaction Procedure. Land Use Policy 25(2), 286–297 (2008)
29. Liu, B., Hsu, W., Ma, Y.: Integrating Classification and Association Rule Mining. In: Proceedings of the Fourth International Conference on Knowledge Discovery and Data Mining, New York, NY, USA, pp. 80–86. AAAI Press, Menlo Park (1998)
30. Macal, C.M., North, M.J.: Agent-Based Modeling and Simulation: Desktop ABMS. In: Proceedings of the 2007 Winter Simulation Conference, Washington, DC, USA, pp. 95–106. IEEE Computer Society Press, Los Alamitos (2007)
31. Midgley, D., Marks, R., Kunchamwar, D.: Building and Assurance of Agent-based Models: An Example and Challenge to the Field. Journal of Business Research 60(8), 884–893 (2007)
32. Miller, H.J., Han, J.: Geographic Data Mining and Knowledge Discovery. Taylor & Francis, Abington (2001)
33. Mirkin, B., Mirkin, B.G.: Clustering for Data Mining: A Data Recovery Approach. Chapman & Hall/CRC, Boca Raton (2005)
34. North, M.J., Macal, C.M.: Managing Business Complexity: Discovering Strategic Solutions With Agent-Based Modeling And Simulation. Oxford University Press, Oxford (2007)
35. Piatetsky-Shapiro, G.: Knowledge Discovery in Databases: 10 Years After. ACM SIGKDD Explorations 1(2), 59–61 (2000)
36. Polhill, J., Parker, D., Brown, D., Grimm, V.: Using the ODD Protocol for Describing Three Agent-based Social Simulation Models of Land-use Change. JASSS 11(2) (2008)
37. Quinlan, J.R.: C4.5: Programs for Machine Learning. Morgan Kaufmann Publishers, San Francisco (1993)
38. Raghavan, S.N.R.: Data Mining in E-commerce: A Survey. Sadhana 30(2&3), 275–289 (2005)
39. Remondino, M., Correndo, G.: Data Mining Applied to Agent Based Simulation. In: Proceedings of the 19th European Conference on Modelling and Simulation, Riga, Latvia (2005)
40. Remondino, M., Correndo, G.: MABS Validation through Repeated Executing and Data Mining Analysis. International Journal of Simulation Systems, Science & Technology 7(6), 10–21 (2006)
41. Rypielski, C., Wang, J.-C., Yen, D.C.: Data Mining techniques for customer Relationship Management. Technology in Society 24, 483–502 (2002)
42. Shen, W., Norrie, D.H.: Agent-based Systems for Intelligent Manufacturing: A State-of-the-art Survey. Knowledge and Information Systems 1, 129–156 (1999)
43. Shervais, S., Wakeland, W., Raffo, D.: Evolutionary Verification and Validation of Software Process Simulation Models. In: The 5th International Workshop on Software Process Simulation and Modelling, Edinburgh, Scotland (2004)
44. Smith, G.D.: Meta-heuristics in the KDD Process. In: Proceedings of the First UK Knowledge Discovery in Data Symposium, Liverpool, UK (2005)

45. Wang, J.T.L., Zaki, M.J., Toivonen, H.T.T., Shasha, D.E.: Data Mining in Bioinformatics. Springer, London (2005)
46. Wang, W., Yang, J.: Mining Sequential Patterns from Large Data Sets. Springer, New York (2005)
47. Weiss, S.M., Indurkha, N.: Predictive Data Mining: A Practical Guide. Morgan Kaufmann Publishers, San Francisco (1997)
48. Windrum, P., Gagiolo, G., Moneta, A.: Empirical Validation of Agent-based Models: Alternatives and Prospects. Journal of Artificial Societies and Social Simulation 10(2), 8 (2007)
49. Wooldridge, M.J., Jennings, N.R.: Intelligent Agent: Theory and Practice. Knowledge Engineering Review 10(2), 115–152 (1995)
50. Wooldridge, M.J.: An Introduction to MultiAgent Systems. John Wiley & Sons, Chichester (2002)
51. Yilmaz, L.: Validation and Verification of Social Processes within Agent-based Computational Organization Models. Computational & Mathematical Organization Theory 12(4), 283–312 (2006)
52. Zhang, D., Zhou, L.: Discovering Golden Nuggets: Data Mining in Financial Application. IEEE Transaction on Systems, Man, and Cybernetics – Part C: Applications and Reviews 34(4), 513–552 (2004)
53. Zhang, G.P.: Neural Networks for Classification: A Survey. IEEE Transactions on Systems, Man, and Cybernetics – Part C: Applications and Reviews 30(4), 451–462 (2000)

Combining Multidimensional Scaling and Computational Intelligence for Industrial Monitoring

António Dourado[1], Sara Silva[1], Lara Aires[1], and João Araújo[2]

[1] Centro de Informática e Sistemas da Universidade de Coimbra
Department of Informatics Engineering
Universidade de Coimbra – Polo II, 3030-290 Coimbra, Portugal
{dourado,sara,laires}@dei.uc.pt
[2] Galp Energia, Refinaria de Sines, Ap. 15, 7520-952 Sines, Portugal
joao.araujo@galpenergia.com

Abstract. Large industrial complexes with hundreds of variables must be tightly monitored for safety, quality and resources optimization. Multidimensional scaling and computational intelligence are proposed in this work as effective tools for building classifiers of the operating state of the industrial process into normal / abnormal working regions. The VisRed, Visualization by Data Reduction computational framework, is extended with techniques from computational intelligence, such as neural networks (several architectures), support vector machines and neuro-fuzzy systems (in an evolving adaptive implementation) to build such classifiers. The Visbreaker plant of an oil refinery is taken as case study and some scenarios show the potentiality of the combined approach.

Keywords: computational intelligence, multidimensional scaling, neural networks, support vector machines, neuro-fuzzy systems, industrial monitoring.

1 Introduction

Technological complexity, physical dimension, and scale of production, make monitoring of industrial processes a more and more important task, but also more and more challenging. The same can be said about diagnosis in medical problems, decision support in finance and services. All these are tasks that can profit from mining of the numerical data available in today's information systems. Multidimensional scaling [1] has been adopted here for these purposes in order to make information easily represented and interpreted by humans [2].

In previous works [3] [4], VisRed and VisRedII were proposed as GNU free platforms for multidimensional scaling, based on Matlab [5]. In multidimensional scaling practical applications, the number of optimization variables easily reaches very high numbers, and this is a hard problem to solve the local minima drawbacks. Powerful optimization techniques, including the metaheuristics of Genetic Algorithms and Simulated Annealing, can be used with that aim and are available in VisRedII. Now the authors have extended its capabilities to VisRedIII with neural networks of several architectures, including support vector machines, and evolving neuro-fuzzy systems (Fig. 1).

P. Perner (Ed.): ICDM 2009, LNAI 5633, pp. 232–246, 2009.

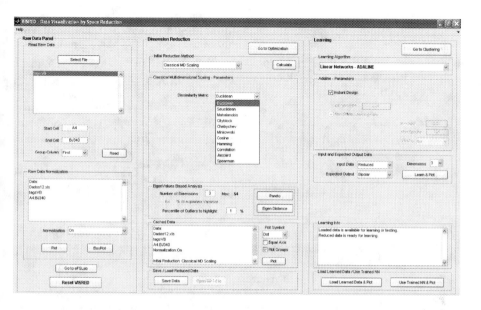

Fig. 1. Main screen of VisRedIII, with Fuzzy systems and Neural Networks

After defining the optimization problem at hand, in Section 2, and the classification problem associated with it for industrial diagnosis and monitoring, the different capabilities of VisRedIII are presented in Section 3 (dimensionality reduction), Section 4 (neural networks), and Section 5 (neuro-fuzzy systems). Section 6 presents one application, an industrial process (visbreaker process of an oil refinery) for which some scenarios have been worked out and some interesting results have been obtained. Conclusions are settled in Section 7.

2   The Optimization and the Classification Problems Involved

In large industrial complexes, such as a refinery, there are thousands of variables, acquired by computerized instrumentation and control systems, that must be used in an effective way to monitor the plant, improving its safety (preventing accidents, namely environmental ones), the quality of output products and the economy of resources. The state of the process is represented by these variables, in high dimensional spaces. For the sake of human visibility, one may reduce that dimension to two or three. For that purpose multidimensional scaling is considered to be an effective technique, since it reduces the dimension of information space from n to two or three dimensions, preserving, as much as possible, the information content.

Let us consider a set of p points in an original n dimensional space. Some measure of dissimilarity between each pair of them can be defined. Dissimilarity may be a distance (such as Euclidian), a ranking, or an order between any two of the p points. A squared symmetric $p \times p$ dissimilarity matrix is constructed with these dissimilarities.

Using for example Euclidean distances, the dissimilarity matrix Δ_n in the original n-dimensional space is constructed. Now multidimensional scaling will find a distribution of the same number of p points into a reduced m-dimensional space, with $m \ll n$. In this reduced space a similarity matrix Δ_m (with the same dimension as Δ_n) is constructed in the same way.

The Euclidian distances between the dissimilarity matrix Δ_n and the dissimilarity matrix Δ_m are then minimized in the least squares sense. The matrix distance (1) is defined as the sum of the Euclidian distances between every point in one matrix and the corresponding point in the other matrix.

$$ J = \left\| \Delta_n - \Delta_m \right\| = \sum_{i=1}^{p} \sum_{j=1}^{p} \left\| (\Delta_n)_{ij} - (\Delta_m)_{ij} \right\| \tag{1} $$

Multidimensional scaling is an optimization process aiming at minimizing (1). If the final distance was zero then the points in the reduced space would be a perfect reduced view of the points in the original high dimensional space. In this situation all the information content expressed by the positions of the points would be perfectly preserved when the dimension reduction was performed. This may be considered a topology preserving method. However, in practical problems that distance J is never zero and its minimization is the goal.

In practical problems one has hundreds or thousands of points in dozens of dimensions. For example, 2000 points in 50 dimensions give 100 000 optimization variables, defining a large scale optimization problem, non-convex. Convergence to a global minimum is not an easy task. Several optimization techniques are implemented in VisRedIII (Fig. 2), including metaheuristics such us genetic algorithms and simulated annealing to ease this aim.

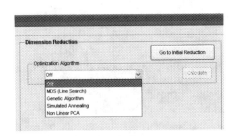

Fig. 2. The optimization techniques implemented in VisRedIII   **Fig. 3.** Clustering techniques in VisRedIII

Multidimensional scaling is preferable to other techniques such as Andrews plots [25] because it allows a more suggestive representation in three dimensions of data for industrial engineers and operators.

After dimension reduction, classification may be made in the reduced space. Clustering techniques may be used and several algorithms are available in VisRedIII

(Fig. 3). However, computational intelligence techniques, such as neural networks, support vector machines and neuro-fuzzy systems can be used as classifiers in some problems such as the industrial monitoring, where in general there are two classes in data: normal state and abnormal state.

3 Dimensionality Reduction Implemented in VisRedIII

Besides the dimensionality reduction techniques implemented in VisRed (Linear Principal Component Analysis, Non Linear Principal Component Analysis, Classical Multidimensional Scaling, Multidimensional Scaling) the following have been introduced (Fig. 4). Some freely available code from [6] has been used.

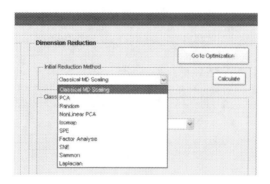

Fig. 4. Methods for dimension reduction implemented in VisRedIII

1. Isomap
This method is an extension of multidimensional scaling. Instead of using Euclidean or other common metrics, it calculates pairwise geodesic distances on a weighted graph, where each point is connected to all its neighbors. Each geodesic distance is calculated as the sum of edge weights along the shortest path between two nodes [7].

2. Stochastic Proximity Embedding (SPE)
This method minimizes the MDS raw stress function. It differs from MDS in the computationally-efficient rule it uses to update the current estimate of the low-dimensional data representation [8].

3. Factor Analysis [9]
This method models observed variables as linear combinations of fewer unobserved variables called factors, plus error terms. It is related to PCA, and becomes essentially the same if the errors are all assumed to have the same variance.

4. Stochastic Neighbor Embedding (SNE)
This method, also related to MDS, converts each high-dimensional similarity into the probability that one data point will pick the other data point as its neighbor [10].

5. Sammon Mapping

This method is a variant of MDS where more emphasis is put on retaining distances that were originally small [1].

6. Laplacian Eigenmaps

This method finds a low-dimensional data representation by preserving local properties, based on the pairwise distances between near neighbors [11].

4   Neural Networks and Support Vector Machines

Ten supervised learning neural networks were implemented (Fig. 5), basically falling into these categories: Adaptive Linear Filters (ADALINE) [12], Feedforward Networks [13], both static and dynamic [14], Radial Basis Networks [15], and Support Vector Machines [16]. They can be trained in a data set and later tested in a different data set, they can use the original high-dimensional or the reduced low-dimensional data, and they can use the original class numbers or their binary or bipolar representation. The implementation of SVMs is based on a Matlab/MEX Interface to SVMlight by Tom Briggs [16]. The SVMlight code is the intellectual property of Thorsten Joachims [18].

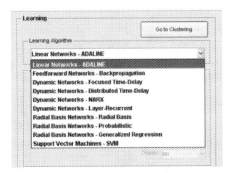

Fig. 5. Neural Networks available in VisRedIII

1. Linear Networks – ADALINE

This is a linear network and, as such, can only solve linearly separable problems. When given a set of input vectors, the network returns a set of output vectors, one for each input. The difference between an output vector and its target, or expected vector, is the error. The goal is to find values for the network weights and biases such that the sum of the squares of the errors is minimized or below a specific value.

2. Feedforward Networks – Backpropagation

Backpropagation networks is the name usually given to feedforward neural networks trained with any variant of the backpropagation algorithm. Backpropagation networks with biases, a sigmoid layer, and a linear output layer are capable of approximating any function with a finite number of discontinuities.

3. Dynamic Networks – Focused Time-Delay

This is a dynamic network. Unlike static feedforward networks, dynamic networks can have feedback elements and contain delays. A Focused Time-Delay neural network is the most straightforward dynamic network, consisting of a feedforward network with a tapped delay line at the input.

4. Dynamic Networks – Distributed Time-Delay

This is another dynamic network. While the Focused Time-Delay network has the tapped delay line memory only at the input to the first layer, the Distributed Time-Delay can have the tapped delay lines distributed throughout the network.

5. Dynamic Networks – NARX

The NARX (nonlinear autoregressive network with exogenous inputs) is a recurrent dynamic network, with feedback connections enclosing several layers of the network.

6. Dynamic Networks – Layer-Recurrent

First introduced by Elman in a simplified form, the Layer-Recurrent network contains a feedback loop, with a single delay, around each layer of the network except for the last layer.

7. Radial Basis Networks – Radial Basis

In Radial Basis networks each output neuron receives the vector distance between its weight vector and the training input vector, multiplied by the bias, and applies a radial basis transfer function to the result.

8. Radial Basis Networks – Probabilistic

Probabilistic networks are radial basis networks containing two layers: the first is a radial basis layer that calculates distances between each input vector and previously seen training vectors; the second is a competitive layer that uses this information to produce a vector containing the probabilities of the input vector belonging to each training class. A final element selects the most probable class.

9. Radial Basis Networks – Generalized Regression

Generalized Regression networks are also built with a radial basis layer, followed by a special linear layer. They perform regression, not classification like the Probabilistic networks. The output of a Generalized Regression network is simply a weighted average of the target values of training samples close to the given input.

10. Support Vector Machines - SVM

SVMs operate by finding a hyperplane in the space of training inputs. This hyperplane will attempt to split the positive examples from the negative examples, such that the distance between the hyperplane to the nearest of the positive and negative examples is the largest possible [19].

5 Evolving Fuzzy System Creation in VisRedIII eFSLab

The computational environment eFSLab is intended to be a complete tool through which a non-expert can create, in an easy way, a zero or first order Takagi-Sugeno

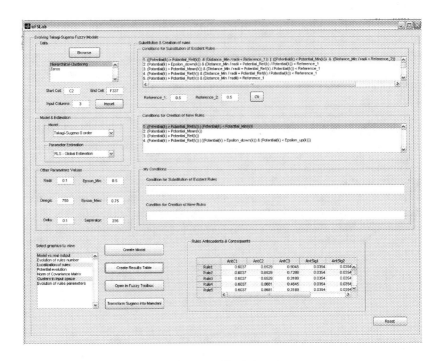

Fig. 6. Main eFSLab interface

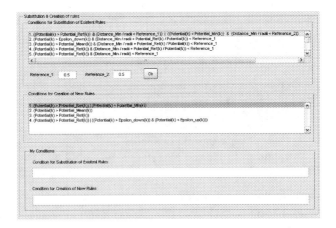

Fig. 7. Panel for creation or modification of rules in eFSLab

(TS) fuzzy system [20] using a data driven approach. To do that, the user only has to choose the data to use in an adequate format, and some relevant parameters. Then, an iterative (evolving) algorithm constructs the fuzzy rules from the data, with the rule base being initially empty. Fig. 6 presents the eFSLab interface called from VisRedIII.

In this process, the first step is to choose the data file to use (it could be an Excel or a Text file) and the data set. eFSLab is also prepared to accept excel files produced by VisRed without being necessary any processing. Then it is necessary to select the model for TS (zero or first order) and for parameter estimation (local or global), and some other specific parameters that try to optimize the rules creation or modification with respect to the new data. Conditions to create or modify rules have been re-searched and implemented; they are available from the main interface. The user can select one of the provided conditions or define a new condition. This can be seen with more detail in Fig. 7.

After all parameters are set, the user can choose whether or not to see the graphical information about the evolution of TS system and, if wanted, which graphics will be presented. Then the TS system is created. Graphical information includes the obtained clusters in input space, the instants of modification or creation of rules, the number of rules, rules parameters evolution, the norm of covariance matrix, the potential evolution of the data samples and the model output versus real output for validation data. One example of these graphics is shown in Fig. 8. It represents the collection of points from the used data set in combination with the clusters obtained at the end of the learning phase of the eFS algorithm, which is based on recursive subtractive cluster-ing [20], as will be explained in the next section.

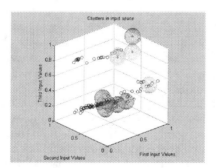

Figure 8. Clusters in input space after training. The dimensions of the spheres depend on the parameters of the clusters in a subtractive clustering environment.

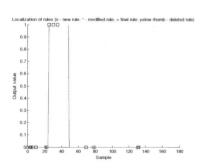

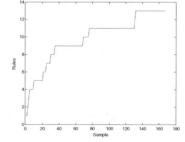

Fig. 9. Sampling localization of rule changes (deleted, created, or modified)

Fig. 10. Evolution of the number of rules

Another example is the graphical representation of the created, modified and deleted rules and the instant when that happens, as in Fig. 9. The evolution of the number of rules during the process can also be visualized, as in Fig. 10.

After the TS system is produced, its characteristics can be explored in various ways using the Matlab Fuzzy Logic Toolbox (analysing antecedents and consequents fuzzy sets, viewing the graphics, etc). A diagnosis text file is created by eFSLab to support the user. A table of the rules, shown in the interface as the one presented in Fig. 11, for the case of five rules, Gaussian membership functions and two antecedents.

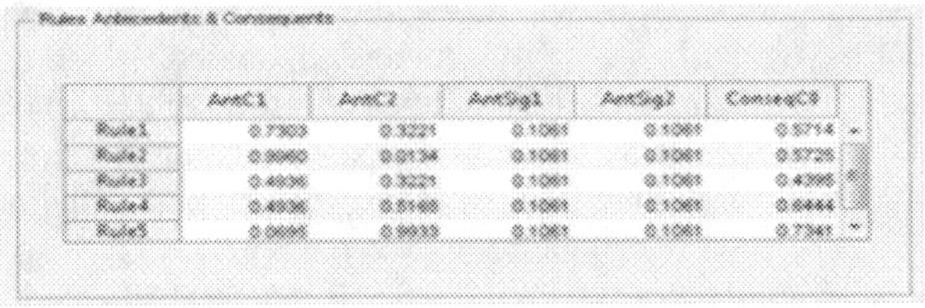

Fig. 11. Table containing the rules created by eFS procedure and antecedents and consequents for each of them

Another capability of eFSLab is to transform a generated zero order TS system into a Mamdani system, by using the *SugenoToMamdani* interface shown in Fig. 12. If needed, Matlab Fuzzy Logic Toolbox with its functionalities provides a deeper analysis of the Mamdani fuzzy system produced. To create TS fuzzy systems the eFS algorithm is used. This algorithm is based on the work of Angelov et al. regarding online learning [21], and improved by [22] [23].

Fuzzy models are created by a data-driven approach and can be zero or first order Takagi-Sugeno with global or local recursive estimation parameters. The algorithm progresses through main stages in an iterative process. The first step is initialization of rule-base structure based on the first data sample that makes the first rule. Then, the next data sample is read. In the next stage the potential of each new data sample is recursively calculated and then the potential of the centers (focal points) of the existing rules (clusters) are recursively updated.

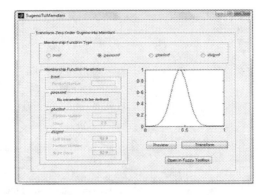

Fig. 12. The *SugenoToMamdani* interface in eFSLab

Then, the rule-base is modified or upgraded by comparing the potential of the new data sample with the updated potential of existing centers. If the new data has a potential higher than all the existent centers, it becomes a new rule, or, if it is too near an existent center, it replaces this neighbor center. In the next stage the parameters of consequence are globally or locally recursively updated using Recursive Least Squares (RLS) or weighted RLS, respectively. Finally the output of the system is predicted and the process returns to stage 2 by reading the next data sample at the next time step.

The algorithm intends to approach the subtractive clustering technique applied in a batch onset. Since it is to be applied in real-time environments, the data appears successively and at each time the new available data is used to improve the fuzzy system, in such a way that the new informative content of data should be reflected in the fuzzy rules, whether by changing some of them, whether by creating new rules if there is sufficiently novelty in the process behavior. There are here some scientific challenges. The main one is how to measure the informative novelty in the new data. The second is how to change/create the rules consequently. The proposed techniques by [20] and [21] are mainly based on heuristics and experimentation and as a consequence are case dependent. In the present work the same approach is followed, but more conditions for rule update/create have been researched and implemented in the computational framework eFSLab.

In the case of the construction of a Mamdani system from a TS one, only the zero order TS models can be considered. In order to do that, the same antecedents of the TS system are used and the fuzzy sets of Mamdani consequents are centred in the Takagi-Sugeno constant consequents.

6 Application to the Visbreaker Process

To test the performance of VisRedIII with neural networks and neuro-fuzzy systems, the detection of failures in sensors of the Visbreaker process has been considered. The Visbreaker Unit is intended to reduce the viscosity of the residual coming from the vacuum column, in an oil refinery. The great economical advantage of the visbreaker process is that it produces a residual with a lower viscosity that the load feed. This way, it is possible to produce fuel oil using a lower quantity of "cutterstocks" (some of them of high benefit). Fig. 13 [23] presents the Visbreaker Unit flow sheet. It can be noticed that it is composed of several interconnected sub-processes.

6.1 Neuro-Fuzzy Systems

Data averaged for each hour, between the 1st of January and the 5th of June of 2008, was collected. Original data comes from 160 tags. After correlation analysis and process expertise, only 59 tags were selected as the most representative of the process. Then, multidimensional scaling is applied to those 59 dimensions and a three dimensional representation was obtained by using VisRed. This three dimensional data is then fed to eFSLab that finds a zero-order TS fuzzy model. For diagnosis purpose a binary output (normal-abnormal) is desired. Experience has shown that the most adequate models are TS zero-order ones.

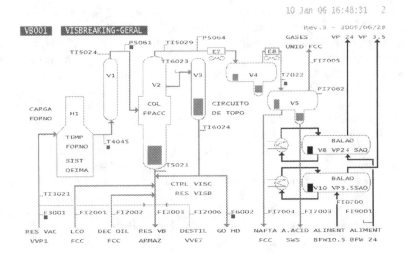

Fig. 13. The Visbreaker Process [24]

Consider now two faults in two sensors in a time horizon of two weeks. This has been done by introducing into the data between the 22nd of May and the 4th of June 2008 a 20% reduction in data from two temperature sensors in the kiln, during the day 23rd of May and the day 4th of June.

In eFSLab it was chosen to construct a zero order TS fuzzy model with Global parameter estimation. The other parameters were set with its default values, except radii, which value was fixed in 0.1. The training set is the first week and the testing set the second one. To improve the classification performance, the default conditions for creation and modification of rules were chosen again, due to the fact that these ones have the capability to produce rules in regions with less data points.

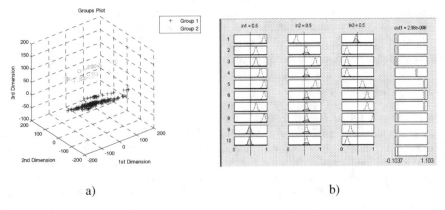

a)   b)

Fig. 14. a) The three-dimensional state space representation of the Visbreaker process Group 1 are normal, Group 2 are abnormal points. b) The 10 fuzzy rules produced.

The first step was dimension reduction of data VisRed. This was processed with Euclidean dissimilarity metric and without normalization. A three dimensional representation of the reduced data is in Fig. 14a). With eFSLab, a 10 rules zero order TS system was obtained, with Radii of 0.1. The rules can be seen in Matlab Fuzzy Toolbox, as in Fig. 14b).

In order to realize the performance of the TS fuzzy system created, the real output and the fuzzy system output can be compared, for the test data set, as in Fig. 15a). In this case the classification is successful in 98.8% of the points. Note that the centres of the spheres in Fig. 15b), projected in each dimension, define the centres of the membership functions of the antecedents of the TS fuzzy rules. Since the method is interactive, processing the information as it arrives, the final results express mainly the more recent data. That is why there are significant regions in the information space where there are no rules. This question however merits further research.

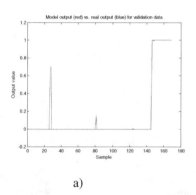
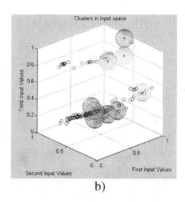

a)   b)

Fig. 15. a) Real output (blue) and simulation output (red) obtained with fuzzy system created. b) The created centers that define the rules of the fuzzy system (same as Fig. 8).

6.2 Neural Networks and Support Vector Machines

The same data sets used for the neuro-fuzzy systems were also used for training different neural networks (ADALINE, feedforward with 1, 5 and 10 neurons on the hidden layer, and radial basis networks) and support vector machines (with different kernel functions). The expected output was represented as 1 for normal points and 2 for abnormal points, codified as binary vectors (1,0) and (0,1). The results show that this is an easy problem for ADALINE and feedforward networks. Both can learn and generalize with 100% accuracy, as shown in Fig. 16 (example only for ADALINE).

Radial basis networks were also able to learn with perfect accuracy, but not able to generalize: none of the abnormal points was detected on the test set, as shown in Fig. 17. The results of the support vector machines depended on the kernel function used. For example, with a polynomial function all the results were 100% accurate, but when using a radial basis function the generalization ability was completely lost, and once again none of the abnormal points was detected on the test set. This fact stresses the inconvenience of the locality property of the radial basis functions.

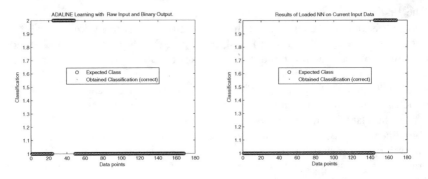

Fig. 16. Learning (left) and generalization (right) results for ADALINE networks

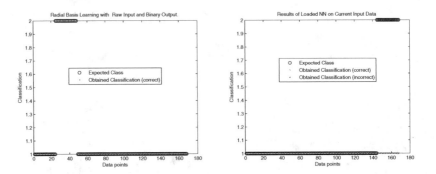

Fig. 17. Learning (left) and generalization (right) results for radial basis networks

7 Conclusions

The decision maker in industrial complexes needs advanced tools to support his needs to maintain operation in high level conditions of safety, quality, and efficiency. Reducing the dimension of the available information enables two or three dimensional synthetic views of the process where essential information, such as normal/abnormal conditions, can be evidenced. If classification techniques are applied, an alarming system may be built based on the classification results of normal / abnormal present state. Advanced computational platforms, integrating, in a friendly way, the multiplicity of methods and techniques available are an important step towards that objective. VisRedIII, presented in this paper, was developed with that aim. It is presently GNU open source software, downloadable from the following address: http://eden.dei.uc.pt/~dourado/VisRedIII.zip.

Integrating multidimensional scaling with artificial neural networks, support vector machines, and fuzzy systems, equipped with a multiplicity of techniques for dimensionality reduction and clustering, it can be used in a flexible way, allowing a rapid prototyping of application not only in industry, but in general quantitative data-mining in all areas.

The case study of the Visbreaker process of the Galp Sines Refinery, in Portugal, shows the potential of this methodology. Sensor faults can have serious dangers, if not monitored in proper time. Kiln temperature distribution profile, if disturbed, processed in VisRed, has the potentiality to give the process operator the information that something must be checked and improved.

Improvement of the eFSLab included in VisRed, an ongoing work, may lead to better fuzzy rules, with a more adequate adaptation to changes in process dynamics. Interpretable fuzzy rules will be the goal of the researched progresses.

Acknowledgments. This work was supported by Project CLASSE POSC/EIA/58162/2004 including EU FEDER support.

References

1. Borg, I., Groenen, P.: Modern Multidimensional Scaling, Theory and Applications, 2nd edn. Springer, Heidelberg (2005)
2. de Oliveira, M.C.F., Levkowitz, H.: From visual data exploration to visual data mining: A survey. IEEE Trans on Visualization and Computer Graphics 9(3), 378–394 (2003)
3. Dourado, A., Ferreira, E., Barbeiro, P.: VISRED – Numerical Data Mining with Linear and Nonlinear Techniques. In: Perner, P. (ed.) ICDM 2007. LNCS, vol. 4597, pp. 92–106. Springer, Heidelberg (2007)
4. Dourado, A., Ferreira, E., Barbeiro, P.: VisRedII-Introducing Meteheuristics in Multidimensional Scaling, Short Paper. In: Perner, P. (ed.) ICDM 2008. LNCS, vol. 5077. Springer, Heidelberg (2008)
5. The Mathworks, Inc., http://www.mathworks.com/
6. van der Maaten, L.J.P.: An Introduction to Dimensionality Reduction Using Matlab. Technical Report MICC 07-07. Maastricht University, Maastricht, The Netherlands (2007), http://ticc.uvt.nl/~lvdrmaaten/Laurens_van_der_Maaten/ Matlab_Toolbox_for_Dimensionality_Reduction_files/ Report_final.pdf
7. Tenenbaum, J.B., de Silva, V., Langford, J.C.: A Global Geometric Framework for Nonlinear Dimensionality Reduction. Science 290(5500), 2319–2323 (2000)
8. Agrafiotis, D.K.: Stochastic Proximity Embedding,, Int. J. Comput. Chem. 24: 1215–1221, Wiley Periodicals (2003)
9. Hill, T., Lewicki, P.: STATISTICS Methods and Applications. StatSoft, Tulsa, OK (Printed Version) (2007) Electronic textbook, http://www.statsoft.com/textbook/stfacan.html#index
10. Geoffrey Hinton and Sam Roweis Department of Computer Science, University of Toronto 10 King's College Road, Toronto, M5S 3G5 Canada fhinton,roweisg@cs.toronto.edu
11. Belkin1, M., Niyogi, P.: Laplacian Eigenmaps for Dimensionality Reduction and Data Representation. Neural Computation 15(6), 1373–1396 (2003)
12. Widrow, B., Sterns, S.D.: Adaptive Signal Processing. Prentice-Hall, New York (1985)
13. Rumelhart, D.E., Hinton, G.E., Williams, R.J.: Learning internal representations by error propagation. In: Rumelhart, D.E., McClelland, J. (eds.) Parallel Data Processing, vol. 1, ch. 8, pp. 318–362. MIT Press, Cambridge (1986)
14. Hagan, M.T., Demuth, H.B., Beale, M.: Neural Network Design. PWS Publishing (1995)

15. Chen, S., Cowan, C.F.N., Grant, P.M.: Orthogonal Least Squares Learning Algorithm for Radial Basis Function Networks. IEEE Transactions on Neural Networks 2(2), 302–309 (1991)
16. Vapnik, V.: Statistical Learning Theory. Wiley Interscience, New York (1998)
17. Briggs, Tom, http://webspace.ship.edu/thbrig/mexsvm/
18. Joachims, Thorsten, http://svmlight.joachims.org/
19. Chen, P.H., Lin, C.J., Schölkopf, B.: A tutorial on v-support vector machines. Appl. Stoch. Models. Bus. Ind. 21, 111–136 (2005)
20. Ross, T.: Fuzzy Logic With Engineering Applications, 2nd edn. McGraw Hill, New York (2004)
21. Angelov, P., Filev, D.: An Approach to Online Identification of Takagi-Sugeno Fuzzy Models. IEEE Transactions on Systems, Man, and Cybernetics - Part B 34 (2004)
22. Victor, J., Dourado, A.: Evolving Takagi-Sugeno fuzzy models. Technical Report, CISUC (September 2003), http://cisuc.dei.uc.pt/acg/view_pub.php?id_p=760
23. Ramos, J.V., Dourado, A.: On line interpretability by rule base simplification and reduction. In: Eunite Symposium, Aachen (2004)
24. Galp data, Sines Refinery (2006)
25. Andrews, D.F.: Plots of high dimensional data. Biometrics 28, 125–136 (1972)

A Case of Using Formal Concept Analysis in Combination with Emergent Self Organizing Maps for Detecting Domestic Violence

Jonas Poelmans[1], Paul Elzinga[3], Stijn Viaene[1,2], and Guido Dedene[1,4]

[1] K.U.Leuven, Faculty of Business and Economics, Naamsestraat 69,
3000 Leuven, Belgium
[2] Vlerick Leuven Gent Management School, Vlamingenstraat 83,
3000 Leuven, Belgium
[3] Amsterdam-Amstelland Police, James Wattstraat 84,
1000 CG Amsterdam, The Netherlands
[4] Universiteit van Amsterdam Business School, Roetersstraat 11
1018 WB Amsterdam, The Netherlands
{Jonas.Poelmans,Stijn.Viaene,Guido.Dedene}@econ.kuleuven.be
Paul.Elzinga@amsterdam.politie.nl

Abstract. In this paper, we propose a framework for iterative knowledge discovery from unstructured text using Formal Concept Analysis and Emergent Self Organizing Maps. We apply the framework to a real life case study using data from the Amsterdam-Amstelland police. The case zooms in on the problem of distilling concepts for domestic violence from the unstructured text in police reports. Our human-centered framework facilitates the exploration of the data and allows for an efficient incorporation of prior expert knowledge to steer the discovery process. This exploration resulted in the discovery of faulty case labellings, common classification errors made by police officers, confusing situations, missing values in police reports, etc. The framework was also used for iteratively expanding a domain-specific thesaurus. Furthermore, we showed how the presented method was used to develop a highly accurate and comprehensible classification model that automatically assigns a domestic or non-domestic violence label to police reports.

Keywords: Formal Concept Analysis, Emergent Self Organizing Map, text mining, actionable knowledge discovery, domestic violence.

1 Introduction

In this paper, we propose a framework for knowledge discovery from unstructured text based on the synergistic combination of two visually appealing discovery techniques known as, Formal Concept Analysis (FCA) [10, 11] and Emergent Self Organizing Maps (ESOM) [2, 5]. The framework recognizes the important role of the domain expert in mining real-world enterprise applications and makes efficient use of specific domain knowledge, including human intelligence and domain-specific constraints.

P. Perner (Ed.): ICDM 2009, LNAI 5633, pp. 247–260, 2009.
© Springer-Verlag Berlin Heidelberg 2009

FCA arose twenty-five years ago as a mathematical theory [10, 15] and has over the years grown into a powerful tool for data analysis, data visualization and information retrieval [12, 13]. This paper is a continuation of the work presented last year [20]. The major contributions of this paper are as follows. We complement the knowledge discovery based on FCA with a special type of topographic map known as ESOM. The ESOM functions as a catalyst for the FCA based discovery. Another key contribution of this paper is that we ground the knowledge discovery approach based on FCA and ESOM in the theoretical foundation of C-K theory. C-K theory is used to give a clear structure to the discovery process based on FCA and ESOM.

We apply the presented framework to a real life case study using data from the Amsterdam-Amstelland police. The case zooms in on the problem of distilling concepts for domestic violence from the unstructured text in police reports. These concepts are used to iteratively enrich the actionable knowledge available to police officers for recognizing cases of domestic violence. Domestic violence is one of the top priorities of the Amsterdam-Amstelland police force [19]. Unfortunately, in the past intensive audits of the police databases related to filed reports established that many police reports tended to be wrongly labelled as domestic or as non-domestic violence cases.

In this paper, we demonstrate that by applying the discovery framework to the unstructured text in police reports we can obtain essential knowledge for upgrading the definition and understanding of the domestic violence phenomenon, and for improving its management. In addition, we show how early detection of domestic violence could be automated. Many new results were obtained in comparison to our previous paper and these are discussed in detail.

The remainder of this paper is composed as follows. In section 2, we introduce the data exploration techniques of FCA and ESOM. In section 3, we discuss the dataset. In section 4, we elaborate on the knowledge discovery process, apply it to the data set at hand, and report results. Section 5 concludes the paper.

2 Exploration Techniques

According to R.S. Brachman and T. Anand [17] much attention and effort has been focused on the development of data mining techniques, but only minor effort has been devoted to the development of tools that support the analyst in the overall discovery task. The authors argue for a more human-centered approach. Human-centered KDD refers to the constitutive character of human interpretation for the discovery of knowledge, and stresses the complex, interactive process of KDD as being led by human thought. This can only be achieved by tools that offer highly interactive user interfaces that continuously engage human control over the information seeking process [16]. According to Brachman et al. [18] this is best embedded into a knowledge discovery support environment. FCA and ESOM are particularly suited for exploratory data analysis because of their human-centeredness. FCA and ESOM offer the user intuitive visual displays of different types of structures available in the dataset and guide the user in its exploration.

2.1 Formal Concept Analysis

The starting point of the analysis is a database table consisting of rows M (i.e. objects), columns F (i.e. attributes) and crosses $T \subseteq M \times F$ (i.e. relationships between objects and attributes). The mathematical structure used to reference such a cross table is called a formal context (M, F, T). In our case, reports of domestic violence (i.e. the objects) are related (i.e. the crosses) to a number of terms (i.e. the attributes); here a report is related to a term if the report contains this term. Given a formal context, FCA then derives all concepts from this context and orders them according to a subconcept-superconcept relation. This results in a line diagram (a.k.a. lattice).

The notion of concept is central to FCA. The way FCA looks at concepts is in line with the international standard ISO 704, that formulates the following definition: A concept is considered to be a unit of thought constituted of two parts: its extension and its intension, [10, 11]. The extension consists of all objects belonging to the concept, while the intension comprises all attributes shared by those objects. For a set of objects $O \subseteq M$, the common features can be identified, written $\sigma(O)$, via

$$A = \sigma(O) = \{ f \in F \mid \forall o \in O : (o, f) \in T \} .$$

Take the attributes that describe a report from the dataset used in this paper, for example. By collecting all reports of this context that share these attributes, we get to a set $O \subseteq M$ consisting of reports. This set O of objects is closely connected to set A consisting of attributes

$$O = \tau(A) = \{ i \in M \mid \forall f \in A : (i, f) \in T \} .$$

In other words, O is the set of all objects sharing all attributes of A, and A is the set of all attributes that are valid descriptions for all the objects contained in O. Each such pair (O, A) is called a formal concept (or concept) of the given context. The set $A = \sigma(O)$ is called the intent, while $O = \tau(A)$ is called the extent of the concept (O, A).

There is a natural hierarchical ordering relation between the concepts of a given context that is called the subconcept-superconcept relation. A concept $d = (O_1, A_1)$ is called a subconcept of a concept $e = (O_2, A_2)$ (or equivalently, e is called a superconcept of a concept d) if and only if the extent of d is a subset of the extent of e (or equivalently, if and only if the intent of d is a superset of the intent of e), or

$(O_1, A_1) \subseteq (O_2, A_2) \Leftrightarrow (O_1 \subseteq O_2 \Leftrightarrow A_2 \subseteq A_1)$.

The set of all concepts of a formal context combined with the subconcept-superconcept relation defined for these concepts gives rise to the mathematical structure of a complete lattice, called the concept lattice of the context. The latter is made accessible to human reasoning by using the representation of a (labelled) line diagram.

In section 4 of this paper, a line diagram is displayed. We use the following conventions. The circles or nodes in the line diagram represent the formal concepts. The diagram displays only concepts that describe objects and is therefore a subpart of the concept lattice. The shaded boxes (upward) linked to a node represent the attributes used to name the concept. The non-shaded boxes (downward) linked to the node represent the objects used to name the concept. The information contained in the concept lattice can be distilled from the line diagram by applying the following reading rule: An object "g" is described by an attribute "m" if and only if there is an ascending path from the node named by "g" to the node named by "m".

Retrieving the extension of a formal concept from a line diagram implies collecting all objects on all paths leading down from the corresponding node. To retrieve the intension of a formal concept one traces all paths leading up from the corresponding node in order to collect all attributes. The top and bottom concepts in the lattice are special. The top concept contains all objects in its extension. The bottom concept contains all attributes in its intension. A concept is a subconcept of all concepts that can be reached by travelling upward. This concept will inherit all attributes associated with these superconcepts.

2.2   Emergent Self Organizing Map

Emergent Self Organizing Maps (ESOM) [2] are a special class of topographic maps [7]. ESOM is argued to be especially useful for visualizing sparse, high-dimensional datasets, yielding an intuitive overview of its structure [6]. Topographic maps perform a non-linear mapping of the high-dimensional data space to a low-dimensional one, usually a two-dimensional space, which enables the visualization and exploration of the data [4]. ESOM is a more recent type of topographic map. According to Ultsch [5], "emergence is the ability of a system to produce a phenomenon on a new, higher level". In order to achieve emergence, the existence and cooperation of a large number of elementary processes is necessary. An emergent SOM differs from a traditional SOM in that a very large number of neurons (at least a few thousands) are used [3]. In the traditional SOM, the number of nodes is too small to show emergence.

The ESOM map is composed of a set of neurons I, arranged in a hex-grid map structure. A neuron $i \in I$ is a tuple (w_i, p_i) consisting of a weight vector $w_i \in W$ and a position $p_i \in P$ in the map. The input space $D \subset R^n$ is a metric subspace of R^n.

Consider a training set $E = \{x_1, ..., x_k\}$ with $x_1, ..., x_k \in R^n$ representing the input vectors for ESOM training. The training algorithm used is the online training algorithm in which the best match for an input vector is searched and the neighborhood of the map is updated immediately. When an input vector x_i is supplied to the training algorithm, the weight of a neuron $n_1 = (w_1, p_1)$ is modified as follows. Let $\eta \in [0,1]$, then

$$\Delta w_1 = r \times h \times (bm_i, n_1, r) \times (x_i - w_1)$$

where

$$D \rightarrow I : bm_i = bm(x_i)$$

represents the best-matching neuron of an input vector , being the neuron $n_b \in I$ having the smallest Euclidean distance to x_i , or

$$n_b = bm(x_i) \Leftrightarrow d(x_i, w_b) \leq d(x_i, w_b) \forall w_b \in W ,$$

where $d(x_i, w_j)$ stands for the Euclidean distance between input vector x_i and weight vector w_j.

The neighborhood of a neuron

$$N_i = N(n_i) = \{n_j \in M \mid h_{ij}(r) \neq 0\}$$

is the set of neurons surrounding neuron n_i and determined by the neighborhood function h. The neighborhood defines a lattice of neurons in the map space K, while r is called the neighborhood radius.

The produced map maintains the neighborhood relationships that are present in the input space and can be used to visually detect clusters. It also provides the analyst with an idea of the complexity of the dataset, the distribution of the dataset (e.g. spherical), and the amount of overlap between different classes of objects. Finally, only a minimal amount of expert knowledge is required for an analyst to start use it effectively for exploratory data analysis. An additional advantage of an ESOM is that it can be trained directly on the available dataset without first having to perform a feature selection procedure [5]. ESOM maps can be created and used for data analysis by means of the publicly available Databionics ESOM Tool [8].

3 Dataset

Our dataset consists of a selection of 4814 police reports describing a whole range of violent incidents from the year 2007. All domestic violence cases from that period are a subset of this dataset. The selection came about amongst others by filtering out police reports that did not contain the reporting of a crime by a victim, which is necessary for establishing domestic violence. This happens, for example, when a police officer is sent to investigate an incident, afterwards writes a report in which he/she mentions his/her findings, but the victim ends up never making an official statement to the police. The follow-up reports referring to previous cases were also removed. Of those 4814 reports, 1657 were classified by police officers as domestic violence.

We also used a validation data set for our experiment. It consists of a selection of 4738 cases describing a whole range of violent incidents from the year 2006 where the victim made a statement to the police. Again, the follow-up reports were removed. 1734 of these 4738 cases were classified as domestic violence by police officers.

4 Iterative Knowledge Discovery with FCA and ESOM

In this section, we elaborate on the applied process for knowledge discovery based on the synergistic combination of the visually appealing discovery techniques presented in section 2. In this setup, FCA is used as a concept generation engine, distilling formal concepts from the unstructured text documents described in section 3. We complement this knowledge discovery with the capabilities of ESOM, which functions as a catalyst for the FCA based knowledge extraction. To frame our approach to knowledge discovery we make use of C-K theory.

C-K theory is a unified design theory that defines design reasoning dynamics as a joint expansion of the Concept (C) and Knowledge (K) spaces through a series of continuous transformations within and between the two spaces [1]. The theory makes a formal distinction between concepts and knowledge: the K space consists of propositions with logical status (i.e. either true or false) for a designer, and the C space consists of propositions without logical status in the knowledge space. According to Hatchuel et al. [1], concepts are candidates to be transformed into propositions of K but are not themselves elements of K.

From the point of knowledge discovery in data, the K space could be viewed as composed of actionable information. It contains the existing knowledge used to operate and steer in the action environment. The C space, on the other hand, is considered as the design space. Whereas K is used as the basis for action and decision making, C puts this actionability under scrutiny for potential improvement and learning. The transformations within and between the C and K spaces are accomplished by the application of four transformation operators: $C \rightarrow K$, $K \rightarrow C$, $C \rightarrow C$, and $K \rightarrow K$. These transformations form what is referred to as the design square, which fundamentally provides structure to the design process. At the basis of the knowledge discovery process are iterations through the design square. The knowledge discovery documented in this paper was driven by a data analyst and a domain expert, who collaborated intensely and continuously interacted with the visual exploration tools. They immersed themselves in a process of iterating through the transformation operators of the design square. This process is summarized in Figure 1.

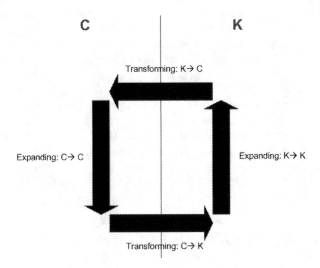

Fig. 1. Design square in action

In this case, the process in Figure 1 works as follows:

- The process starts with the data analyst constructing an initial FCA lattice and ESOM map using the police reports contained in the dataset, and the terms contained in a thesaurus (i.e. $K \rightarrow C$). The thesaurus was refined after each iteration of re-indexing the reports and visualizing and analyzing the data with the FCA lattice and ESOM maps.
- The FCA lattice and the ESOM map provide a reduced search space to the domain expert, who then visually inspects and analyses the lattice and

map (i.e. C → C). In other words, the FCA lattice and ESOM map are used in the capacity of information browser.

- Based on anomalies and counter-intuitive elements found by analyzing the lattice, or using ESOM to pinpoint outliers, clusters and areas of the map containing a mixtures of case types, police reports can be selected for in-depth manual inspection (i.e. C → K).

- These police reports are in turn used, for example, to discover new referential terms to improve the thesaurus, to enrich and validate prior domain knowledge, to discover new classification rules, and for operational validation (i.e. K → K).

The obtained results, together with the relevant prior knowledge of the domain expert are then incorporated into the existing visual representation, resulting in a new lattice and ESOM map (i.e. K → C) for starting a new iteration of the design square.

In the remainder of this section we limit ourselves to illustrating how ESOM and FCA help to operationalise the four design cube operators that constitute the knowledge discovery process.

4.1 Transforming: K→C

The definition of domestic violence employed by the police organization of the Netherlands is as follows: "Domestic violence can be characterized as serious acts of violence committed by someone in the domestic sphere of the victim. Violence includes all forms of physical assault. The domestic sphere includes all partners, ex-partners, family members, relatives and family friends of the victim. The notion of family friend includes persons that have a friendly relationship with the victim and (regularly) meet with the victim in his/her home [16]."

The lattice structure in Figure 2 was fundamentally influenced by this domestic violence definition. Prior to the analysis with FCA, certain terms were clustered in term clusters based on this definition and added to a thesaurus. For example, to verify whether a criminal offence has occurred, reports were searched for terms such as "hit", "stab" and "kick". These terms were grouped into the term cluster "acts of violence". Another term cluster, for example, verified whether a person from the domestic sphere of the victim was involved in the crime. In this case, reports were searched for terms such as "my dad", "my ex-boyfriend", and "my uncle". These terms were grouped into the term cluster "persons of domestic sphere".

Using the reference definition of domestic violence employed by the police was but one way to come up with term clusters to structure the lattice in Figure 2. Term clusters also emerged from in-depth scanning of certain reports highlighted during a knowledge iteration cycle. This is the way, for example, in which the term cluster "relational problems" was created. Terms in that cluster made reference to a broken relationship. In-depth scanning of reports is also the way in which we found that many cases did not have a formally labelled suspect. Thus, we incorporated the attribute "no suspect" into the lattice. Reports that were assigned the label "domestic violence" had been classified as such by police officers. The remaining reports were classified as non-domestic violence.

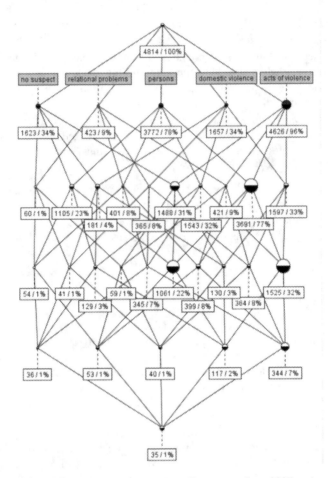

Fig. 2. Lattice based on the police reports from 2007

Some 123 elements were collected into an initial thesaurus. This gathering process was not only influenced by the above domestic violence definition, but also incorporated information from other prior knowledge sources such as expert advice. Indexing the set of 4814 reports from 2007 with this thesaurus resulted in a data set (i.e. reports are objects and thesaurus elements are object attributes) for training a toroidal ESOM map. The latter is presented in Figure 3. The green squares refer to neurons that dominantly contain non-domestic violence cases, while the red squares refer to neurons that dominantly contain domestic violence cases.

4.2 Expanding: C→C

The notion of expansion plays a key role in C-K theory. An analysts ability to recognize an expansion opportunity depends on his sensitivity to these opportunities, his training, and the knowledge at his disposal. We argue here that FCA and ESOM help analysts recognize and exploit these opportunities. Basically, C space expansion is

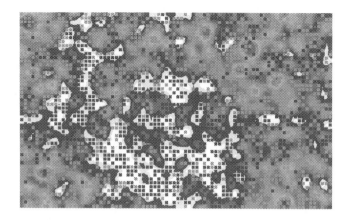

Fig. 3. ESOM map based on the police reports from 2007

driven by the analyst spotting and investigating anomalies, outliers and concept gaps from these visual exploration tools. For example, the FCA lattice in Figure 2 allowed us to make some interesting observations associated with the data at hand. We use the numbers in Table 1 to illustrate.

Table 1. Interesting observations from the lattice in Figure 2

	Non-domestic violence	Domestic violence
No "acts of violence"	128	60
No "acts of violence" and "persons of domestic sphere"	63	18
"acts of violence" and no "persons of domestic sphere"	863	72
"relational problems"	58	365

From Table 1 it is clear that a total of 60 domestic violence cases did not contain a term from the "acts of violence" term cluster. Of these 60 cases, 18 contained a term from the cluster containing terms referring to a person in the domestic sphere of the victim. Interestingly, some 28% (i.e. 863) of the non-domestic violence reports only contain terms from the "acts of violence" cluster, while there are only 72 domestic violence reports in the dataset that share that characteristic. Apparently, some cases that were labelled as domestic violence did not fit with the definition of domestic violence that was used to start this discovery exercise in the first place. The associated reports were therefore selected for in-depth investigation.

Zooming in onto the term cluster "relational problems" we observed some more interesting facts. Apparently, only 58 non-domestic violence reports contained one or more terms from that cluster. We concluded that the presence of at least one of the terms from this cluster in a police report seemed to be a strong indication for domestic violence. This was enough evidence to warrant manual inspection of these 58 police reports.

Visual inspection of the patterns laid down by the ESOM map in Figure 3 also allowed us to make interesting observations. For example, using colour coding made it easy to spot outlying observations, that is, red squares located in the middle of a large group of green squares, and vice versa. For inspection we made use of the ESOM tool's functionality to select neurons for displaying the cases that had these neurons as best match. We thought that these neurons were associated with cases that might have been wrongly classified by police officers.

4.3 Transforming: C→K

The C → K operator transforms concepts in C into logical questions in K. An answer to such a question is in our case found by manually inspecting selected police reports. We refer to this manual analysis as the validation of concept gaps, giving rise to multiple types of discoveries: confusing situations, new referential terms, faulty case labellings, niche cases, and data quality problems.

For example, and with reference to Table 1, the 18 domestic violence cases that contained a term from the cluster "persons of domestic sphere" but no violence term were selected for manual investigation. In-depth analysis showed that these reports contained violence related terms, such as "abduction", "strangle" and "deprivation of liberty", that were originally lacking from the initial thesaurus. Another example, are the 42 cases that did not contain a violence term or a term referring to a person of the domestic sphere. These cases turned out to be wrongly classified as domestic violence.

Table 1 also indicates that there were 58 police reports that were classified as non-domestic violence while containing a term from the "relational problems" cluster. Investigation revealed that a startling 95% of these cases had been wrongly labelled as non-domestic violence. Moreover, about 70% of these cases had in common that a third person made a statement to the police for someone else. Moreover, analysis of the remaining 30% led to the discovery of an important new concept that turned out to be lacking from the domain expert's initial understanding of domestic violence. Many of the reports included expressions such as "I was attacked by the ex boyfriend of my girlfriend" and "I was maltreated by the ex girlfriend of my boyfriend". This gave rise to a term cluster "attack by ex-person against new friend" that would be added to the thesaurus.

We also went after novel and potentially interesting classification attributes. The fact of not making mention of a suspect in a report is such an attribute. This can be inferred from the FCA lattice in Figure 2. It shows how some 34% of the reports (1623 cases) did not mention a suspect. However, going back to our original definition of domestic violence, which talked about a perpetrator belonging to the domestic sphere of the victim, an offender thus had to be known for a valid labelling of domestic violence. However, the lattice in Figure 1 shows that 181 of these "no suspect" cases were actually labelled as domestic violence, which led us to further investigate these reports.

Some of the ESOM outliers also helped us to enrich the K space. Inspection of these outlier cases helped us to enrich the thesaurus. Some of the features that constituted this enrichment are mentioned in Table 2.

Table 2. Thesaurus features discovered by exploring ESOM outliers

pepper spray
homosexual relationship, lesbian relationship
sexual abuse, incest
alternative spelling of some words (e.g. ex-boyfriend, exboyfriend, ex boyfriend)
weapons lacking in the thesaurus: belt, kitchen knife, baseball bat, etc.
terms referring to persons: partner, fiancée, mistress, concubine, man next door, etc.
terms referring to relationships: love affair, marriage problems, divorce proceedings, etc.
reception centers: woman's refuge center, home for battered woman, etc.
gender of the perpetrator: mostly male
gender of the victim: mostly female
age of the perpetrator: mostly older than 18 years and younger than 45 years
age of the victim: mostly older than 18 years and younger than 45 years
terms referring to an extra marital affair: I have an another man, lover, I am unfaithful, etc.

Many of the reports also contained confusing situations that upon disclosure were used to refine the notion of domestic violence.

4.4 Expanding: K→K

This expansion of the K space constitutes validation or testing of the proposed expansion with the ultimate goal of producing actionable intelligence. K-validation of a concept comes down to a confrontation of the output from the C→K transformation with a selection of knowledge sources available to the K space (e.g. cross-checking with other databases, setting up field experiments, soliciting expert advice).

For example, analysis of the misclassified reports of which we made mention in the previous section showed that apparently, for some unknown reason, police officers regularly misclassified burglary, car theft, bicycle theft and street robbery cases as domestic violence. To consolidate the agreement on this kind of mistake a new term cluster was introduced that would not only influence subsequent iterating through the design square, but police reporting itself by addressing it during police training. The latter would also elaborate on what seemed to be confusing situations to police officers in terms of labelling cases correctly (e.g. when third persons make statements for someone else).

In the previous section, we also described how the analysis of police reports revealed interesting cases in which the ex-boyfriend attacked the new boyfriend. We presented these doubtful cases to the board members responsible for the domestic violence policy. Police officers and policy makers confirmed that this type of situation was indeed to be seen as domestic violence, mainly because the perpetrator often aims at emotionally hurting the ex-partner. Consequently, the expectation was for the terms contained in this cluster to relatively frequently occur in cases labelled as domestic violence reports. However, this assumption turned out to be incorrect when scrutinizing the data. Police officers clearly had trouble allocating a correct label to these kind of cases, which anew gave rise to a need for training

The "no suspect" cases were yet another example of the potential for knowledge expansion. Remember that some of the cases that were labelled as domestic violence by police officers did not make mention of a suspect, which was weird. Studying the way in which police officers registered victim reports helped us uncover some haphazard behaviour in the process. Apparently, while some officers immediately registered a suspect at the moment the victim mentioned this person as a suspect, others preferred to first interrogate these people before casting the label of suspect. In the latter cases, the person that was mentioned would then be added to the list of persons who were said to be involved in or to have witnessed the crime. These lists tended to account for a rather diverse and extensive set of people. Suspects easily got lost in these lists. When we inquired about the proper policy regarding the labelling of suspects, we were told there simply was none. This analysis made a strong case for such a policy.

In the process of digging up evidence and confronting the different stakeholders with that evidence we exposed a serious mismatch between the management's conception of domestic violence and that of police officers. We found that management employed a much broader definition of domestic violence than most police officers.

4.5 Actionable Intelligence

Several iterations through the design cube resulted in truly valuable upgrades of the K space from the point of improving action in the field, that is, on the street. Some of the most important achievements of our work are the following:

- We were able to upgrade the definition for domestic violence that would act as a principle guideline for labelling cases.
- Several types of niche cases were identified as valid exceptions to the general definition, and advice, grounded in evidence, was formulated for policy redesign.
- We ended up extracting a set of 22 domestic violence and 15 non-domestic violence classification rules. Using these rules, 75% of cases from the year 2007 could be labelled automatically and correctly as either domestic or non-domestic violence. Validation of these results by applying these rules to the police reports from the year 2006 allowed us to obtain a similar performance at 72%. These rules have been incorporated in an early case filter to identifying cases that warrant in-depth manual inspection. Before, all claims had to be manually checked.
- The set of identified classification rules did not just allow the police to classify newly incoming cases. The rules could also be usefully employed to reclassify cases from the past to provide for more correct performance management and reporting over time. Domestic violence cases that were not recognized as such in the past might also be re-opened for investigation.

5 Conclusion

In this paper, we proposed an approach to knowledge discovery from unstructured text using FCA and ESOM. The approach was framed and illustrated with reference to C-K theory (i.e. the design square) to provide for a deep understanding of the nature of the exploration; an exploration that essentially is human-centered. With this paper we argued for the discovery capabilities of FCA and ESOM, acting as information browsers in the hands of human analysts. The tools were shown to help analysts progress with knowledge expansion by progressively looping through the design square in an effective way. We showcased the framework using a real life case study with data from the Amsterdam-Amstelland police. The case zoomed in on the problem of distilling concepts for domestic violence from the unstructured text in police reports. The data exploration for this case study resulted in several improvements related to the operational reporting on and the handling of domestic violence cases. This included the implementation of an effective early case filter to identify cases that truly warrant in-depth manual inspection.

Acknowledgments

The authors would like to thank the Amsterdam-Amstelland police for providing them with the necessary degrees of freedom to conduct and publish this research. In particular, we are most grateful to Deputy Police Chief Reinder Doeleman and Police Chief Hans Schönfeld for their continued and relentless support. Jonas Poelmans is aspirant of the Research Foundation – Flanders.

References

[1] Hatchuel, A., Weil, B.: A new approach of innovative design: an introduction to C – K theory. In: Proc. of ICED 2003, August 2003, Stockholm, Sweden, p. 14 (2003)
[2] Ultsch, A., Moerchen, F.: ESOM-Maps: Tools for clustering, visualization, and classification with Emergent SOM. Technical Report Dept. of Mathematics and Computer Science, University of Marburg, Germany, No. 46 (2005)
[3] Ultsch, A., Hermann, L.: Architecture of emergent self-organizing maps to reduce projection errors. In: Proc. ESANN 2005, pp. 1–6 (2005)
[4] Ultsch, A.: Density Estimation and Visualization for Data containing Clusters of unknown Structure. In: Proc. GFKI 2004 Dortmund, pp. 232–239 (2004)
[5] Ultsch, A.: Maps for visualization of high-dimensional Data Spaces. In: Proc. WSOM 2003, Kyushu, Japan, pp. 225–230 (2003)
[6] Ultsch, A., Siemon, H.P.: Kohonen's Self Organizing Feature Maps for Exploratory Data Analysis. In: Proc. Intl. Neural Networks Conf., pp. 305–308 (1990)
[7] Van Hulle, M.: Faithful Representations and Topographic Maps from distortion based to information based Self-Organization. Wiley, New York (2000)
[8] http://databionic-esom.sourceforge.net/
[9] Ultsch, A.: Data mining and knowledge discovery with Emergent SOFMS for multivariate Time Series. In: Kohonen Maps, pp. 33–46 (1999)
[10] Ganter, B., Wille, R.: Formal Concept Analysis: Mathematical Foundations. Springer, Heidelberg (1999)
[11] Wille, R.: Restructuring lattice theory: an approach based on hierarchies of concepts. In: Rival, I. (ed.) Ordered sets, pp. 445–470. Reidel, Dordrecht (1982)
[12] Priss, U.: Formal Concept Analysis in Information Science. In: Cronin, B. (ed.) Annual Review of Information Science and Technology, ASIST, vol. 40 (2005)
[13] Wille, R.: Why can concept lattices support knowledge discovery in databases? Journal of Experimental & Theoretical Artificial Intelligence 14(2), 81–92 (2002)

[14] Stumme, G., Wille, R., Wille, U.: Conceptual Knowledge Discovery in Databases Using Formal Concept Analysis Methods. In: Żytkow, J.M. (ed.) PKDD 1998. LNCS (LNAI), vol. 1510, pp. 450–458. Springer, Heidelberg (1998)

[15] Stumme, G.: Formal Concept Analysis on its Way from Mathematics to Computer Science. In: Priss, U., Corbett, D.R., Angelova, G. (eds.) ICCS 2002. LNCS, vol. 2393, p. 2. Springer, Heidelberg (2002)

[16] van Dijk, T.: Huiselijk geweld, aard, omvang en hulpverlening. Ministerie van Justitie, Dienst Preventie, Jeugd-bescherming en Reclassering (1997)

[17] Brachman, R., Anand, T.: The process of knowledge discovery in databases: a human-centered approach. In: Fayyad, U., Piatetsky-Shapiro, G., Smyth, P., Uthurusamy, R. (eds.) Advances in knowledge discovery and data mining. AAAI/MIT Press (1996)

[18] Brachman, R.J., Selfridge, P.G., Terveen, L.G., Altman, B., Borgida, A., Halper, F., Kirk, T., Lazar, A., Mc Guinnes, D.L., Resnick, L.A.: Integrated support for data archaeology. International Journal of Intelligent and Cooperative Information Systems 2, 159–185 (1993)

[19] http://www.politie-amsterdam-amstelland.nl/get.cfm?id=86

[20] Poelmans, J., Elzinga, P., Viaene, S., Dedene, G.: An exploration into the power of Formal Concept Analysis for domestic violence analysis. In: Perner, P. (ed.) ICDM 2008. LNCS, vol. 5077, pp. 404–416. Springer, Heidelberg (2008)

Ordinal Evaluation: A New Perspective on Country Images

Marko Robnik-Šikonja[1], Kris Brijs[2], and Koen Vanhoof[2]

[1] University of Ljubljana,
Faculty of Computer and Information Science,
Tržaška 25, 1001 Ljubljana, Slovenia
Marko.Robnik@fri.uni-lj.si
[2] University of Hasselt, Department of Economics,
Universitaire Campus, 3590 Diepenbeek, Belgium
{kris.brijs,koen.vanhoof}@uhasselt.be

Abstract. We present a novel use of ordinal evaluation (OrdEval) algorithm as a promising technique to study various marketing phenomena. OrdEval algorithm has originated in data mining and is a general tool to analyze data with ordinal attributes, including surveys. Its many favorable features, including context sensitivity, ability to exploit meaning of ordered features and ordered response, and robustness to noise and missing values in the data, offer marketing practitioners a perspective, not available with classical analytical toolbox.

We present a case study applying OrdEval algorithm on country-of-origin (COO) information. We demonstrate some interesting advantages it has to offer and show how to extract and interpret new insights allowing marketing practitioners to further optimize the management of products abroad.

Data for the empirical study was gathered by means of 1225 questionnaires. Results indicate that, contrary to the classical view on COO-effects, the processing of country-related cognitions, affects and conations is a non-linear and asymmetric phenomenon. The practical implications of this finding for marketers are discussed more in detail.

1 Introduction

In recent years we can observe large changes in economy in general and marketing in particular as a result of internet expansion, globalization, and ubiquitous information availability. One of the scientific fields which gained momentum as a result of this was data analysis under various names: statistics, data mining, machine learning, intelligent data analysis, knowledge discovery. Many new data analysis techniques emerged which exploit availability of more and different data from several sources, and increased computational power of nowadays computers. Some examples of these techniques are support vector machines, text analytics, association rules, ensemble techniques, subgroup discovery, etc. These techniques have been accepted into analytics' standard toolbox in many disciplines: genetics, engineering, medicine, finance, vision, statistics, marketing, etc.

The *OrdEval* algorithm [11] is a novel analytical tool which emerged in data mining context with purpose to evaluate the importance and impact of various factors in the

P. Perner (Ed.): ICDM 2009, LNAI 5633, pp. 261–275, 2009.
© Springer-Verlag Berlin Heidelberg 2009

given data (e.g., survey). For example, in the analysis of customer satisfaction data for a particular product/service, OrdEval can determine the importance of each product's feature to the overall customer's satisfaction, and also indicate the thresholds where satisfaction with individual feature starts having strong positive or negative impact on the overall satisfaction. The output of OrdEval are probabilistic factors indicating the probability that increase/decrease in the individual feature or the feature's value will have impact on the dependent variable. The intuition behind this approach is to approximate the inner workings of the decision process taking place in each individual respondent, which forms relationship between the features and the response. If such brain introspection would be possible one could observe a causal effect the change of a feature's value has on the response value. By measuring such an effect we could reason about the importance of the feature's values and the type of the attribute. Also, we could determine which values are thresholds for change of behavior. While this is impossible, OrdEval algorithm uses the data sample and approximates this reasoning. For each respondent it selects its most similar respondents and does inference based on them. For example, to evaluate the effect an increase in certain feature value would have on overall satisfaction, the algorithm computes the probability for such an effect from the similar respondents with increased value of that feature. To get statistically valid and practically interesting results the overall process is repeated for large enough number of respondents, and weighted with large enough number of similar respondents. The motivation and contribution of this paper is to demonstrate how OrdEval works in a marketing context, how its output can be visualized and adapted to include information relevant for marketing practitioners, and the new insights into the country-of-origin (COO) problem, which we used as our case study.

We continue with a brief overview of the literature on COO-effects. Without going into the details, it situates the broader (marketing) context in which the technical contributions of this paper are to be seen. Although the accent of this work is on the use and adaptations of OrdEval algorithm in marketing, some background information on what COO-research is all about is necessary for understanding concepts and ideas contained by the present paper. Besides that, we motivate a short excursion into the COO-field by signaling that the contribution of this paper not only lays in technical aspects but also in developing superior consumer information with managerial relevance.

2   Country-Of-Origin Research

Research on COO-effects is mainly concerned with the effect information about a product's source country exerts on consumer's attitude towards the product. Empirical data overwhelmingly indicates that people indeed are guided somehow by the COO-cue when they are confronted with products coming from abroad. For instance, cars or washing machines 'Made in' Germany are usually preferred over Korean- or Chinese-made models. Below, we briefly discuss how both country image (CI) and product attitude are to be understood.

A product's COO is often explicitly mentioned in order to favorably influence people's reaction towards the product. Communicating a product's COO can be done in several manners like for instance by means of a 'Made in' label, a foreign sounding brand name or visual imagery. Confronted with these COO-stimuli, an internally stored schema is automatically activated in the consumer's mind. Generally, these memory networks are referred to as country images.

The literature reveals 9 environmental conditions as core aspects of the country image construct. These are cultural identity, political climate, language, national history, natural landscape, meteorological climate, level of technological and economic development, religion, and people's character [9].

Most scholars working within the COO-field operationalize country images as three-component attitude constructs. That is, most existing scales for measuring country image capture respondents' beliefs, feelings and intentions towards a country's environmental conditions. These three components which we use throughout this paper as well are normally labeled respectively as the country image's cognitive (mental), affective (emotional) and conative (intentional) dimensions.

Common to each of these three types of COO-effects is that they are theoretically explained as processes where the attitude towards a product is (un)favorably 'biased' in function of how consumers are thinking, feeling or (morally) oriented towards the product's COO. Technically, this biasing effect is believed to be a symmetric and linear mechanism where the scores for product-related beliefs, affects and purchase intentions increase or decrease proportionately to variations in the scores obtained for the country image's constituent components. These processes are better known as upward and downward reinforcement.

However, this traditional view on the functioning of marketing phenomena is challenged by some authors [1, 6, 7, 4]. In particular, Prospect Theory and its assumptions of loss aversion and diminishing sensitivity has brought several marketing scholars to alter their views on classic decision making mechanisms such as upward and downward reinforcement. For instance, within the field of customer satisfaction, a number of studies were capable of demonstrating that the relationship between attribute levels and overall satisfaction is rather asymmetric and S-shaped [7, 8, 11]. In contrast with research on customer satisfaction, the COO-literature about asymmetric and non-linear mechanisms is rather scarce. To the best of our knowledge, there are only two trying to explain the asymmetric nature of COO-effects [12, 5]. As such, this topic is still widely open for further exploration.

In light of these ideas, we intend to reconsider the classical explanations for the technical mechanisms in support of COO-effects. In our attempt to address this issue, we first demonstrate a technique that is capable of capturing the type of processes we would like to investigate.

3   OrdEval

Only recently, Robnik-Šikonja and Vanhoof [11] introduced the Ordered Evaluation (OrdEval) algorithm. It can be used for analysis of any data where the dependent variable has ordered values, meaning that is suitable also for surveys where answers are given in the graded manner. The methodology uses conditional probabilities called 're-inforcement factors' as they approximate upward and downward reinforcement of the feature value. For each value of the feature we obtain estimates of two conditional probabilities: the probability that the response value increases given the increase of the feature value (upward reinforcement), and the probability that response value decreases given the decrease of the feature value (downward reinforcement). To take the context of other features into account, these probabilities are computed in the local context, from the most similar instances. The visualization of these factors gives clear clues about the role of each feature, the importance of each value and the threshold values. To understand the idea of OrdEval algorithm, the feature should not be treated as a whole. Rather we shall observe the effect a single value of the feature may have. Assume for a moment that we could observe the inner workings of the decision process which forms relationship between the features and the response. In other words, suppose that we can observe a causal effect the change of a feature's value has on the response value. By measuring such an effect we could reason about the importance of the feature's values. Also, we could determine which values are thresholds for change of behavior and we could characterize the behaviors. This is of course impossible, but OrdEval algorithm uses the data sample and approximates this reasoning.

To explain the idea of the approach we need some definitions. Let R be a randomly selected observation and S the observation most similar to it. Let j be the value of the feature A_i at observation R. We observe the necessary changes of response value and features (A_i with value j in particular) which would change S to R. If these changes are positive (increase of response and/or feature values), let us define the following probabilities.

- $P(C_{i,j}^p)$ is a probability that the response value of R is larger than the response value of its most similar observation S. $P(C_{i,j}^p)$ is therefore the probability that the positive change in similar observation's response value is needed to get from S to R (note that for R the value of attribute A_i is j).
- $P(A_{i,j}^p)$ is a probability that j (the value of feature A_i at R) is larger than the value of feature A_i at its most similar observation S. By estimating $P(A_{i,j}^p)$ we gather evidence of the probability that the similar observation S has lower value of A_i and the change of S to R is positive.
- $P(C^pA_{i,j}^p)$ is a probability that both the response and j (the value of feature A_i at R) are larger than the response and feature value of its most similar observation S. With $P(C^pA_{i,j}^p)$ we estimate the probability that positive change in both the response and A_i value of similar instance S is needed to get the values of R.

Similarly, for negative changes which would turn S into R (decrease of response and/or feature values), we define $P(C_{i,j}^n)$, $P(A_{i,j}^n)$, and $P(C^nA_{i,j}^n)$. The output of the algorithm are two factors, upward and downward reinforcement, computed for each value of each feature. These factors measure the upward/downward trends exhibited in the data. The upward reinforcement of the i-th feature's value j is defined as

$$U_{i,j} = P(C_{i,j}^p | A_{i,j}^p) = \frac{P(C^pA_{i,j}^p)}{P(A_{i,j}^p)} \tag{1}$$

This factor reports the probability that a positive response change is caused by the positive feature change. This intuitively corresponds to the effect the positive change in feature's value has on the response. Similarly the downward reinforcement is defined as

$$D_{i,j} = P(C_{i,j}^n | A_{i,j}^n) = \frac{P(C^n A_{i,j}^n)}{P(A_{i,j}^n)} \tag{2}$$

and reports the effect the decrease of attribute's value has on the decrease of the class' value. The $U_{i,j}$ and $D_{i,j}$ factors are efficiently estimated by the OrdEval algorithm which we present in Fig. 1 in a simplified form intended for easier comprehension.

Input: for each respondent a vector of feature values and the overall score
Output: reinforcements $U_{i,j}$ and $D_{i,j}$ for all features i and their values j

1 **for** all features i ant their values j **do**
2 | initialize $A_{i,j}^p, C^p A_{i,j}^p, A_{i,j}^n, C^n A_{i,j}^n$ to 0

3 **for** pre-specified number of respondents **do**
4 | randomly select a respondent R
5 | find k nearest respondents closest to R
6 | **for** each closest respondent S and each features i **do**
7 | | **update** weights of $A_{i,j}^p, C^p A_{i,j}^p, A_{i,j}^n, C^n A_{i,j}^n$ as follows
8 | | | **if** feature value of S is lower than j **then** increment $A_{i,j}^p$
9 | | | **if** both feature value and overall score value of S are lower than j **then**
10 | | | | increment $C^p A_{i,j}^p$
11 | | | **if** feature value of S is higher than j **then** increment $A_{i,j}^n$
12 | | | **if** both feature value and overall score value of S are higher than j **then**
13 | | | | increment $C^n A_{i,j}^n$
14

15 **for** all features i ant their values j **do**
16 | let $U_{i,j} = C^p A_{i,j}^p / A_{i,j}^p$ and $D_{i,j} = C^n A_{i,j}^n / A_{i,j}^n$

Fig. 1. Pseudo code of OrdEval algorithm

The algorithm assumes that the cause of the differences in overall score are the differences in the attributes' values and gives these values some credit for that, but only if the sign of the differences in class and attribute is the same. It first sets counters of (co)occurring changes to zero (lines 1 and 2). Than it randomly selects a respondent R (line 4) and searches for its k nearest respondents (line 5). For each of these most similar respondents it updates the counters for all the features depending on the overall scores and feature values of the randomly selected respondent and the near respondents (lines 7 - 14): if the feature value of the near instance is lower than the value of the random instance (line 8) then the change is positive and we update $A_{i,j}^p$ for the value j of the given feature i (j is the value of feature i for respondent R). If additionally the overall score of the similar respondent is lower than the score of the random respondent (line 9) then the change in both overall score and feature is positive and we update $C^p A_{i,j}^p$ for given feature i and its value of random respondent j (line 10). Similarly we do for negative changes in feature and overall score (lines 11-13). We repeat the whole process (lines 3 - 14) for a pre-specified number of iterations. Conservatively we can set this number to be equal to the number of respondents, but we get useful results even if we run only a few iterations (e.g., logarithm of the number of respondents). Finally the upward and downward enforcement factors for all the values of attributes are computed as conditional probabilities (lines 15-16).

To show the behavior and usability of the algorithm we first define a simple artificial problem which is motivated by the Behavioral Decision Theory, stating that there are several distinct manners according to which marketing stimuli (like the COO) can be used during the formation of product attitude [4].

Our data set is described by six important and two irrelevant features. The important features correspond to different feature types from the marketing theory: two basic features (B_{weak} and B_{strong}), two performance features (P_{weak} and P_{strong}), two excitement features (E_{weak} and E_{strong}), and two irrelevant features ($I_{uniform}$ and I_{normal}). The values of all features are randomly generated integer values from 1 to 5, indicating for example score assigned to each of the features by the survey's respondent. The dependent variable for each instance (class) is the sum of its features' effects, which we scale to the uniform distribution of integers 1-5, indicating, for example, an overall score assigned by the respondent.

$$C = b_w(B_{weak}) + b_s(B_{strong}) + p_w(P_{weak}) + p_s(P_{strong}) + e_w(E_{weak}) + e_s(E_{strong})$$

The effects of attributes are as follows.

- Basic features are taken for granted by customers; high score in these features does not significantly increase the overall score, while a low score has a decreasing effect on dependent variable. We define two variants of basic features, one with weaker and another with stronger negative impact:

$$b_w(A) = \left\{ \begin{array}{l} -2; A <= 2 \\ 0; A >= 3 \end{array} \right\}, \quad b_s(A) = \left\{ \begin{array}{l} -4; A <= 3 \\ -2; A = 4 \\ -0; A = 5 \end{array} \right\}.$$

- Performance features have positive correlation with overall score: the higher the value of the attribute the bigger the effect on the overall score. We define the performance effects as

$$p_w(A) = \left\{ \begin{array}{l} -3; A = 1 \\ -2; A = 2 \\ -0; A = 3 \\ 2; A = 4 \\ 3; A = 5 \end{array} \right\}, \quad p_s(A) = \left\{ \begin{array}{l} -5; A = 1 \\ -3; A = 2 \\ -0; A = 3 \\ 3; A = 4 \\ 5; A = 5 \end{array} \right\}.$$

- Excitement features describe properties of product/service which are normally not very important to the users, but can cause excitement if the score is very high. We define two grades of excitement effect as

$$e_w(A) = \left\{ \begin{array}{l} 0; A <= 4 \\ 1; A = 5 \end{array} \right\}, \quad e_s(A) = \left\{ \begin{array}{l} 0; A <= 4 \\ 4; A = 5 \end{array} \right\}.$$

We generated 1000 instances for this data set. While the value distribution and the independence of features are unrealistic, note that we have experimented also with more realistic distributions as well as with different types of correlation, but the results and conclusions remain unchanged. Table 1 shows the upward and downward reinforcement factors the OrdEval algorithm returned for this data set. The direct interpretation and analysis of these numbers is of course possible, but visualization makes it easier.

Table 1. Upward and downward reinforcement factors for the pedagogical data with different types of features

	P_{weak}	P_{strong}	B_{weak}	B_{strong}	E_{weak}	E_{strong}	$I_{uniform}$	I_{normal}
$U_{..1}$	0	0	0	0	0	0	0	0
$U_{..2}$	0.42	0.45	0.29	0.30	0.33	0.33	0.33	0.32
$U_{..3}$	0.50	0.69	0.51	0.26	0.30	0.29	0.29	0.30
$U_{..4}$	0.53	0.73	0.30	0.54	0.31	0.24	0.32	0.30
$U_{..5}$	0.39	0.45	0.33	0.47	0.39	0.72	0.29	0.33
$D_{..1}$	0.38	0.44	0.26	0.28	0.31	0.32	0.30 0	.29
$D_{..2}$	0.46	0.66	0.49	0.25	0.27	0.29	0.27	0.28
$D_{..3}$	0.51	0.69	0.28	0.52	0.29	0.24	0.28	0.28
$D_{..4}$	0.36	0.43	0.30	0.45	0.36	0.71	0.26	0.28
$D_{..5}$	0	0	0	0	0	0	0	0

The slope visualization proposed by [11] (upward and downward reinforcement are represented with the steepness of the line segment between two consecutive feature values) is unusual for marketing research practitioners and, as we argue below, does not convey all the information necessary for this specific field. We therefore propose a marketing friendly visualization of the OrdEval results on Fig. 2, which contains results for each feature separately.

The 8 graphs are a sort of bar charts with addition of confidence intervals. For each graph a left-hand side with blue bars contains downwards reinforcements for each feature score separately. Upwards reinforcement factors for all the scores are represented with red bars on the right-hand side of each graph. Before we explain the results let us give a motivation for grey box-and-whiskers graphs on top of each reinforcement bar.

There are two problems with reinforcement factors when used in marketing:

- Imbalanced value distribution: it is quite common that for certain features some scores are almost non-existent (e.g., extremely low score of a basic feature is very rare - such a customer, would probably change the supplier), and also the reverse might be true, namely on a scale 1-5 it is not uncommon that almost all the scores are 4 and 5. Such imbalance also has consequences for reinforcement factors, since the probability of the increased/decreased overall score might be an artefact of the skewed distribution of values.
- Lack of information about significance of the reinforcement factors: the user does not know what expected range of a certain reinforcement factor is and weather the computed score is significantly different from the uninformative feature.

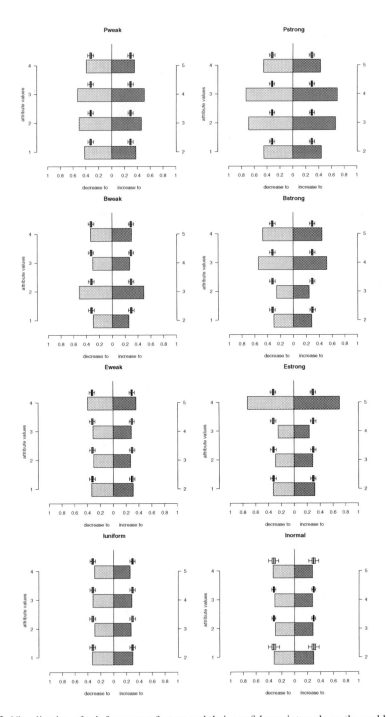

Fig. 2. Visualization of reinforcement factors and their confidence intervals on the problem with different types of features

To solve both problems we compute confidence intervals for each reinforcement factor. Since we cannot assume any parametric distribution and have to take the context of similar respondent into account we proceed as follows:

1. for each feature we construct e.g., $n = 200$ features with bootstrap sampled values from the original feature (alternatively the values can be randomly shuffled), we call these features normalizing features,
2. when searching similar respondents we only take original features into account, but we estimate also the reinforcement factors of randomly constructed features,
3. for each reinforcement factor $U_{i,j}$ and $D_{i,j}$ (upward and downward reinforcement for each value of each feature) we perform a statistical testing based on bootstrap estimates [3].

   (a) the null hypothesis states that the reinforcement factor is uninformative, i.e., it is equal to the median of its random normalizing features
   (b) the alternative hypothesis is one-sided, as we are interested if the reinforcement of the original feature is larger than the random normalizing reinforcement
   (c) set fixed confidence level, e.g. $\alpha = 0.05$
   (d) sort the reinforcement factors of random normalizing features in ascending order
   (e) if the reinforcement factor of the original feature is larger than $n(1 - \alpha)$th sorted factor we can reject the null hypothesis, and assume that the computed reinforcement contains significant information

4. the sorted reinforcement factors are the source of information for box-and-whiskers plot: the box is constructed from the 1st and 3rd quartile, middle line is median, while the whiskers are $100\alpha/2$ and $100(1 - \alpha/2$ percentiles (e.g. 2.5 and 97.5 percentiles) giving the borders of confidence interval (e.g., 95% confidence interval).

On Fig. 2 reinforcement factors (blue and red bars) reaching beyond the box-and-whiskers therefore contain significant information. Since the way we construct confidence intervals is not sensitive to the number of instances, these intervals are valid even for low number of scores. We can observe that the algorithm has captured the important landmarks of the features:

– for performance features P_{weak} and P_{strong} (two graphs in the top row) all the upward and downward reinforcements are significant, and the relative length of the bars is roughly proportional to the difference between impacts of the values,
– for basic feature B_{weak} (left-hand graph in the second row) the thresholds at values 2 and 3 (increasing feature from 2 to 3 strongly increases the overall score, and decreasing this feature from 3 to 2 strongly decreases the overall score).
– for basic attribute B_{strong} (right-hand graph in the second row) the upward thresholds at values 3 and 4 and downward reinforcement thresholds 4 and 5,
– for excitement features E_{weak} and E_{strong} (third row graphs) the jump from 4 to 5 and back is detected, in upward and downward enforcement, respectively. The reinforcements are larger for E_{strong} as expected.
– irrelevant random features $I_{uniform}$ and I_{normal} have no significant values (bottom row).

Note that only the reinforcement for the thresholds we have defined, are significantly larger than the boundaries of confidence intervals defined by the normalization features.

The properties of the used approach relevant to our study in particular, and in more general terms, to the COO-field at large, are manifold. Firstly, there is substantial *context sensitivity*. Typically the features are highly conditionally dependent upon the response and have to be evaluated in the context of other features. OrdEval is intrinsically contextual and assumes neither independence nor some fixed distribution of the features. The context of other features is handled through the distance. By using different distance measures and different features in the calculation of the distance, we are even in a position to use different contexts, e.g., we could use some background socio-economic information to calculate the similarity of respondents. Secondly, there is the *ability to handle ordered features and ordered response* and to use the information the ordering contains. The order of the attribute's values contains information which is comparable but not the same as values of numerical features, e.g., values poor, good, very good, excellent are ordered in expressing certain attitude but this ordering is not necessarily linear. Thirdly, we have *awareness of the meaning implied by the ordering* of the answers and the positive (negative) correlation of changes between feature values and the response (e.g., if the value of the feature increases from poor to good, we have to be able to detect both positive and negative correlation to the change of the overall response value). Fourthly, OrdEval has the *ability to handle each value of the feature separately*, e.g., for some features the value of good and very good have identical neutral impact on the response, value poor may have strong negative, and value excellent highly positive impact. We are able to observe and quantify each feature's values separately and thereby identify important thresholds. Next to that, *visualization of the output* allows experts to use it as a powerful exploratory data analysis tool, e.g., to identify type of features and the impact of their individual values. Also, the *output is in the form of probabilities*. Probability theory is commonly used and therefore the results in form of probabilities are comprehensible and interpretable by a large audience and can also be used operationally. Finally, we have *fast computation and robustness to noise and missing values*. A study of the family of the algorithms similar to OrdEval has shown that feature evaluation is possible and reliable even for extremely noisy data [10].

The following section describes the methodological design of an empirical study we conducted to test the OrdEval algorithm and to investigate the true nature of the mechanisms supporting COO-effects.

4   Case Study: COO

We demonstrate the analysis possible with OrdEval on a specific marketing-related issue, i.e., the functioning of COO-effects. A complete description of the data set, methodological design of the survey and questionnaire are given in [2]. Here we give only the relevant selection.

The products selected for the study were DVD-players (utilitarian) and beer (hedonic). The countries-of-origin selected were Spain and Denmark. Both countries were sufficiently familiar to respondents and mutually different on a number of country-specific aspects. Data was gathered by means of two surveys (one for Spain/Spanish products and one for Denmark/Danish products). These were distributed to respectively 616 and 609 graduate students of Belgian nationality. Several meta-analyses report that there are no significant differences in the estimates of COO-effects sizes between student and non-student samples [13].

The questionnaire consisted of 7 sections. First, subjects indicated sex and age. For all the remaining questions the subjects evaluated each of the country characteristics on a 7 point scale. Subjects were asked on their cognitions about nine environmental conditions (cultural identity, political climate, language, history, landscape, climate, technological and economic development, religion and people's character). Ten questions measured subjects' feelings towards COO following by four items retrieving respondents' behavioral intentions towards Spain or Denmark (I would like to shop, work, buy products, do business with and invest in country X). Next reliability, durability, performance and easiness of use were queried for DVD-players and taste, naturalness, aroma and prestige for beer. Subjects also evaluated DVD-players and beer (overall quality, likeability, appeal). Finally, purchase intentions were assessed with three questions (i.e., I would be willing to buy, consider buying, there is a chance I would buy product X). The 7 point scale was later turned into the 3 point scale (low, medium, high) for all variables as follows: $1,2 \rightarrow$ low; $3,4,5 \rightarrow$ medium; $6,7 \rightarrow$ high. Additional summarization over relevant variables was performed to get an overall score for the cognitive, emotional, and conative dimension which we used in our study.

4.1 Results

For each combination of product (Beer, or DVD-player) and country (Spain or Denmark) we computed reinforcement factors with OrdEval and generated a graph with visualization of reinforcement factors. The most interesting results we report below.

Reinforcement factors for Spanish beer (see Fig. 3) are followed by the results for Spanish DVD-players (Fig. 4). Each of the visual outputs can be read in a similar fashion. The three columns contain graphs for each CI dimension (feelings, conative, and cognitive) serving as an overall score (dependent variable). The three rows contain country-product score (product evaluation, beliefs, and purchase intentions) serving as independent variable. While these graph look like showing one dimensional dependencies, note that this is not so, as the context of all independent variables is taken into account through similarity of instances. Each of these independent variables can have three different values (low, medium and high). Upward/downward reinforcement is indicated with red/blue bars on right/left hand side of each of the nine graphs.

Several things can be learned from these graphs, like (non)linearity, (a)symmetry, threshold values and significance of reinforcements. For linear features one would expect that the stronger the preference for certain COO-dimension the better the country image. For example on Fig. 3 all the features show such tendency for the positive feelings toward Spain (first column). To detect symmetry, comparison of the red and blue bars on the same level allows us to determine what the valence of country image effects is like, namely the visual outputs on the same levels allow comparison between upward and downward reinforcement. In most cases on Fig. 3 there is a clear tendency towards asymmetry. Threshold values are detected by observing for which value the reinforcement factors (red and blue bars) become significant and exceed in length the whiskers

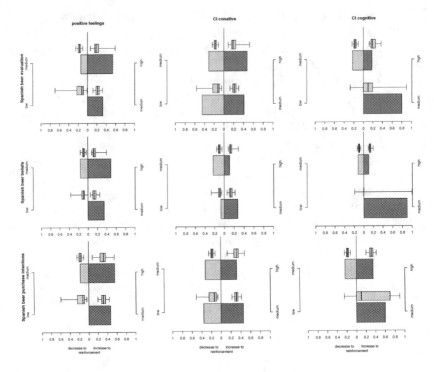

Fig. 3. Visualization of OrdEval results for Spanish beer

of gray box and whiskers graph above it. In the same way we detect significant reinforcements - they exceed whiskers of confidence interval. Since confidence intervals
do not assume any distribution and also take number of instances into account we get
practically useful information.

Within the Spanish survey, results obtained for beer (Fig. 3) and DVD-players
(Fig. 4) are very consistent. Contrary to the traditionally supported linear and symmetric perspective towards COO-effects, application of the OrdEval algorithm indicates
that country image (CI) is operating in a non-linear and asymmetric manner. The non-
linearity implies that the probability of increase or decrease in the dependent variable
(i.e., product attitude) varies in function of what the precise value obtained for the independent variable (i.e., the CI) is like. The asymmetry implies that the probability of
increase or decrease in the dependent variable differs between upward and downward
reinforcement. If we apply this overall finding to each of CI's three constituent components taken separately, we can come to the following more detailed conclusions.

The finding of non-linearity implies that the CI's cognitive component probably will
cause an effect on product attitude only in those specific cases where high scores for the
CI's cognitive component change into moderate scores. More in particular, the critical
threshold for the ci's cognitive component to exert an effect on product attitude is situated at the transition from high to medium. The finding of asymmetry means that the
probability of an effect on product attitude to occur can only be supported for downward reinforcement effects. Put differently, while a decrease in the value for the CI's

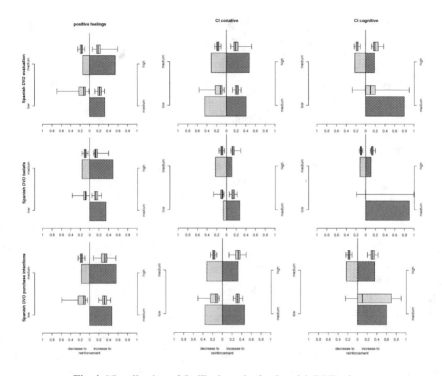

Fig. 4. Visualization of OrdEval results for Spanish DVD-players

cognitive component (from high to moderate) can be expected to (negatively) affect product attitude, the opposite scenario, i.e., an eventual increase in the value for the CI's cognitive component will fail to produce a (positive) effect on product attitude.

Non-linearity for the CI's affective component manifests itself somewhat differently compared with the CI's cognitive component in that the probability of CI's affective component to exert an effect on product attitude can be established for both changes from low into moderate and from moderate into high values. However, in general, the probability of an effect to occur is higher for changes from moderate to high values than for changes from low into moderate values. Thus, the critical threshold for the CI's affective component can be situated at the transition from medium to high. In line with the outcome for the CI's cognitive component, the CI's affective component operates asymmetrically in that it is expected to influence product attitude almost exclusively in cases where the values for the CI's affective component increase.

Finally, results obtained for the CI's conative component indicate that subjects' product attitude seems to be sensitive to changes in values for the CI's conative component in a different manner compared to how their product attitude reacts to changes in values obtained for the CI's cognitive and affective components. As already discussed, changes in values for the CI's cognitive and affective component influence product attitude only in one single direction (i.e., a downward effect for the cognitive component and an upward effect for the affective component). For the CI's conative component, probability scores indicate that both upward and downward effects can be expected to occur. However, comparable to the findings for the two other components of CI, the effects generated by the CI's conative component are supported by a similar mechanism, i.e., non-linear and asymmetric. Both upward and downward reinforcement effects triggered by the CI's conative component have a specific threshold value above or underneath which no further change in product attitude is to be expected. For upward reinforcement, this critical value lies at the transition from a scalar value of low to medium, for downward reinforcement the critical value is to be situated at the transition from high to medium. The asymmetry is reflected in that an increase in value from low to medium causes upward reinforcement without the opposite process of a decrease in value from medium to low resulting in a downward reinforcement effect. In line with this, downward reinforcement for the CI's conative component is generated in case there is a decrease in value from high to medium without upward reinforcement taking place for an increase in value from medium to high.

Results for the Danish survey are very much alike those for the Spanish survey, left aside a few minor differences. Analysis of the data by means of the OrdEval algorithm again shows that the CI's three basic components are operating in a non-linear and asymmetric manner.

5   Conclusions

We separate our conclusions into two groups, first COO related relevant for marketers, and the second a methodological one, relevant for data miners.

From a marketing point of view, we can state that this study clearly raises some issues with regard to the traditional view on COO-effects as a linear and symmetric phenomenon. Different from previous publications on COO-effects, we consider COO-effects as a non-linear phenomenon. As a consequence, marketers, in order to deal effectively with a product's COO, should be knowledgeable about the very precise critical or threshold value of CI. Next to that, we see COO-effects as an asymmetric phenomenon. This implies that marketers should be aware of the fact that upward and downward reinforcement effects in product attitude operate in full independence from each other.

From a data mining point of view the paper has adapted a general methodology for analysis of ordered data to the specifics of marketing. OrdEval algorithm possesses several favorable properties like context sensitivity, ability to handle ordered features and ordered response, awareness of the meaning implied by the ordering, ability to handle each value of the feature separately, output is in the form of probabilities, and fast computation and robustness to noise and missing values. In this paper we proposed an algorithm to compute confidence intervals for reinforcement factors thereby solving important problems which prevented its practical use. For example imbalanced value distribution is quite a common phenomenon but it has severe consequences for reinforcement factors, since the probability of the increased/decreased overall score might be an artefact of the skewed distribution of values. Another such obstacle is information about significance of the reinforcement factors: the user does not know what expected range of a certain reinforcement factor is and weather computed score is significantly different from the uninformative feature. By computing distribution independent confidence intervals we provide information on reliability of the reinforcement scores which give them practical importance and enables confident decision making. Additionally the proposed visualization of the reinforcement factors enables detection of (non)linearity, (a)symmetry, threshold values and significance of the results.

References

[1] Anderson, E.W., Sullivan, M.: The antecedents and consequences of customer satisfaction for firms. Science 12(2), 125–143 (1993)

[2] Brijs, K.: Unravelling country-of-origin: Semiotics as a theoretical basis for a meaning-centred approach towards country-of-origin effects. PhD thesis, University of Hasselt (2006), http://hdl.handle.net/1942/1819

[3] Efron, B., Tibshirani, R.J.: An introduction to bootstrap. Chapman & Hall, New York (1993)

[4] Einhorn, H.J., Hogarth, R.M.: Behavioral decision theory: Processes of judgement and choice. Annual Review of Psychology 32, 53–88 (1981)

[5] Maheswaran, D., Chen, C.Y.: Nation equity: Incidental emotions in country-of-origin effects. Journal of Consumer Research 33(3), 370–376 (2006)

[6] Matzler, K.E., Bailom, F., Hinterhuber, H., Renzl, B., Pichler, J.: A reconsideration of the importance-performance analysisthe asymmetric relationship between attribute-level performance and overall customer satisfaction: A reconsideration of the importance-performance analysis. Industrial Marketing Management 33, 271–277 (2004)

[7] Mittal, V., Kamakura, W.: Satisfaction, repurchase intention, and repurchase behavior: Investigating the moderating effect of customer characteristics. Journal of Marketing Research 38, 131–142 (2001)

[8] Peeters, G., Czapinski, J.: Positive-negative asymmetry in evaluations: The distinction between affective and informational negativity effect. European Review of Social Psychology 1, 33–60 (1990)

[9] Pharr, J.: Synthesizing country-of-origin research from the last decade: is the concept still salient in the era of global brands? Journal of Marketing Practice and Theory 13(4), 34–45 (2006)

[10] Robnik-Šikonja, M., Kononenko, I.: Theoretical and empirical analysis of Relief and RReliefF. Machine Learning Journal 53, 23–69 (2003)

[11] Robnik-Šikonja, M., Vanhoof, K.: Evaluation of ordinal attributes at value level. Data Mining and Knowledge Discovery 14, 225–243 (2007)

[12] Story, J.W.: The effects of perceived technological incongruence on perceptions of fit between countries, products, and attributes. Journal of Business Research 58(10), 1310–1319 (2005)

[13] Verlegh, P.W.J., Steenkamp, J.-B.E.M.: A review and meta-analysis of country-of-origin research. Journal of Economic Psychology 20(5), 521–546 (1999)

Evaluation of Fusion for Similarity Searching in Online Handwritten Documents

Sascha Schimke[1], Maik Schott[1], Claus Vielhauer[2], and Jana Dittmann[1]

[1] Otto-von-Guericke University of Magdeburg
[2] University of Applied Sciences Brandenburg
{sascha.schimke,maik.schott,
jana.dittmann}@iti.cs.uni-magdeburg.de,
claus.vielhauer@fh-brandenburg.de

Abstract. With the spread of TabletPCs handwriting raises in its significance and importance in the digital domain. Also there exist other devices with pen-based inputs like PDAs, digitizer tablets and pads specially prepared with sensors. The advantage of handwritten input methods is their possibility of an ad hoc creation of technical sketches and drawings alongside with text and that keyboards may be in some cases and environments bothersome. Therefore the amount of handwritten documents is likely to increase. But a great problem is a proper full text search on such documents. This paper discusses the effects of multi-sample and multi-algorithm fusion approaches, known from biometrics to increase the performance. The tests are done by using three different devices (Logitech ioPen, Pegasus PC NotesMaker, ACE CAD DigiMemo Digital) and five different feature extraction methods (square grid, triangular grid, slope, curvature and slant of writing) and show that fusion can improve the retrieval performance in terms of precision and recall from 0.903 and 0.935 without fusion to 0.958 and 0.943 with fusion, respectively.

1 Introduction

Although pen-based devices for computer input are not new – the first being the RAND tablet from 1956 [15] – they have only recently broken into the customer market. As more and more devices with pens are used, electronically handwritten documents increases too. Therefore there is a need to analyze handwritten documents for latter processing and to search in such documents for retrieval purposes.

For a classification of this work, handwritten data can be categorized in 3 ways. First, how were they recorded. Off-line approaches often use scans of physically handwritten documents and are thus image-based. In on-line approaches the samples of the handwriting are acquired during the actual writing process of the user and thus based on the pen movement

The second categorization is by their type. Handwritten data can be either textual and use some specially developed simple alphabets, a standard alphabet of a language or a cursive. Other types are gestures to give commands to the computer, symbols and drawings, or signatures, which are used in biometrics for the authentication of persons [12].

P. Perner (Ed.): ICDM 2009, LNAI 5633, pp. 276–290, 2009.
© Springer-Verlag Berlin Heidelberg 2009

The third and last categorization is by their time of processing, i.e. are they immediately recognized as needed for interactive scenarios or is a deferred recognition used as in the automatic evaluation of paper forms.

This work is based on [14] with its focus on texts, symbols and drawings, recorded online with immediate recognition. In this paper we presented a novel algorithm for searching in online handwritten documents without actually knowing what letters and symbols were written, using four methods for feature extraction: Freeman grids [3], slope (direction of writing), curvature and slant.

As there is an continuously growing amount of born-digital documents – existing only in digital form –, it can be assumed that online handwritten documents also occur more often. This type of documents may then become subject of digital preservation. As digital preservation is not just only about storing data, but also accessing and discovering them this work is also suited for this domain. Especially since with this approach works from the same author and thus his or her complete works in the digital preservation system can be found even if they are not tagged as such.

In many domains such as pattern recognition the system performance is improved by combining the results single subsystems or experts. Another domain where such combination known as fusion is applied is biometrics. Influenced by this, our work discusses the possibilities to adapt the strategies from biometrical fusion for similarity search in online handwritten documents.

Therefore, this work first presents in section 2 preliminary reflections, describing related work and the used features. Section 3 deals with the actual fusion, foundations as well as how it is concretely used. This is followed by section 4 with the evaluation of the various fusion approaches and finally this works concludes with an outlook for future work in section 5.

2 Preliminary Reflections

First, in subsection 2.1 related work for searching in online handwritten documents is discussed. The next subsection 2.2 presents the features used for searching in our database.

2.1 Related Work

An important approach on handwriting recognition was the single stroke recognizer from Dean Rubine [13]. This algorithm only supports the recognition of single stroke gestures defined by 13 different features, but is even widely used today. This approach was also used by Lopresti and Tomkins in their ScriptSearch algorithm [10] to search for handwritten texts and drawings. Like our approach their objective was to search for texts in a user-dependent manner. For this they separated the text into strokes, applied Rubine's algorithm and clustering and performed a search using the edit distance [6]. Another approach for search in handwritten documents was done by Perrone et al. [11]. They performed an automatic character recognition and used the n-best lists, that assign the handwritten words, the assumed semantic and the probability of this assignment. Regarding all these approaches, Rubine's algorithm is only usable for single stroke gestures, the ScriptSearch and many related algorithms are content independent and thus allow the retrieval of words, symbols, and bad handwriting. But as they rely on a strict word separation they reach their limit on syllabification/hyphenation. On the other hand this is not a problem for a text recognition approach. However, the later is only usable for texts in a supported script, and therefore may not work for documents with Cyrillic and Chinese characters. Our system includes all the advantages of ScriptSearch but also supports syllabification/ hyphenation.

2.2 Features

The input data are the spatial and temporal quantized horizontal x and vertical positions y of a pen and the pressure. Some devices have only binary pressure values, representing if the pen is on the writing surface or not. At first this data is preprocessed. The first preprocessing is noise removal, like duplicates of samples, i.e. successive samples with the same pen positions. This may happen with a high sampling rate and slow or no pen movement. Some devices do not associate a sample with a time stamp, like the DigiMemo Digital Notepad from ACE CAD Enterprises [17]. For these devices, gaps cannot be directly recognized as such. In general gaps provide hints for the beginning and end of a stroke. Strokes are used in this work as the input for feature extraction. The stroke segmentation considers a sample with a non-zero pressure and the immediate next sample with a non-zero pressure as the start and end points of a stroke, respectively.

For a fixed sampling rate the pair-wise spatial distance between two samples becomes larger with a higher writing speed. This occurs even more for devices with a variable sampling rate, e.g. the Logitech io Personal Digital Pen [18] between 20 and 50 Hz. To eliminate the influence of the writing speed on the samples, the last preprocessing step performs an equidistant resampling with a sample distance ws. This is done with the common approach of using cubic spline interpolation, separately for the x and y values [1].

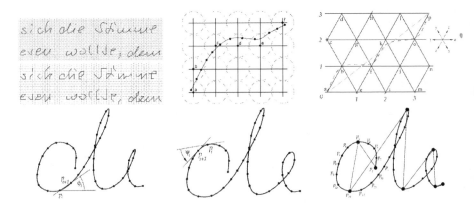

Fig. 1. Quantization of a text (top-left) and circular neighborhoods of a square grid (top-center), triangular gird (top-right), slope (bottom-left), curvature (bottom-right), slant (bottom-right)

The actual features used in this work are square grids, triangular grids, slope, curvature and slant.

The *square grid* feature extractor is based on a coding of line drawings first introduced by Freeman in [3]. The input is overlaid with a square grid of width *wg*. Each line formed by the input samples is approximated by a sequence of lines running through directly neighboring grid edges in a horizontal, vertical or diagonal manner (Fig. 1, top-left). As these lines can only have angles of $0°, 45°, 90°, 135°, 180°, 225°, 270°$ or $315°$, the input lines can be represented by a code string with the alphabet $A_\square=\{0, 1, 2, 3, 4, 5, 6, 7\}$. For actually selecting the appropriate grid vertices the circular quantization [5] (Fig. 1, top-center) is used. This method assumes a circular neighborhood of diameter *wg* for each grid vertex. If an input line runs through a circular neighborhood the corresponding edge vertex is saved in a vertex list. These vertices of this list are then connected in the mentioned manner. As only two vertices can be connected that are at the same grid block, a modified Bresenham algorithm is used to interpolate missing vertices.

The *triangular grid* is similar to the square grid approach with the difference that the grid consists of triangles (Fig. 1, top-right) instead of square blocks. The alphabet of the resulting code string is $A_\triangle=\{0, 1, 2, 3, 4, 5\}$.

The *slope* is the angle φ_i between two neighboring samples p_i and p_{i+1} (Fig. 1, bottom-left). For efficiency this slope is quantized w.r.t. to a certain quantization *q* to an integer codomain $[0, q-1]$ – corresponding to an alphabet $A_q=\{0, ..., q-1\}$ – according to the following formula:

$$\text{quant}_q(\varphi_i) = \left\lfloor \frac{q \cdot \varphi_i}{2 \cdot \pi} \right\rfloor \tag{1}$$

The *curvature* ψ_i of writing (Fig. 1, bottom-center) is also used as a feature by Shomaker and Segers in [9] or Jain and Namboodiri in [4]. A simple way of calculation is to use the difference between two neighboring slopes, i.e. $\psi_i = \varphi_i - \varphi_{i+1}$.

The last used feature is the *slant* σ_i as used by Denoue and Chui in [2]. The slant is the angle between a local maximum point and the following local minimum point and vice versa (Fig. 1, bottom-right). The slant is calculated in the same manner as the slope, including the quantization, but using only neighboring extremum points.

The actual search is done with one of these features. Because of quantization errors and as two writings of the same text or symbol are never the same, the feature strings can also differ. Therefore not exact matches are searched, but similar matches, i.e. fuzzy string searching. This is accomplished by using the approximate string matching approach of Sellers [8] which in turn uses the edit distance. The edit distance measures the minimum count of needed operations (insertion, deletion and replacing) to transform one string into another. The result of the approximate string matching is a $(|r|+1)\times(|s|+1)$ matrix D, with r being the feature string of the query, s being the feature string of the document. Each element $D_{i,j}$ is an edit distance between the query feature substring $r_{1...i}$ and the document feature substring $s_{j-k+1...j}$ with the a priori unknown length k. The last row $D_{|r|}$ thus contains the edit distances between the complete query string r and each part of the document string ending at document position j. Therefore a similarity vector $S = (S_1, S_2, ..., S_{|s|})^T$ between the query and the document can be calculated with:

$$S_j = 1 - \frac{D_{|r|,j}}{|r|} \text{ with } 1 \le j \le |s| \tag{2}$$

The approximate string search tries to find such a j, describing the tail substrings $s_{j-k+1...j}$, whose similarity S_j does not falls below a given threshold τ. In the ε neighborhood of a document position j with an optimal similarity $S_j=1$ further positions j' with $j-\varepsilon \le j' \le j+ \varepsilon$ are found with a gradually lowering similarity $S_{j'}$ but nevertheless well above the threshold. Therefore only hits are considered that do not have overlapping document positions with other hits and saved into a matching set H.

The next section describes the type of used fusions and how they are employed.

3 Fusion

This section treats the various fusion strategies usable for similarity searching for online handwritten texts. Although biometric authentication systems and handwriting search systems share similarities in their general structure and data processing, there exist differences, e.g. in biometrics reference data is enrollment, whereas in the handwriting search system the documents are the references but need not to be enrolled. In biometric systems an important processing step is the matching between the test feature set and the reference feature set of the same person for verification or a set of reference feature sets for identification. This means a repeated matching between similar feature sets is performed. In our system instead the feature sequences have different sizes. Whereas some feature sequences are significantly shorter than the search query, other feature sequences are significantly longer than whole documents. For this cause no direct matching can be performed.

On decision level, in biometric authentication a class is assigned to test feature set or the probabilities that the test feature set is a member of the respective classes. In our system the result of a query is an unknown count of matches and their degree of similarity with the test feature sequence.

The next sub-sections therefore discuss new approaches how the biometric techniques of multi-sample fusion (3.1) and multi-algorithmic fusion (3.2) on feature extraction, matching score and decision level can be used for searching for handwritten texts.

3.1 Multi-sample Fusion

In biometrics multi-sample fusion a single biometric trait is presented to the systems multiple times for feature extraction. Dependent on the fusion level either the feature sets are combined or the matching is performed multiple times and the results are aggregated.

In the case of searching for handwritten texts an analogous approach would be that the query string is written multiple times. This time-consuming task for the user is circumvented by an alternative approach. Instead of multiple writing of the query string, some results of the query are used in further iterations as query strings themselves, as seen in Fig. 2.

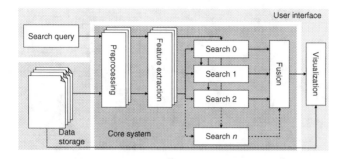

Fig. 2. Chain of execution for multi-sample fusion

Thus an important parameter is the count of iterations. This parameter is defined by using the *coi* best matches w.r.t. a threshold for the maximum dissimilarity between the match and the query string. Therefore either matching level fusion or decision level fusion is applicable. To fuse the matching sets H_0, H_1, ..., H_{coi} of the single queries on *decision level*, we either use intersection or union of these sets:.

$$H_\cap = \bigcap_{i=0}^{n} H_i \text{ and } H_\cup = \bigcup_{i=0}^{n} H_i \text{ with } H_\cap, H_\cup, H_0, ..., H_{coi} \notin H \qquad (3)$$

H_0 is the matching set of the original query string, H_1, ... , H_{coi} of the *coi* iterations, $H_\cap$ and $H_\cup$ of the fused sets, respectively. A fusion of the similarity values, i.e. on *matching score level*, is performed by using the distance matrices of each query. The distance matrix D of a query contains the edit distances between parts of the query and document strings. As each similarity vector – described in section 2.2 – has the same length, the results of each query can thus be fused by aggregating the similarity vectors S_i, i.e. by using the k-th power mean:

$$\overline{S}(k) = \sqrt[k]{\frac{1}{coi} \sum_{i=0}^{coi} S_i} \qquad (4)$$

The actual evaluations will be done using the arithmetic mean ($k=1$), minimum ($k \to +\infty$) and maximum ($k \to -\infty$).

3.2 Multi-algorithmic Fusion

This sub-section considers the multi-algorithmic fusion on pattern, matching score and decision level. Fig. 3 exemplarily illustrates the chain of execution for multi-algorithm decision level fusion.

A *feature extraction level fusion* in the context of handwritten data relates to the fusion of the alphabet used for encoding the local properties of the handwriting in the form of feature sequences. To fuse sequences with different feature types from the different search algorithms, an element-wise mapping between the feature sequences

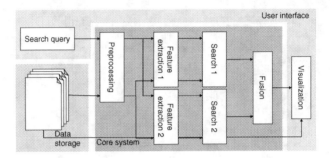

Fig. 3. Chain of execution for multi-algorithm decision level fusion

is necessary. Let $s_1 \in A_1^*(s_1=s_{1,1}, \ldots, s_{1,n}), \ldots, s_k \in A_k^*(s_k=s_{k,1}, \ldots, s_{k,n})$ be k sequences of the alphabets $A_1, \ldots, A_k$. If there exists a element-wise mapping $(s_{1,i}, \ldots, s_{k,i})$ with $1 \leq i \leq n$ between $s_1, \ldots, s_k$, a fusion of these sequences can be realized by defining an alphabet A_{fusion}:

$$s_{fusion} = \left(\left(\left(s_1 \cdot |A_2| + s_2\right) \cdot |A_3| + s_3\right)\ldots\right) \cdot s_k + |A_k| \text{ with } s_{fusion} \in A_{fusion}$$

$$= \begin{cases} s_k & \text{for } k = 1 \\ s_{k-1} \cdot |A_k| + s_k & \text{otherwise} \end{cases} \tag{5}$$

The size of the resulting alphabet is thus the cumulative size of the alphabets $A_1, \ldots, A_k$.

The needed unique mapping is only possible for features where the sample points were used, i.e. the features φ of the slant and the features ψ of the curvature. The size quantization parameter q can be freely chosen, whereas the parameter ws for the equidistant resampling must be same for both. As the slant does not use neighboring points, it is not affected by the sampling width ws. For a fusion of the slant features with $q_\varphi = 16$ with the curvature features with $q_\psi = 8$, the length of the resulting alphabet is $|A_{fusion}|=16\cdot8$.

A multi-algorithm *decision level fusion* is identical to the multi-sample decision level fusion described in the last subsection.

For *matching score fusion* an element-wise mapping is required too, as during query one similarity value per symbol of the feature sequences of the documents is calculated. The actual matching score fusion is performed by using the k-th power mean of the similarity values like the multi-sample matching score fusion.

4   Evaluation

This section evaluates the performance of the various fusion approaches previously introduced.

First in subsection 4.1 we present the test setup and measure and in the subsection 4.2 the actual results.

4.1 Test Setup

The general performance measures in information retrieval are precision and recall. Unlike most retrieval systems, the performance of our system depends heavily on the writer of the documents. Our system therefore has traits of a biometrical authentication system, where an evaluation should be performed person-wise.

Therefore the precision and recall are first measured locally by each person p and denoted as $precision_p(\tau)$ and $recall_p(\tau)$, respectively, with τ being the threshold for the query as described in section 2.2. The global precision and recall rates $precision(\tau)$ and $recall(\tau)$ are the average of all persons precision and recall rates.

The intersection of the graphs of both functions is similar to the Equal Error Rate (EER) of biometrical systems [16], with the difference that in biometric systems the EER should be 0, whereas the precision and recall should be 1 at the intersection.

For visualization, the precision can also be plotted as the function of the recall, to obtain a precision-recall graph, similar to a Receiver Operating Characteristic (ROC) curve. The intersection of the main diagonal with the curve matches the intersection of the precision curve with the recall curve. Each point of the precision-recall graph matches a threshold τ. The calculation of the intersection of the main diagonal of the precision-recall graph with the curve thus cannot be measured exactly. An approximation can be done by calculating the F_1 measure of the precision-recall pair of the curve which lies nearest to the main diagonal. The F_1 measure was first introduced in [7] which sets both retrieval performance measures in relation to each other by using the harmonic mean:

$$F_1 = \frac{2 \cdot precision(\tau_{EER}) \cdot recall(\tau_{EER})}{precision(\tau_{EER}) + recall(\tau_{EER})} \tag{6}$$

The intersection is then:

$$\tau_{EER} = \arg\min_{\tau} |precision(\tau) - recall(\tau)| \tag{7}$$

In summary, we use $precision(\tau_{EER})$, $recall(\tau_{EER})$ and F_1 as performance measures for evaluation.

We used three devices for online handwriting: the Logitech io Personal Digital Pen [18] – working optically by using special dotted paper –, the Pegasus PC Notes Maker [19] – using the propagation delay of ultrasound – and the ACE CAD Enterprise DigiMemo Digital Notepad [17] – using an electromagnetic field and a special pad.

The total count of writers, documents, words, symbols and queries in our test database is given in Table 1. From the 181 documents 83 were in German, 51 in English and 41 in Urdu.

All correct matches of each query were beforehand manually annotated, to obtain the a priori knowledge for evaluation. For this 158 word groups in 1.632 instances were annotated and used as query.

The tests were performed using different grid widths wg and resampling widths ws defined in section 2.2 A smaller width increases the geometric precision of the extracted features, but also increases the extracted feature vector and thus the calculation

Table 1. Overview of the test database

	ioPen	DigiMemo	NotesTaker	Total
Count of writers	13	4	9	23
Count of documents	72	22	87	181
Count of words	10.607	2.176	9.111	21.894
Count of symbols	173	0	0	173
Count of queries	527	113	992	1.632

duration. Therefore $wg=ws=7$ was selected as lower bound – 0.14 till 0.26 mm depending on the device resolution – and $wg=20$ as the upper bound – 0.4 till 0.74 mm depending on the device resolution.

Multi-sample fusion

For the multi-sample decision level fusion the count of iterations coi is also included as evaluation parameter. The actual query time is increased by $coi+1$ times compared to using no fusion. Therefore coi is below or equal to 3. The result sets of each iteration are fused by using the union.

For the multi-sample matching score level fusion the aggregation function of the similarity vectors is an additional parameter, whereas the functions are minimum, average (arithmetic mean) and maximum.

Multi-algorithm fusion

The multi-algorithm fusion is additionally to the matching score and decision level also performed on the lower feature extraction level. But as shown in section 3.2 the only usable features for feature extraction level fusion are the slope and curvature. For both others exemplary ROC curves are given.

4.2   Test Results

In this subsection first in Table 2 the test results (F_1 measure) for each device and algorithm without any kind of fusion are given. This also done as it is not feasible to present the results for all these combinations with multi-sample fusion and its 3 kinds of aggregation functions. The results of the multi-sample fusion are then given based on the best algorithm. Also the performances with different grid widths wg for the grid features and resampling widths ws for the slope and curvature are given, excluding the slant feature which is independent of a certain grid width. But the slant, as well as the slope and curvature are affected by a quantization parameter q, as described in section 2.2, which is therefore given in the right side of the table.

It can be seen that the results differ greatly between the devices, with the DigiMemo in general having the highest performance and the NotesTaker having the lowest one. The cause for this may be the different sampling precisions. Although the NotesTaker with 1200 ppi has the highest resolution it has a smaller precision as not the position of the pen tip is measured but the position of its ultrasound emitter, whose relative position changes with the inclination of the pen. Also the database is inhomogeneous w.r.t.

Table 2. F_1 measures for each device, algorithm and different wg and q, where applicable (best performer for each device in bold, overall best performer with gray background)

$wg/$ ws	ioPen	Digi Memo	Notes Taker	q	ioPen	Digi Memo	Notes Taker
Feature: square grid							
7	**0.859**	0.914	**0.741**				
10	0.829	**0.918**	0.731	n.a.			
15	0.786	0.902	0.713				
20	0.748	0.864	0.683				
Feature: triangular grid							
7	0.781	0.904	0.630				
10	0.723	0.902	0.609	n.a.			
15	0.721	0.916	0.599				
20	0.651	0.866	0.582				
Feature: slope (left: $q=8$, right: $ws=7$)							
7	0.806	0.902	0.675	3	0.709	0.824	0.615
10	0.813	0.912	0.681	8	**0.806**	**0.902**	**0.675**
15	0.793	0.906	0.675	12	0.801	0.896	0.670
20	0.750	0.896	0.660	16	0.789	0.847	0.631
Feature: curvature (left: $q=8$, right: $ws=10$)							
7	0.409	0.590	0.300	3	0.412	0.532	0.258
10	0.424	0.612	0.286	8	0.424	0.612	0.286
15	0.399	0.615	0.299	12	0.394	0.627	0.323
20	0.410	0.645	0.292	16	0.418	0.633	0.331
Feature: slant							
				3	0.353	0.450	0.220
n.a.				8	0.432	0.594	0.289
				12	0.420	0.634	0.307
				16	0.390	0.557	0.287

writers and document types. However, in [4] an IBM CrossPad – similar to the Digi-Memo – yielded similar results of a precision of 0.923 and a recall of 0.900 and thus an F_1 of 0.911. Comparing the features, the square grid performs the best and more than 2 times better than the worst feature (slant). Furthermore, the results show that with an increasing grid size wg or resampling width ws the performance decreases. Therefore in the following tables the used values for wg and ws are {7, 10, 15} only. For the quantization instead no such clear trend can be given. The best results of table are 0.859 for the ioPen and $wg=7$, 0.918 for the DigiMemo and $wg=10$ and 0.741 for the NotesTaker and $wg=7$.

Multi-sample fusion

As the square grid feature was shown to perform the best, the multi-sample fusion is done using this feature. The precise test results are given in Table 3.

Table 3. F_1 measures of multi-sample matching score level fusion (best performer for each device in bold, overall best performer with gray background)

wg	coi	Average (F_1)			Minimum (F_1)			Maximum (F_1)		
		ioPen	Digi Memo	Notes Taker	ioPen	Digi Memo	Notes Taker	ioPen	Digi Memo	Notes Taker
7	0	0.859	0.914	0.741	**0.859**	0.914	0.741	0.859	0.914	0.741
	1	0.878	0.934	0.766	0.854	0.924	**0.751**	0.879	0.931	0.762
	2	**0.887**	0.939	0.778	0.821	0.921	0.746	**0.881**	**0.946**	**0.763**
	3	0.885	0.936	**0.786**	0.754	0.911	0.735	0.880	0.944	0.761
10	0	0.829	0.918	0.731	0.829	0.918	0.731	0.829	0.918	0.731
	1	0.861	0.938	0.762	0.833	**0.931**	0.742	0.853	0.934	0.752
	2	0.864	0.944	0.776	0.788	0.925	0.739	0.855	0.945	0.755
	3	0.871	**0.950**	0.783	0.687	0.901	0.725	0.855	0.942	0.749
15	0	0.786	0.902	0.713	0.786	0.902	0.713	0.786	0.902	0.713
	1	0.832	0.922	0.740	0.802	0.912	0.724	0.811	0.918	0.733
	2	0.833	0.926	0.757	0.737	0.913	0.714	0.831	0.930	0.735
	3	0.838	0.939	0.771	0.650	0.914	0.706	0.832	0.939	0.733

In almost all cases a fusion ($coi>0$) increases the F_1 measure compared to not using a fusion ($coi=0$). In general, with each iteration coi a better performance is gained. Even then, there are only minor differences between each iteration count. The overall best performances for each device are 0.887 (precision: 0.887, recall: 0.886) for the ioPen, 0.950 (precision: 0.958, recall: 0.943) for the DigiMemo and 0.786 (precision: 0.776, recall: 0.797) for the NotesTaker, each using the average aggregate function. In contrast, the minimum aggregation function performs the worst.

Table 4 shows the result for the multi-sample decision level fusion for the three devices.

Table 4. F_1 measures of multi-sample decision level fusion (best performer for each device in bold, overall best performer with gray background)

wg	coi	ioPen (F_1)	DigiMemo (F_1)	NotesTaker (F_1)
7	0	0.859	0.914	0.741
	1	0.875	0.936	0.757
	2	0.874	**0.947**	**0.758**
	3	**0.884**	0.942	0.754
10	0	0.829	0.918	0.731
	1	0.859	0.915	0.748
	2	0.856	0.939	0.746
	3	0.860	0.941	0.743
15	0	0.786	0.902	0.713
	1	0.819	0.916	0.727
	2	0.822	**0.947**	0.728
	3	0.824	0.937	0.724

The results of the decision level fusion follow the trend of the matching score fusion, i.e. only small differences between $coi=1$, $coi=2$ and $coi=3$ but significant differences between those 3 and $coi=0$. The best performances for each device are 0.884 (precision: 0.892, recall: 0.875) for the ioPen, 0.947 (precision: 0.953, recall: 0.943) for the DigiMemo and 0.758 (precision: 0.751, recall: 0.764) for the NotesTaker. These are a little less than for the matching level fusion, which also applies if all results are considered. Thus for multi-sample fusion the matching score level should be preferred over the decision level.

Multi-algorithm fusion
The multi-algorithm fusion is performed on feature extraction, matching score and decision level. As there are many possibilities which feature extraction algorithm to take for fusion and with which parameter set only exemplary results are shown. Nevertheless, these represent the general trend of the performance changes using the multi-algorithm fusion for each of the three different levels.

As for *feature extraction level* Fig. 4 shows exemplarily for the ioPen the precision-recall graphs of several fusions of the curvature and slope algorithms with a fixed resampling width $ws=10$. The dashed lines represent the results of the slope algorithm without fusion with the curvature algorithm. For the other devices and their parameters the graphs follow the same trend.

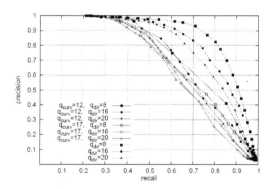

Fig. 4. Precision-recall graphs for multi-algorithm feature level fusion for ioPen using the curvature and slope algorithms

In all used combinations, the fusion has worse results than the slope algorithm alone. A possible cause is that curvature algorithm in general performs only half as good as the slope algorithm and therefore the former cannot improve the latter.

On the *matching score level* the scores can be either fused by the minimum, average or maximum aggregation function. Fig. 5 shows the precision-recall graphs of the fusion between the curvature and slope algorithms.

Like the fusion on the feature extraction level, no improvements are gained compared to not using fusion, probably because of the same reasons. In comparison to the multi-sample matching score level fusion the average aggregation function in general performs better than the minimum and maximum aggregation functions. However, for the multi-algorithm fusion there is no definite performance difference between the last both as it was the case for the multi-sample fusion.

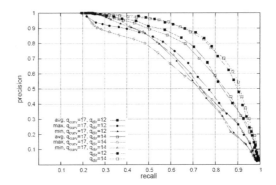

Fig. 5. Precision-recall graphs for multi-algorithm matching score level fusion for ioPen using the curvature and slope algorithms

In contrast to the other levels, for a fusion on the *decision level* the parameters of the algorithms can be freely chosen. However, there exists no single similarity value for which the precision and recall can be calculated, but one similarity value per algorithm. Therefore, no precision-recall graph can be given for the fused system. Table 5 shows the results for the fusion of the square grid algorithm with the other algorithms by using the union the fuse the single test results. The parameters for the algorithms are $wg=7$ for square and triangular grid, $ws=7$ and $q=12$ for the slope, $ws=7$ and $q=14$ for the curvature and $q=17$ for the slant.

Table 5. F_1 measures of multi-algorithm decision level fusion (best performer for each device in bold, overall best performer with gray background)

	F_1 ioPen	F_1 DigiMemo	F_1 NotesTaker
Square grid	0.859	0.914	0.741
Square grid U triangular grid	0.871	0.928	0.758
Square grid U slope	**0.876**	**0.930**	**0.763**
Square grid U curvature	0.859	0.922	0.730
Square grid U slant	0.864	0.922	0.751

In contrast to the fusion on the other both levels, for the decision level fusion there is a performance gain up to 2.2% for the NotesTaker device.

Comparison with other works

This section summarizes the results of our different feature extraction algorithms with and without fusion and relates their performance to other works. These are the ScriptSearch algorithm Lopresti et al. [10], the Word Spotting system of Jain et al. [4] and a pattern recognition approach of Perrone et al. [11]. For the last the results were only given as ROC curves, so that the actual precision(τ_{EER}), recall(τ_{EER}) and thus F_1 can only be estimated. Also the results are given dependently on the query length.

Table 6. Comparison of our feature extraction algorithms with and w/o fusion with other works (overall best performer with gray background)

Description	Precision	Recall	F_1
Lopresti et al.	0.572	0.643	0.605
Jain et al.	0.923	0.900	0.911
Perrone et al. (single word query)	≈ 0.7	≈ 0.7	≈ 0.7
Perrone et al. (multi-word query)	≈ 0.9	≈ 0.9	≈ 0.9
ioPen (square grid, wg=7)	0.848	0.870	0.859
DigiMemo (square grid, wg=10)	0.903	0.935	0.918
NotesTaker (square grid, wg=7)	0.728	0.756	0.741
Multi-sample matching score level fusion			
ioPen (square grid, wg=7, n=2)	0.887	0.886	0.887
DigiMemo (wg=10, n=3)	0.958	0.943	0.950
NotesTaker (wg=7, n=2)	0.776	0.797	0.786

The table shows that the DigiMemo results using square grid features equals the one in [4], where a similar device was used. However with using a multi-sample matching score level fusion our system clearly surpasses Jain et al.'s.

5 Conclusion and Future Work

This paper presented five types of features for online handwriting: square grid, triangular grid, slope, curvature and slant. It was shown that the square grid approach performed best with F_1 values for the DigiMemo device of over 0.9. The other approaches therefore seem less feasible for similarity in searching in online handwritten documents, although for example the curvature is successfully used in the biometrics domain [9].

This paper has also shown that a similarity searching system in online handwritten documents benefits from using fusion, with performance increases from 0.859 to 0.887 for the ioPen, from 0.918 to 0.950 for the DigiMemo and from 0.741 to 0.786. Using a multi-sample approach the system benefited in almost all cases, especially on the matching score level. The multi-algorithm approach only led to some slight improvements on the decision level, whereas on the other levels even worse results than without any fusion at all were obtained. This contrasts to the biometrics domain, where also for multi-algorithm approaches improved results occur [12].

Future work may include the investigation of person-independent features, features with much smaller feature strings to speed up the search process and better algorithms for a fuzzy search as currently implemented algorithms for this case have a quadratic worst-case complexity. Regarding the fusion, multi-modal fusion was not evaluated in this work, as well as weighted fusion. A possible modality would be speech, or more precisely, the spoken texts.

Acknowledgements

The work in this paper has been supported in part by the European Commission through the IST Programme under the EU Network of Excellence SIMILAR

(FP6–507609) and the application in the context of digital long-term preservation through the FP7 ICT Programme under Contract FP7-ICT-216736 SHAMAN. The information in this document is provided as is, and no guarantee or warranty is given or implied that the information is fit for any particular purpose. The user thereof uses the information at its sole risk and liability.

References

1. de Boor, C.: A Practical Guide to Splines. Springer, Heidelberg (1978)
2. Denoue, L., Chiu, P.: Ink Completion. In: Graphics Interface 2005 – Posters and Demos, Victoria, British Columbia, Canada (May 2005)
3. Freeman, H.: Computer Processing of Line-Drawing Images. ACM Computing Surveys 6(1), 57–97 (1974)
4. Jain, A.K., Namboodiri, A.M.: Indexing and Retrieval of On-line Handwritten Documents. In: Proceedings of the International Conference on Document Analysis and Recognition, pp. 655–659. IEEE Computer Society Press, Los Alamitos (2003)
5. Koplowitz, J., Toussaint, G.T.: A unified theory of coding schemes for the efficient transmission of line drawings. In: Proceedings of the 1976 IEEE Conference on Communications and Power, October 1976, pp. 205–208 (1976)
6. Levenshtein, V.I.: Binary Codes Capable of Correcting Deletions, Insertions and Reversals. Soviet Physics Doklady 10(8), 707–710 (1966)
7. van Rijsbergen, C.J.: Information Retrieval. Department of Computer Science, University of Glasgow (1979)
8. Sellers, P.H.: The Theory and Computation of Evolutionary Distances: Pattern Recognition. Journal of Algorithms 1(4), 359–373 (1980)
9. Schomaker, L., Segers, E.: Finding Features used in the Human Reading of Cursive Handwriting. International Journal on Document Analysis and Recognition 2, 13–18 (1999)
10. Lopresti, D.P., Tomkins, A.: On the Searchability of Electronic Ink. In: Proc. of International Workshop on Frontiers in Handwriting Recognition, December 1994, pp. 156–165 (1994)
11. Perrone, M.P., Russell, G.F., Ziq, A.: Machine Learning in a Multimedia Document Retrieval Framework. IBM Systems Journal 41(3), 494–503 (2002)
12. Scheidat, T., Vielhauer, C., Dittmann, J.: Handwriting verification – Comparison of a multi-algorithmic and a multi-semantic approach. In: Image and Vision Computing (2007)
13. Rubine, D.: Specifying Gestures by Example. In: Proceedings of the 18th annual conference on Computer graphics and interactive techniques, pp. 329–337. ACM Press, New York (1991)
14. Schimke, S., Vielhauer, C.: Similarity Searching for On-line Handwritten Documents. Journal on Multimodal User Interfaces 1(2) (2007)
15. Davis, M.R., Ellis, T.O.: The RAND Tablet: A Man-Machine Graphical Communication Device. Memorandum rm-4122-arpa, RAND Corporation (August 1964)
16. Vielhauer, C.: Biometric User Authentication for IT Security: From Fundamentals to Handwriting. Springer, New York (2006)
17. ACE CAD Enterprise Co., Ltd.: DigiMemo Digital Notepad, http://www.acecad.com.tw/
18. Logitech, Inc.: io Personal Digital Pen, http://www.logitech.com/
19. Pegasus Technologies Ltd.: PC Notes Taker, http://www.pegatech.com/

Self-training Strategies for Handwriting Word Recognition

Volkmar Frinken and Horst Bunke

Institute of Computer Science and Applied Mathematics, University of Bern,
Neubrückstrasse 10, CH-3012 Bern, Switzerland
{frinken,bunke}@iam.unibe.ch

Abstract. Handwriting recognition is an emerging subfield of human-computer interaction that has many potential industrial applications, e.g. in postal automation, bank check processing, and automatic form reading. Training a recognizer, however, requires a substantial amount of training examples together with their corresponding ground truth, which needs to be created by humans. A promising way to significantly reduce this effort, and hence cut system development costs, is offered by semi-supervised learning, in which both text with and text without transcription is used for training. However, until today there is no straightforward and established way of semi-supervised learning, particularly not for handwriting recognition. In the self-training approach, an initially trained recognition system creates a new training set from un-labeled data. Using this set, a new recognizer is created. The creation of the training set is done by selecting elements from the unlabeled set, according to their recognition confidence. The success of self-training depends crucially on the data selected. In this paper, we test and compare different rules used to select new training data for single word recognition with and without additional language information in the form of a dictionary. We demonstrate that it is possible to substantially increase the recognition accuracy for both systems.

1 Introduction

Offline handwriting recognition has received a substantial amount of attention in intelligent information processing in the past and has important applications, e.g. in postal address identification [1], Bank check processing [2], prescreening of handwritten notes [3], and the creation of digital libraries of historical documents [4]. After several decades of research and emerging applications, however, it is still considered a difficult problem that has only been partially solved [5,6].

One of the key issues in handwriting recognition is the fact that the writing style varies greatly between different persons. Hence a large amount of training data is necessary to obtain a well performing recognizer. Training data usually consist of samples of handwritten words along with their correct transcriptions,

P. Perner (Ed.): ICDM 2009, LNAI 5633, pp. 291–300, 2009.

also called *labels* or *ground truth*. The creation of this ground truth has to be done manually, which makes it costly and time consuming, while the collection of handwritten samples itself can be done very efficiently.

It has been shown that in various classification scenarios unlabeled examples can significantly improve the recognition accuracy, when applying semi-supervised learning [7]. This raises the question whether of it can be helpful for handwriting recognition as well. Most of the existing works, however, deal with the standard classification scenario where a single point in a feature space has to be mapped into the label space [8,9]. In the current paper, a more general problem is considered in the sense that a (possibly long) sequence of feature vectors has to be mapped to a (usually much shorter) sequence of labels, or characters. Some research has been done for sequential semi-supervised learning, mostly with Hidden Markov Models, but only moderate success has been achieved following this approach [10,11]. Only few publications exist that deal specifically with semi-supervised learning for handwritten word recognition. In [12] the authors adapt a recognition system to a single person by using unlabeled data. This system is highly specialized after the adaptation and thus not suitable for general handwriting recognition. The task of unconstrained, writer independent recognition was addressed in [13]. Nevertheless, the setup used in that paper did not take any information about the language into account and therefore allows only limited conclusions to be drawn about self-learning. In this paper, we extend the approach proposed in [13] by integrating a dictionary into the recognizer. In this way, we obtain a more robust system with a higher performance.

Self-training is a semi-supervised learning methodology that can be widely used due to its general and abstract formulation. It states that a recognizer initially should be trained with all available labeled data. Afterwards, the recognizer assigns pseudo-labels to the unlabeled data set by recognizing it. The best recognized elements are then included in a new training set that is used to generate a new recognizer. These steps are repeated until some criterion is met, e.g. the convergence of the recognition accuracy.

However, there is no straightforward way to tell which elements should be used for retraining. On the one hand, using only few, but confidently recognized samples might not change the recognizer at all. On the other hand, using too many elements might impede the recognizer, because the more data with pseudo-labels are added, the more misclassified samples enter the training set. We therefore investigate different retraining rules that determine which data should be used for retraining. In this work we compare the effects of different retraining rules for two types of word recognizers, one with and the other without the use of a dictionary.

The rest of the paper is structured as follows. In Section 2, details of the task of handwritten word recognition and the recognizer are presented. Semi-supervised learning and self-training are discussed in Section 3. The experiments are presented in Section 4 and the paper concludes with Section 5.

2   Neural Networks Based Recognizer

2.1   Preprocessing

In offline handwriting recognition, the computer tries to recognize words from scanned documents. We follow the common approach to create a sequence of feature vectors from the scanned image of a word and use a recognizer suited for sequential data. To transform an input word into a sequence of feature vectors, several preprocessing steps have to be applied. The words used in the experiments come from the IAM database [14]. They are extracted from pages of handwritten texts, which were scanned and separated into individual text lines. After binarizing the image with a threshold on the grey scale value, the slant and skew of each textline was corrected and the width and height were normalized. Next, text lines are split into individual words. Then features are extracted using a horizontally sliding window. A window with a width of one pixel is used to extract nine geometric features at each position, three global and six local ones. The global features are the 0th, 1st and 2nd moment of the black pixels' distribution within the window. The local features are the position of the top-most and that of the bottom-most black pixel, the inclination of the top and bottom contour of the word at that position, the number of vertical black/white transitions and the average grey scale value between the top-most and bottom-most black pixel. For details on these steps, we refer to [15].

2.2   Bidirectional Long Short-Term Memory Neural Network

The recognizer used in this paper is a recently developed recurrent neural network, termed *bidirectional long short-term memory* (BLSTM) neural network [16]. A hidden layer is made up of so-called *long short-term memory* blocks instead of simple nodes. These memory blocks are specifically designed to address the *vanishing gradient problem*, which describes the exponential increase or decay of values as they cycle through recurrent network layers. This is done by nodes that control the information flow into and out of each memory block. For details about BLSTM networks, we refer to [16,17].

The network is made up of two separate input layers, two separate recurrent hidden layers, and one output layer. Each input layer is connected to one hidden layer. The hidden layers are both connected to the output layer. The network is *bidirectional*, i.e. a sequence is fed into the network in both forward and backward modes. The input layers consist of one node for each feature. One input and one hidden layer deal with the forward sequence, the other input and hidden layer with the backward sequence. At each position p of the input sequence of length l, the output layer sums up the values coming from the hidden layer that has processed positions 1 to p and the hidden layer that has processed positions l down to p. The output layer contains one node for each possible character in the sequence plus a special ε node, to indicate "no character". At each position, the output activations of the nodes are normalized so that they sum up to 1, and are treated as probabilities that the node's corresponding character occurs at this position.

The output of the network is therefore a sequence of output activations for the nodes in the output layer. Afterwards, this sequence is transformed into the word that the recognizer outputs. In this step, external language information might be helpful [16]. A dictionary containing the most likely words can guide the recognition, but it also prevents the recognition of words that are not contained in the dictionary. Also, a dictionary might not always be available. Depending on the type of recognition, with or without a dictionary, different forms of post-processing take place.

If no dictionary guides the recognition, only the nodes with the highest probability are considered and all others are discarded. This results in a sequence of one output activation at each time step. Eliminating repeated activations of the same node and of the ε node (in that order), yields the desired character sequence.

If a dictionary is given, the Connectionist Temporal Classification (CTC) algorithm computes a solution for the most likely words among all words in the dictionary [17]. For a given word, a path is constructed through the output sequence, such that the nodes along this path yield the word's characters when omitting repeated occurrences of the same node and all occurrences of the ε node. The multiplication of all output activations along the path results in a probability value and the maximum value is used as the word's probability. Finally, the word with the highest probability is returned.

2.3   Confidence Values

The probability values that the network outputs turn out to be unreliable and should not be used as a confidence value for the recognition. We therefore performed the following steps to obtain better confidence values.

First, several separately trained neural networks recognize the word independently and vote on the output. A simple count of how many networks agree gives a preliminary confidence value n for that output y. To break a tie in the voting, the network with the best performance on an independent validation set is used. In step two, the recognition accuracy on the validation set with a preliminary confidence value n is computed separately for all values of n and stored as $p(n)$. In the third step, the output itself is additionally taken into account. The recognition accuracy for each different output y and each value of n is computed on the validation set and stored as $p(y, n)$ if enough samples exist to estimate it robustly. Finally, the recognition's confidence value of the word's output is set to $p(y, n)$ if that value exists, otherwise to $p(n)$. For more details on these confidence values, we refer to [18].

3   Self Learning

3.1   General Methodology

The basic idea of semi-supervised learning is to enhance a recognizer's accuracy by using both labeled and unlabeled data for training. This is done by incorporating information about the a priori data distribution which can be inferred from the unlabeled examples [9]. However, in contrast with most of the scenarios where semi-supervised learning was used in the past, however, handwritten words are usually represented by sequences, rather than individual feature vectors.

Nevertheless, two frameworks exist that are general enough to deal with sequential information. These are co-training, introduced in [19], and self-training, which goes back as far as [20] and was put into context to co-training in [21]. Both are forms of the EM algorithm [9,22], which iteratively adapts the model to increase the likelihood of the data. In the current paper we focus on self-training.

Self-training as introduced in [21] uses only the most confidently recognized elements to retrain the recognizer, but that might not always be the best choice [13]. A conservative approach would be to ensure that only correctly labeled data are used for retraining. However, if retraining is done with only those elements whose correctness can be guaranteed, the retraining set does not change significantly and the classifier may remain nearly the same. Enlarging the retraining set, on the other hand, is only possible at the cost of increasing noise, i.e. adding mislabeled words to the training set. With only few correctly recognized words and large amounts of possible misrecognitions, the challenge of successful self-training lies in finding the optimal trade-off between data quality and data quantity for retraining.

3.2   Retraining Rules

We compare three deterministic rules to select the elements used for retraining. The three retraining rules each compute a confidence threshold and select all elements recognized with a confidence equal to or above that threshold. Additionally, all three methods select the whole original training set for retraining. The difference between the three rules lies in the procedure to choose the threshold. The most conservative among the three retraining rules tries to ensure that no missclassified samples are included in the set used for retraining. It places its associated confidence threshold to the lowest possible value without misclassifying any of the words in the validation set. Since this is the highest threshold considered by the three rules, it is termed *High Confidence* retraining rule. To obtain a larger retraining set, the *Medium Confidence* retraining rule selects all words that are more likely to be correct than wrong. Consequently its threshold is placed at the lowest value for which the recognition accuracy of those words having this value is above 50% on the validation set. As the third deterministic and most relaxed rule, the *Low Confidence* retraining rule selects all words for retraining, regardless of their recognition confidence.

4   Experiments

4.1   Experimental Setup

To analyze the effect of the retraining rules, the same set of experiments have been performed in two recognition scenarios. In one scenario, the neural network recognizes words without using a dictionary, while in the other scenario a dictionary is used. The words included in the dictionary are the 4,000 most frequently occurring words from the IAM Handwriting Database [14]. The database is split up in a test set (5,342 words), a validation set (5,590 words), and a work set (38,127 words) each with different writers[1]

To simulate the existence of few labeled data and large amounts of unlabeled data, the work set was randomly split up into a labeled set to be used for training the initial recognizer and an unlabeled set to be used for retraining. All retraining rules were tested with 2,000, 4,000, 6,000, 8,000, and 10,000 labeled words.

For self-training without a dictionary, ten neural networks were used in parallel for each setup to calculate the confidences. Self-training with the help of a dictionary was only done with five networks for each setup, which resulted in slightly less precise confidence values, but was necessary to reduce computational costs.

For the purpose of initialization, the networks were trained on the labeled part of the working set. In each self-training iteration, the networks decoded a) the validation set from which the confidence mapping was computed, b) the unlabeled set from which the new training sets were created and c) the test set to compute the recognition accuracy. The new training sets were then used to retrain each network. For all retraining rules, the retraining sets were exactly the same for all networks in each step. Retraining was bounded to 10 backpropagation iterations to keep the computation time within reasonable limits.

4.2   Results

The goal of the experiments is twofold. Firstly we wanted to investigate the behavior of the retraining rules for recognition systems using a dictionary. Secondly, we wanted to compare the results to those reported in [13], where a system without a dictionary was considered. The results of our experiments appear in Fig. 1, where the left- and right-hand parts correspond to the case without and with a dictionary, respectively. Due to the different nature of the recognition tasks, the accuracy values are given by different formulas. In case of recognition without a dictionary, the label accuracy is of interest and given by the number of correctly recognized letters minus the number of added and missing letters, divided by the number of letters in the word's ground truth. In case of recognition with a dictionary, the word accuracy is reported, which is the number of correctly recognized words divided by the entire number of words in the test set.

First of all, it can be observed that the accuracy increases as the size of the labeled set used for the initial training increases. The accuracies of the networks trained with the entire labeled work set are indicated by the horizontal line, which are 80.32% label accuracy for recognition without a dictionary and 84.70% word accuracy in the other case.

In the scenario without a dictionary, the *Low Confidence* retraining rule consistently performs best, followed by the *Medium Confidence* retraining rule. The

---

[1] The splitting has be done according to the standard writer independent recognition task `http://www.iam.unibe.ch/fkiwww/iamDB/tasks/largeWriterIndependentTextLineRecognitionTask.zip`

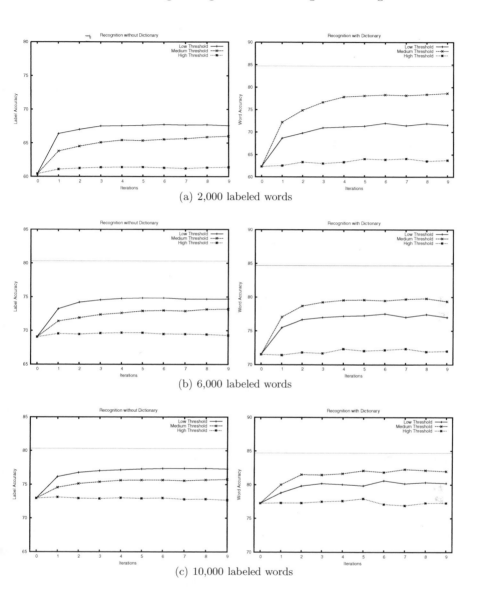

Fig. 1. The averaged accuracies of all networks on the test set for each self-training iteration. Accuracies for recognition without a dictionary are displayed on the left-hand side, while the recognition results with a dictionary are displayed on the right-hand side. Note that on the left-hand side the label accuracy and on the right-hand side the word accuracy is given.

High Confidence retraining rule did not constantly increase the recognition accuracy; it actually resulted in a deterioration of the recognition accuracy in some cases. In the other scenario, where a dictionary was used, the results are different.

Again, the *High Confidence* retraining rule did not increase the recognition accuracy substantially. However, the *Medium Confidence* and *Low Confidence* retraining rules both achieved an increase. Where a dictionary is used, the *Medium Confidence* retraining rule performs better then the *Low Confidence* retraining rule. This is in contrast to the case where no dictionary is used. The increase for the *Low Confidence* and *Medium Confidence* is statistically significant at every position[2]. The difference in accuracy after applying the *High Confidence* retraining rule versus the initial recognition accuracy is significant only at a few positions in both systems.

5   Conclusion

Applying semi-supervised learning to general handwriting recognition is a novel approach to reducing the human effort needed to create a recognizer for handwritten words. We demonstrated that it is possible to achieve a substantial increase in the recognition accuracy with unlabeled data. Furthermore, we underlined the importance of a suitable retraining rule for self-training. Several self-training experiments were performed for two different single word recognition scenarios, viz. recognition without a dictionary and recognition with a dictionary. In both cases, a significant improvement in accuracy can be observed if the retraining rule is suitably chosen. It turns out, however, that the retraining rule with the best performance is not the same in both cases. When self-training is performed in the scenario without a dictionary, the rule that includes all data performes best. This might indicate that for this setup some form of unsupervised learning takes place within the neural network. For recognition with a dictionary, the *Medium Confidence* retraining rule performed best. This indicates that recognition results with a low confidence might be changed too much due to the dictionary use so that their inclusion in the retraining set has an adverse effect on the recognition accuracy.

Future work will therefore investigate whether further improvements are possible with a retraining rule for the recognition with a dictionary that combines these results. All recognition results with a confidence below a certain threshold could be added directly (as the recognized label sequence) to the training set, while those words with a recognition confidence above the threshold are to be added after being corrected according to the dictionary via the CTC algorithm.

Another line of research to be pursued in the future are self-training experiments using Hidden Markov Models [1] and co-training experiments using both recognizers.

Acknowledgments

This work has been supported by the Swiss National Center of Competence in Research (NCCR) on Interactive Multimodal Information Management (IM2).

---

[2] A t-test for paired samples with $\alpha = 0.05$ was used.

References

1. Brakensiek, A., Rigoll, G.: Handwritten Address Recognition Using Hidden Markov Models. In: Dengel, A.R., Junker, M., Weisbecker, A. (eds.) Reading and Learning. LNCS, vol. 2956, pp. 103–122. Springer, Heidelberg (2004)
2. Palacios, R., Gupta, A., Wang, P.S.: Handwritten Bank Check Recognition Of Courtesy Amounts. Int'l Journal of Image and Graphics 4(2), 1–20 (2004)
3. Ye, M., Viola, P.A., Raghupathy, S., Sutanto, H., Li, C.: Learning to Group Text Lines and Regions in Freeform Handwritten Notes. In: Ninth Int'l Conf. on Document Analysis and Recognition, pp. 28–32. IEEE Computer Society Press, Los Alamitos (2007)
4. Govindaraju, V., Xue, H.: Fast Handwriting Recognition for Indexing Historical Documents. In: First Int'l Workshop on Document Image Analysis for Libraries, pp. 314–320. IEEE Computer Society Press, Los Alamitos (2004)
5. Bunke, H.: Recognition of Cursive Roman Handwriting - Past, Present and Future. In: Proc. 7th Int'l Conf. on Document Analysis and Recognition, August 2003, vol. 1, pp. 448–459 (2003)
6. Vinciarelli, A.: A Survey On Off-Line Cursive Word Recognition. Pattern Recognition 35(7), 1433–1446 (2002)
7. Chapelle, O., Schölkopf, B., Zien, A.: Semi-Supervised Learning. MIT Press, Cambridge (2006)
8. Zhu, X.: Semi-Supervised Learning Literature Survey. Technical Report 1530, Computer Science, University of Wisconsin-Madison (2005), http://www.cs.wisc.edu/~jerryzhu/pub/ssl_survey.pdf
9. Seeger, M.: Learning with Labeled and Unlabeled Data. Technical Report, University of Edinburgh, 5 Forest Hill, Edinburgh, EH1 2QL (2002)
10. Ji, S., Watson, L.T., Carin, L.: Semisupervised Learning of Hidden Markov Models via a Homotopy Method. IEEE Transactions on Pattern Analysis and Machine Intelligence 31(2), 275–287 (2009)
11. Inoue, M., Ueda, N.: Exploitation of Unlabeled Sequences in Hidden Markov Models. IEEE Transactions on Pattern Analysis and Machine Intelligence 25(12), 1570–1581 (2003)
12. Ball, G.R., Srihari, S.: Prototype Integration in Off-Line Handwriting Recognition Adaptation. In: Proc. Int'l. Conf. on Frontiers in Handwriting Recognition, pp. 529–534 (2008)
13. Frinken, V., Bunke, H.: Evaluating Retraining Rules for Semi-Supervised Learning in Neural Network Based Cursive Word Recognition. In: 9th Int'l Conference on Document Analysis and Recognition (accepted for publication) (2009)
14. Marti, U.-V., Bunke, H.: The IAM-Database: An English Sentence Database for Offline Handwriting Recognition. Int'l Journal on Document Analysis and Recognition 5, 39–46 (2002)
15. Marti, U.-V., Bunke, H.: Using a Statistical Language Model to Improve the Performance of an HMM-Based Cursive Handwriting Recognition System. Int'l Journal of Pattern Recognition and Artificial Intelligence 15, 65–90 (2001)
16. Graves, A., Liwicki, M., Fernández, S., Bertolami, R., Bunke, H., Schmidhuber, J.: A Novel Connectionist System for Unconstrained Handwriting Recognition. IEEE Transaction on Pattern Analysis and Machine Intelligence (accepted for publication)
17. Graves, A., Fernández, S., Gomez, F., Schmidhuber, J.: Connectionist Temporal Classification: Labelling Unsegmented Sequential Data with Recurrent Neural Networks. In: 23rd Int'l Conf. on Machine Learning, pp. 369–376 (2006)

18. Bertolami, R., Zimmermann, M., Bunke, H.: Rejection Strategies for Off-Line Handwritten Text Line Recognition. Pattern Recognition Letters 27(16), 2005–2012 (2006)
19. Blum, A., Mitchell, T.: Combining Labeled and Unlabeled Data with Co-Training. In: COLT 1998: Proc. of the 11th annual Conference on Computational Learning Theory, pp. 92–100. ACM, New York (1998)
20. Scudder, H.J.: Probability of Error of Some Adaptive Pattern-Recognition Machines. IEEE Transaction on information Theory 11, 363–371 (1965)
21. Nigam, K., Ghani, R.: Analyzing the Effectiveness and Applicability of Co-Training. In: 9th Int'l Conf. on Information and Knowledge Management CIKM, pp. 86–93 (2000)
22. Dempster, A.P., Laird, N.M., Rubin, D.B.: Maximum Likelihood from Incomplete Data via the EM Algorithm. Journal of the Royal Statistical Society, Series B 39(1), 1–38 (1977)

On a New Similarity Analysis in Frequency Domain for Mining Faces within a Complex Background

D.A. Karras

Chalkis Institute of Technology, Automation Dept. and Hellenic Open University,
Rodu 2, Ano Iliupolis, Athens 16342, Greece
dakarras@teihal.gr, dakarras@ieee.org, dakarras@usa.net

Abstract. A novel similarity analysis is presented in this paper for dealing with the problem of mining faces in a complex image background. The proposed approach integrates a robust feature extraction technique based on a specific method of eigenanalysis in the frequency domain of the unique classes identified in the problem at hand, with neural network based classifiers. Such an eigenalysis aims at identifying principal characteristics in the frequency domain of the above mentioned uniquely identified classes. Each unknown image, in the testing phase, is then, analyzed through a sliding window raster scanning procedure to sliding windows identified, through a first stage neural classifier, as belonging to one of the unique classes previously mentioned. After such a sliding window labeling procedure it is reasonable for a second stage neural classifier to be applied to the testing image viewed as a sequence of such labeled sliding windows for obtaining a final decision about whether a face exists within the given test image or not. Although the proposed approach is a hierarchical procedure, its most critical stage is the similarity analysis performed through eigenanalysis in the frequency domain, since, if good identification/ labeling accuracy could be then obtained, it would facilitate final face mining.

Keywords: face detection, eigenanalysis, neural classifiers, 2D-FFT, similarity.

1 Introduction

The main goal of this paper is the study of the problem of automatic detection and recognition of human faces inside a complex background. While a solution of such a problem could lead to many applications like security systems etc. this task is quite difficult to be solved. More specifically, its ultimate goal could be specified as follows. Given an image database with instances of human faces our aim is the development of a methodology which is able to receive as input an image of an unknown human face within a complex background and detect whether it corresponds to a face within the database. System's functioning can be considered as successful if it is independent of any change occurred due to positioning, orientation or any other kind of distortions of the input faces with respect to their stored instances. The achievement of this ultimate goal could be seen as a two stage process.

P. Perner (Ed.): ICDM 2009, LNAI 5633, pp. 301–309, 2009.
© Springer-Verlag Berlin Heidelberg 2009

First,

- Detection of human face existence within a complex background in a given image.

Second,

- Identification of this human face as being one matching another one –or not- stored within a given image database.

The present project has been focused on the solution of the former stage of the complete process. While the second stage problem is the one most cited in the literature the one herein investigated is rarely considered. **The main novelty of this paper is in the methodology developed to tackle it which is based on a type of eigenanalysis, which has not been previously investigated in the literature so far**. Eigenanalysis is considered in a different context for solving the second stage problem of face matching.

There is a multitude of research approaches in the literature for solving the problem of face recognition. A fairly good account of them can be found in [1] as well as in (NATO-ASI for face recognition, 1997). Some of the first algorithms use feature based techniques, like eye area etc., while other are based on template matching. In (Brunelli R. and Poggio T., 1992) there is a detailed consideration and comparison of feature based algorithms with the template based ones. Another approach relies on analyzing a face image according to an eigenface decomposition similar to the Karhunen-Loeve one (e.g Kirby M. and Sirovich L., 1990). This latter method is one of the most prevalent techniques for solving the problem. Other efforts are based on feature extraction methods applied to an image transform domain, like the wavelet domain (von der Malsburg C., 1997). Concerning the classification techniques that have been employed in this task, the most successful ones are discriminant analysis and Artificial Neural Networks (ANN) as indicated in (e.g NATO-ASI for face recognition, 1997).

All the above mentioned methods correspond to the second stage of the human face recognition process from images. On the other hand, the investigations for solving the human face detection problem within a complex background are in a preliminary stage of development. One can mention the multiresolution approach presented by (Huang T.S., 1997), where two contrasting algorithms have been evaluated. The first one was rule-based, having a hierarchy of three resolutions, with the target to detect candidate eye and mouth regions relying on the application of specific rules to the extracted edges. The second algorithm is a three step procedure where template matching with feature based techniques have been combined via hierarchical Markov Random Fields (MRF) and Maximum A posteriori Probability (MAP) estimation. Finally, the approach originated by Poggio T. at 1994 employing ANN (of the Multi-Layer Perceptron- MLP type) trained with sets of raw pixel intensity based patterns formed by applying a sliding window raster scanning the images. The complex background related patterns were randomly injected during the learning process, while the sets containing the patterns associated to face regions were unchanged during the same process. Other similar approaches in this line of research have been presented in [7]-[12].

2   The Proposed Similarity Analysis in the Frequency Domain Using an Eigenalysis Methodology

The main steps of the novel approach developed for the detection of human faces within a complex background can be specified as in the following paragraphs.

- Building of the image database containing human faces under various conditions concerning lighting, face orientation, noise introduced due to scanning operations, resolution level and background complexity. Both natural and artificial background has been considered. The artificial background has been introduced to simple passport photos through using the Adobe Photostyler software package. Therefore, both natural and synthetic photos comprise our face detection database containing 1040 images of dimensions 256 X 256.

In the following figure 1 a sample of the images comprising such a face database is shown

Fig. 1. A sample of the image database outlining the emphasis paid to the existence of a complex background in the phase images instead of the usually used segmented face images in the most common benchmark databases

- 700 images out of the total set of 1040 ones have served to train the proposed system which is based on the eigenanalysis approach, while the rest ones have served to test it. To be more specific, this system relies on the eigenvector analysis of each one of the nine important regions-classes in which a human face, lying within a complex background, can be decomposed: Left and right cheek, chin, eyes, ears, mouth, nose, forehead and background.
- Each database image has been manually decomposed into the above specified regions-classes by applying the Adobe Photostyler software package again. Therefore, every image has been decomposed into subimages of dimensions K X L, each containing only one face region-class. By applying the above specified approach we have formed training/test subimages for each face region-class as follows: Left Cheek (530/300), Right Cheek (590/260),

Chin (680/370), Eyes (1370/740), Ears (500/320), Mouth (710/400), Nose (700/400), Forehead (680/400), Background (4720/800 and 4720/2070). Thus, for the nine face class-regions we have formed a total of 10480/3990 and 10480/5260 training/test subimages.

- For each class-region i the following processing has been performed. Each subimage associated to such a class-region has been raster scanned via sliding windows of 32 X 32. Then, the 2-D FFT has been applied to each such window with 32 X 32 coefficients forming X_{32x32}, the corresponding 2D-FFT matrix.

- The autocorrelation matrix corresponding to that face class-region has, then, been estimated as follows. Each of these sliding windows k can be represented by a vector $X^{(ik)}_{32x32} = (x^{(ik)}_1, x^{(ik)}_2, x^{(ik)}_3, .. x^{(ik)}_{1024})$ and its transpose by $X^{(ik)T}_{32x32} = (x^{(ik)}_1, x^{(ik)}_2, x^{(ik)}_3, .. x^{(ik)}_{1024})^T$. Then, by applying the well known theory of autocorrelation matrices:

$$R^i = E(X^{(ik)}_{(32x32)} * X^{(ik)T}_{(32x32)}) = \frac{1}{N_i} \sum_k X^{(ik)}_{(32x32)} * X^{(ik)T}_{(32x32)} \tag{1}$$

where, R^i is the autocorrelation matrix for class-region i, E is the symbol denoting Expected value of a random variable and N_i is the total number of 32X32 sliding windows within the training set of subimages belonging to the class-region i.

- After defining the face class-region autocorrelation matrices, then, Principal Component Analysis (PCA) is applied to each of them. This process leads to the determination of the 1024 eigenvectors of each autocorrelation matrix. Due to the well known fact from signal processing theory that the eigenvectors corresponding to the largest eigenvalues of a matrix describe characteristics of the specific matrix and not more general traits of it, while the eigenvectors corresponding to its smallest eigenvalues are associated with noise, we have considered the 101 only eigenvectors, in the range 920-1020. Therefore, 909 such eigenvectors have been defined for all the nine face class-regions.

- During the training and test phase of the suggested system the following algorithm is applied. For each class-region i, every subimage of it (either training or test) is raster scanned with sliding windows of 40X40. Let us consider as Y_{40x40} one such window. There are 64 sliding windows Y_{32x32} within it. We can again consider its corresponding autocorrelation matrix R_{Y40x40} according to equation (1), where $N_i= 64$. Again, this analysis is performed as previously in the 2D-FFT domain.

An instance of this procedure is shown in the next figure 2

- For each such autocorrelation matrix R_{Y40x40} we consider its projections onto each one of the 909 previously defined eigenvectors corresponding to the nine face class-region autocorrelation matrices. Let us define $a_m = R_{Y40x40} * e_m$, the projection of R_{Y40x40} onto the eigenvector e_m, where m =1..909. Then, the inner product $< a_m, e_m>$ is a measure of similarity of the vectors a_m and e_m. Finally,

$$\cos\varphi_m = < a_m, e_m>/(\|a_m\|*\|e_m\|)$$

- is a coefficient suitable for expressing the deviation of the projection of R_{Y40x40} from e_m.

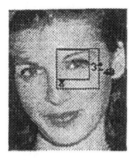

Fig. 2. Outline of a sliding window of 40 X 40 dimensions in the eye-class region. We view how 32 X 32 windows are positioned inside the 40 X 40 window to facilitate calculation of class-region autocorrelation matrix characteristics.

- All the 909 cosines $\cos\varphi_m$ previously defined comprise the feature vector associated with Y_{40x40}, which is input for the classifiers subsequently involved. The output of these classifiers is simply a scalar indicating whether an input vector belongs to a face region or not.
- Concerning the pattern recognition systems used at the classification stage of the proposed methodology several kinds of ANN, of the MLP type, have been employed. To be more specific, On-line Backpropagation, RPROP and Scaled Conjugate Gradient (SCG) are the training algorithms that have been compared [13]. Several different MLP architectures of the type 909-h1-h2-1 have been involved.

3 Experimental Study

Four methodologies have been compared in the experimental study conducted in this section. First, the traditional approach of feeding MLP classifiers with pixel intensity based input vectors. Second, the traditional eigenfaces technique [1] involving again MLP neural networks. Third, the suggested technique of feeding them with input vectors comprised of features obtained by performing the above defined similarity analysis based on eigenanalysis but in the pixel intensity domain. And, finally, the suggested technique of feeding MLP classifiers with input vectors comprised of features obtained by performing the above defined similarity analysis based on eigenanalysis but in the frequency domain as described in section 2.

Concerning the first approach, the best classification accuracy 79.57% for the test set has been obtained by using a 1024-30-30-1 MLP trained with RPROP.

Concerning the second approach, the best classification accuracy 82.12% for the test set has been obtained by using an MLP trained with RPROP.

Regarding the third approach, the best classification accuracy 85.24% for the larger test set and 92% for the smaller test set have been obtained by involving a 909-30-30-1 MLP trained with RPROP and On-line Backpropagation respectively.

Regarding the final approach, the best classification accuracy 87.12% for the larger test set and 96% for the smaller test set have been obtained by involving a 909-30-30-1 MLP trained with RPROP and On-line Backpropagation respectively, again, as in the third approach above.

In the next figures 3, 4, and 5 the evolution of training and test mean square errors is shown for the different learning algorithms involved: Online Backpropagation, Scaled Conjugate Gradient and RPROP [13], when the third and fourth above mentioned approaches are used.

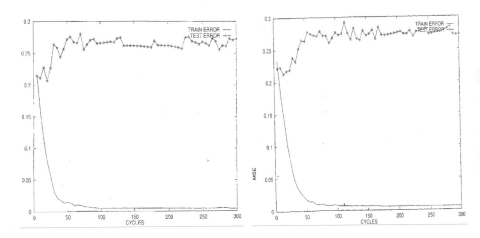

Fig. 3. Evolution of Training and Test set Mean square error versus learning cycles when On line Backpropagation is the training algorithm. In the left graph we see the results obtained by using similarity analysis in the frequency domain (fourth approach) while in the right we see the results obtained by using similarity analysis in the pixel domain (third approach). We can notice how smoother is the generalization error evolution in the case of similarity analysis in the frequency domain, which is the proposed approach.

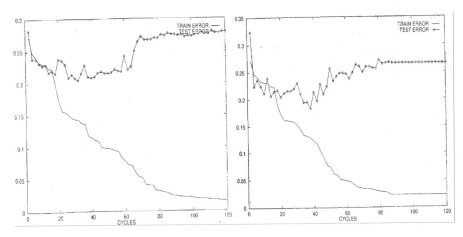

Fig. 4. Evolution of Training and Test set Mean square error versus learning cycles when SCG is the training algorithm. In the left graph we see the results obtained by using similarity analysis in the frequency domain (fourth approach) while in the right we see the results obtained by using similarity analysis in the pixel domain (third approach). We can notice again how smoother is the generalization error evolution in the case of similarity analysis in the frequency domain, which is the proposed approach.

From these figures it is obvious that the test error becomes minimum during the first learning cycles for all the algorithms involved. For instance, regarding the above runs, On line BP gets its minimum mean square error for the test set near 35 learning cycle, SCG near 40 learning cycle and RPROP near 20 learning cycle. Thus, although the task is quite complex due to the large input space, it is computationally feasible to involve neural classifiers to deal with it.

From these graphs we see that SCG is behaved well but not as expected concerning its generalization performance since conjugate gradients is the preferred method when large scale networks or benchmarks are involved. RPROP and Online Back propagation seem to achieve the best results. We can notice from the above graphs that the generalization error evolution is much smoother in the case of similarity analysis in the frequency domain, which is the proposed approach, than in the case of similarity analysis in the pixel domain. These promising results clearly illustrate the efficiency of the proposed face detection within a complex background system.

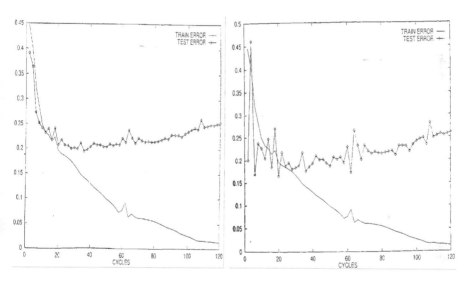

Fig. 5. Evolution of Training and Test set Mean square error versus learning cycles when RPROP is the training algorithm. In the left graph we see the results obtained by using similarity analysis in the frequency domain (fourth approach) while in the right we see the results obtained by using similarity analysis in the pixel domain (third approach). We can notice again how smoother is the generalization error evolution in the case of similarity analysis in the frequency domain, which is the proposed approach.

4   Conclusions and Future Prospects

Automated Face segmentation from natural images is known to be a difficult task, depending on the background complexity. However, without good face segmentation there is difficulty in designing efficient face identification systems performing well in images with random backgrounds. A new methodology is herein presented for dealing with the problem of face segmentation within a complex background. The novelty of the proposed approach lies on the feature extraction technique involved, which is based on similarity analysis using eigenanalysis in the frequency domain of the 9 unique classes identified in the problem at hand after experimentation but, also via the literature. One important class out of them is of course the background while the rest are specific face class regions. The proposed approach integrates this feature extraction methodology with neural network based classifiers. Such an eigenalysis aims at identifying principal characteristics of the above mentioned uniquely identified classes. Each unknown image, in the testing phase, is then, analyzed through a sliding window raster scanning procedure to sliding windows to be labeled, through a first stage neural classifier, as belonging to one of the 9 unique classes previously mentioned.

In order, however, to finally conclude about face existence or non existence in a given image, a second stage neural classifier could be applied to the resulted sequence of such labeled sliding windows. We have experimented with the first stage of such an hierarchical system and we have illustrated that it is a computationally feasible task. In addition, neural classifiers generalization performance is quite promising and moreover, we have illustrated that the proposed feature extraction methodology outperforms more traditional rival techniques. A future research effort should incorporate the second stage classification task above mentioned, in the proposed hierarchical face segmentation system. In addition, we have investigated the performance of several different MLP neural classifiers.

References

[1] Heisele, B., Ho, P., Wu, J., Poggio, T.: Face recognition: component-based versus global approaches. In: Computer Vision and Image Understanding, vol. 91, pp. 6–21. Academic Press, London (2003)

[2] NATO-ASI workshop, "Face Recognition: From Theory to Applications", Scotland, UK, June 23-July 4 (1997)

[3] Brunelli, R., Poggio, T.: Face Recognition: Features versus Templates. Technical report TR 9110-04, Istituto per la Ricerca Scientifica e Technologica (October 1992)

[4] Kirby, M., Sirovich, L.: Application of the Karhunen-Loeve procedure for characterization of human faces. IEEE Trans. On Pattern Analysis and Machine Intelligence 12(1), 103–108 (1990)

[5] von der Malsburg, C.: A Face Recognition System Based on the Morlet Wavelet Transform. In: NATO-ASI workshop, Face Recognition: From Theory to Applications, Scotland, UK, June 23-July 4 (1997)

[6] Huang, T.S.: Face and Facial Feature Detection. In: NATO-ASI workshop, "Face Recognition: From Theory to Applications", Scotland, UK, June 23-July 4 (1997)

[7] Cootes, T.F., Taylor, C.J.: Constrained active appearance models. In: International Conference on Computer Vision, pp. 748–754 (2001)

[8] Fleuret, F., Geman, D.: Coarse-to-fine visual selection. Int. Journal of Computer Vision 41(2), 85–107 (2001)

[9] Rowley, H., Baluja, S., Kanade, T.: Neural network-based face detection. IEEE Transactions on Pattern Analysis and Machine Intelligence 20(1), 23–38 (1998)

[10] Osuna, E., Freund, R., Girosi, F.: Training support vector machines: an application to face detection. In: IEEE Int Conference on Computer Vision and Pattern Recognition, pp. 130–136 (1997)

[11] Rowley, H.A., Baluja, S., Kanade, T.: Human face detection in visual scenes. CMU-CS-95-158R, Carnegie Mellon University (November 1995),
http://www.ri.cmu.edu/pubs/pub_926_text.html
[12] Poggio, T., Sung, K.K.: Example-based learning for view-based human face detection. In: Proc. of the ARPA Image Understanding Workshop, pp. II:843–850 (1994),
http://citeseer.nj.nec.com/poggio94examplebased.html
[13] Haykin, S.: Neural Networks, A Comprehensive Foundation, 2nd edn. Prentice Hall, Englewood Cliffs (1999)

Clustering with Domain Value Dissimilarity for Categorical Data

Jeonghoon Lee[1,*], Yoon-Joon Lee[1,**], and Minho Park[2]

[1] School of EECS, Division of Computer Science,
Korea Advanced Institute of Science and Technology,
Daejeon 350-701, Republic of Korea
{jhoon,yoonjoon.lee}@kaist.edu
http://dbserver.kaist.ac.kr/~leejh
[2] Information Technology Department, The Bank of Korea,
Seoul 135-080, Republic of Korea
minhopark@bok.or.kr

Abstract. Clustering is a representative grouping process to find out hidden information and understand the characteristics of dataset to get a view of the further analysis. The concept of similarity and dissimilarity of objects is a fundamental decisive factor for clustering and the measure of them dominates the quality of results. When attributes of data are categorical, it is not simple to quantify the dissimilarity of data objects that have unimportant attributes or synonymous values. We suggest a new idea to quantify dissimilarity of objects by using distribution information of data correlated to each categorical value. Our method discovers intrinsic relationship of values and measures dissimilarity of objects effectively. Our approach does not couple with a clustering algorithm tightly and so can be applied various algorithms flexibly. Experiments on both synthetic and real datasets show propriety and effectiveness of this method. When our method is applied only to traditional clustering algorithms, the results are considerably improved than those of previous methods.

Keywords: Clustering, Data Mining, Categorical Data, Similarity, Dissimilarity, Domain Value, Co-occurrence.

1 Introduction

The grouping problem has become increasingly important for data mining or knowledge discovery and data analysis in recent days. Large datasets of retail transaction, credit card usage, network stream, and surveillance data embed

---

* Corresponding author.
** Supported by the Korea Science and Engineering Foundation (KOSEF) grant funded by the Korea government (MOST) (No. R01-2007-000-20135-0), and the Ministry of Knowledge Economy, Korea, under the Information Technology Research Center support program supervised by the Institute of Information Technology Advancement (grant number IITA-2009-C1090-0902-0031).

immense information. The grouping methods make it possible to simplify a large set of data for analysis and to find hidden relationship among them [1].

Clustering is a grouping process of identify homogeneous data objects and group them into cluster a.k.a. class based on the values of their attributes [2]. The data objects in the same cluster share similar properties at most and objects in the different clusters do at least. Clustering is called "Unsupervised Grouping and Learning" because it does not use any data to guide how the clusters are built. It plays an important role in a variety of field from discovery of knowledge to prediction of unseen phenomenon and can be a useful preprocess for other purposes like data summarization as well.

Numerical data is the object of traditional clustering in nature of number system. In the late 90s, prosperity of IT technologies caused the expansion of data mining field. In the result, not only numeric data but also categorical data such as text and market basket became a significant subject for knowledge processing.

Domain of categorical attribute is described by a set of descriptive values. These values are dichotomous or discrete. Dichotomous attribute has only two values, for example, the attribute Gender whose values can be male or female. In special case, this attributes can have binary values, 0 and 1. For example, an attribute, which indicates whether a customer bought a specific item or not, can be represent the two cases is 0 (did not buy) and 1 (did buy). Discrete attribute has a certain number of values, for example, the attribute Color where values can be white, blue, red, yellow, and so on, or the variable can only take values 1, 2, 3, or 4 [1].

Similarity of objects is a fundamental decisive factor of clustering and evaluated by the distance between objects in general. The distance measure dominates the quality of clustering. Understanding characteristics of data is essential to select an appropriate distance measure. Based on characteristics of categorical attributes, it is the focus of this paper to find out intrinsic distance for such values.

In this paper, we propose a dissimilarity[1] measure referred to as DVD (Domain Value Dissimilarity) to represent intrinsic difference of categorical values. Our measure is independent on a specific algorithm, and can be applied to both partitioned clustering algorithm and hierarchical clustering algorithm. DVD can be in a wide use even for and information retrieval method using distance measure.

The rest of the paper is structured as follows. In Section 2, we describe problems of traditional similarity measures and our motivation. In Section 3, we describe how to estimate intrinsic difference of values based on the information of data groups related with the values and quantify the Value Differences (VD) of them. Then we present DVD measure for data object with categorical attributes. In Section 4, we show a simple sketch to apply DVD measure based on the Value Difference to clustering process. We empirically evaluate propriety of DVD measure with synthetic data comparing to traditional measure system for

---

[1] We consider dissimilarity as distance and opposite concept of similarity.

categorical values and quality of DVD with real data in Section 5. Finally, in Section 6, we summarize our research results.

2   Motivation and Problems of Traditional Similarity Measure

The distance of numeric data is very descriptive and intuitive based on geometric analogies. A variety of distance measures for numerical data have been researched and used for a long time but in applied area like informational retrieval not only in fundamental area like mathematics and statistics. Generally, the distance of them is considered in the Euclidean space $\mathbb{R}^{n}$[2].

The distance of numeric data is dependent on the semantic quantity of the attribute values. In Fig. 1, the distance between O and A is closer than that between O and B.

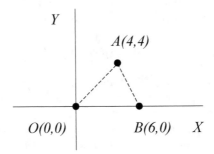

Fig. 1. Distance in a 2 dimensional Euclidean space

The problem of similarity becomes more complicate with categorical attributes, which is referred to attributes of non-numeric value because there is no inherent distance measure between values in them. The research of similarity among objects with categorical values has a long history. Starting with Pearsons *chi-square*[3] in the late 19th century, various similarity measures has been devised. Nevertheless, the previous traditional clustering works with categorical data [3][4][5] generally used the similarity concept of dichotomy, because of its simplicity and convenient for use. They adopted similarity measures like *Simple Matching Coefficient*[4] and *Jaccard Coefficient*[5] [6], which considers values of categorical attributes as binary. Though some research developed applied measures

---

[2] The distance between two objects $x = (x_1, x_2, \cdots, x_n)$ and $y = (y_1, y_2, \cdots, y_n)$ is defined with norm, r, as $\left(\sum_{i=1}^{n} | x_i - y_i |^r \right)^{\frac{1}{r}}$. Especially, 2-norm distance is known as Euclidean distance.

[3] The chi-square test is a statistical test procedure to understand if a relationship exists between two categorical variables.

[4] Simple Matching Coefficient is defined as $\frac{the\ number\ of\ attributes\ with\ matching\ values}{the\ number\ of\ total\ attributes}$.

[5] Jaccard Coefficient is defined as $\frac{the\ number\ of\ matching\ presences}{the\ number\ of\ attributes\ not\ involved\ in\ null\ matches}$.

as "link" of ROCK [4], which is the number of common neighbors between data items, however it is still based on dichotomy. These measures are based on the distinctness of objects such that two objects either have the same value (1) or do not (0). This excessive simplification of the data characteristics makes a loss of the important knowledge of data in the result.

For objects with discrete attributes (i.e., multi-valued categorical attributes), the distance concept is more complicated because information about the scale between values should be taken into account. Assuming that the quality of an item is measured by an attribute with the values {poor, ok, good, excellent}, it seems reasonable that an item I1, that has "excellent" value, would be closer to an item I2, that has "good" value, than it would be to an item I3, that has "ok" value. This observation is generally transformed into quantitative scaled system as follows.

When the values of attribute have order, they are often mapped to a series of integers, e.g., {poor=0, ok=1, good=2, excellent=3}. Then the distance between I1 and I2, $d(I1, I2)$, is $3 - 2 = 1$ whereas that between I1 and I3, $d(I1, I3)$, is $3 - 1 = 2$. When n ordinal values are mapped to integer 0 to $n - 1$, the distance can be normalized with the divider, $n - 1$, to be fallen between 0 and 1.

This approach for similarity and dissimilarity of multi-valued categorical attributes assumes equal intervals, so would seem unnatural. In other words, assumption that differences between adjacent values are all the same is not exactly true because it could not be said that the difference between poor and ok is the same with the difference between good and excellent.

This problem becomes more severe for multi-valued categorical attributes without order. Assuming that a kind of vehicle possessed by customers is one of customers wealth indexes, its values can be various kinds of car from small-sized car to luxury vehicle, for example, {CL550, SL550, E350, CLK550, ... }[6].

In binary mapping system, the similarity between items is represented with 1 (are same) and 0 (are different). In this system, the similarity between CL550 and SL550 is 0 and that between CL550 and E350 is also 0. In other word, it can say that the dissimilarity between CL550 and SL550 is as same as that between CL550 and E350.

In fact, the wealth index of customer who bought CL550 is closer to that of one who bought SL550 than that of one who bought E350. It is true that there is intrinsic order scaled among categorical values which are not represented in binary mapping system. To measure the similarity and dissimilarity among data with them correctly, a method to quantify intrinsic difference among categorical values is required.

Some studies have been suggested to compute distance between values of attributes [7][8]. Their performance, however, are not yet to be good enough in terms of effectiveness. We'll compare the results of [8] with those of our approach by experiments in Section 5.

---

[6] Values are product model numbers of Mercedes Benz[(TM)] company. CL550 and SL550 are available at about $100,000, and E350 and CLK550 are available about $50,000 (http://www.mbusa.com/mercedes/#/vehiclesMenu/)

Recently, another approaches to clustering categorical data can be found in [9][10]. Both of them are adopted concept of Entropy based on Information Theory. Intuitively, well-grouped clusters are to have low entropy in consequence. These methods put relationship information of objects to use assuming attributes are independent on one another tacitly rather than find out intrinsic difference of values specifically.

Some other probabilistic and information-theoretic similarity measures for categorical data were comparably evaluated in a recent research [11]. Each measure compared in [11] adopted its own data-driven characteristics and it was concluded that there is no one best performing similarity measure for all kinds of data. From this point of view, our research focuses on data objects with multi-valued categorical attributes whose values are not ordered.

3   Value Difference of Categorical Attributes

Ideally, experts who have full domain knowledge about categorical attribute could find out the exact difference of values. However, help of expert cannot be always expected practically, so induction from some empirical heuristics is required. We present how to measure difference between two categorical values in this section. To approximate difference of values, our research is based on the following hypothesis.

Hypothesis. If an attribute i is dominant for clustering and two values of the attribute i, m and n have distinguished difference with each other, a data group including the value, m for the attribute i has different characteristics from a data group including the value, n for the attribute i.

Our hypothesis means that the groups may be distinguished by the difference among the values of their dataset. Based on the hypothesis, we try to estimate the difference between two values, called Value Difference (VD). We start with defining attribute domain D, the set of distinct values on a given attribute.

Definition 1. *Let* $\mathcal{D} = \{D_1, D_2, \ldots, D_k\}$ *be a set of bounded domains and* $\mathcal{S} = D_1 \times D_2 \times \cdots \times D_k$ *an k-dimensional non-numerical space. We refer to* $D_1, \ldots, D_k$ *as the attributes of* $\mathcal{S}$*. An attribute domain,* D_i*, is a set of all distinct values occurring in the attribute* i *and is defined as follows.*

$$D_i = \left\{ d_{i1}, d_{i2}, \ldots, d_{i|D_i|} \right\} \ . \tag{1}$$

The dataset consists of a set of k-dimensional objects $U = \{u_1, u_2, \ldots, u_p\}$ where $u_l = \{u_{l1}, u_{l2}, \ldots, u_{lk}\}$. The ith element of u_l is drawn from the attribute domain D_i. The order in an attribute domain is only for itemizing elements of set and has no significance of relative relationship of them.

Let's take an example from Mushroom dataset of UCI Machine Learning Repository[7]. Data has 22 categorical attributes and an attribute, "cap-color" which represents the color of a mushroom cap. There are 10 distinct values in "cap-color": b(buff), c(cinnamon), e(red), g(gray), n(brown), p(pink), r(green), u(purple), w(white), y(yellow), and the value domain of "cap-color" is as follows. $D_{cap-color} = \{$buff, cinnamon, red, gray, brown, pink, green, purple, white, yellow$\}$.

Suppose that there is a group of data objects which have a common value α for an attribute A, we can count the occurrence of values of a specific attribute B in the group. This count-distribution information contains some characteristics of data objects with the value α for the attribute A with respect to the attribute B. We form a vector with this count-distribution information, called a Group Occurrence Vector (GOV) of an attribute A's value V with respect to an attribute B. GOV can be formalized as follows.

Definition 2. *A Group Occurrence Vector (GOV) of a value m in attribute i, d_{im} with respect to an attribute j is a vector whose elements are the occurrence counts of d_{im} and each element in D_j together, respectively. When $g^j_{d_{im}n}$ is the count of simultaneous occurrence of d_{im} and d_{jn} in the group of objects which have a value m for an attribute i,*

$$g^j_{d_{im}} = \left(g^j_{d_{im}1}, g^j_{d_{im}2}, \dots, g^j_{d_{im}|D_j|} \right) \ . \tag{2}$$

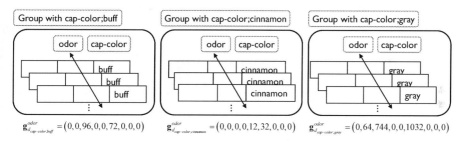

Fig. 2. GOVs of "buff", "cinnamon", "gray" of attribute "cap-color" with respect to "odor"

GOV of a specific value represents the characteristics of the value in dataset using the distribution of the other attribute's values. For example, an attribute "odor" of Mushroom data has 9 values: almond, creosote, foul, anise, musty, none, pungent, spicy and fishy. GOVs of "buff", "cinnamon", "gray" of attribute "cap-color" for "odor" are (0, 0, 96, 0, 0, 72, 0, 0, 0), (0, 0, 0, 0, 12, 32, 0, 0, 0),

---

[7] The UCI Machine Learning Repository is a collection of databases, domain theories, and data generators that are used by the machine learning community for the empirical analysis of machine learning algorithms
(http://archive.ics.uci.edu/ml/index.html).

(0, 64, 744, 0, 0, 1032, 0, 0, 0) respectively. As shown in Fig. 2, GOV has feature information about the group of objects sharing a common value and this information can be used to estimate the characteristics of the value shared.

The difference between two GOVs can represent that of two values in an attribute by our hypothesis. Observing the characteristics of GOV, it is sparse since it has relatively few non-zero attributes. The sparsity can be removed by normalization and the similarity of GOV should not depend on the number of shared 0 values. We adopt Cosine Similarity, which is independent on sparsity and handles non-binary vectors. Using Cosine Similarity, Value Difference of two values in a value domain of an attribute could be figured out.

Definition 3. *Suppose the GOVs of two value of an attribute i, d_{im} and d_{in} with respect to an attribute j are $\boldsymbol{g}^j_{d_{im}}$ and $\boldsymbol{g}^j_{d_{in}}$ respectively, The Value Difference (VD) between the two values withe respect to j can be defined as follows. When $\cdot$ indicates the vector dot product, $\boldsymbol{g}^j_{d_{im}} \cdot \boldsymbol{g}^j_{d_{in}} = \sum_{\alpha=1}^{|D_j|} g^j_{d_{im}\alpha} \times g^j_{d_{in}\alpha}$, and $\|\boldsymbol{g}^j_{d_{im}}\|$ is the magnitude of vector $\boldsymbol{g}^j_{d_{im}}$, $\|\boldsymbol{g}^j_{d_{im}}\| = \sqrt{\sum_{\alpha=1}^{|D_j|} \left(g^j_{d_{im}\alpha}\right)^2}$,*

$$v^j_{d_{im},d_{in}} = 1 - \frac{\boldsymbol{g}^j_{d_{im}} \cdot \boldsymbol{g}^j_{d_{in}}}{\|\boldsymbol{g}^j_{d_{im}}\| \|\boldsymbol{g}^j_{d_{in}}\|} \ . \tag{3}$$

The value difference $v^j_{d_{im},d_{in}}$ has value between 0 and 1. This shows how much the two distributions of the attribute j's values with respect to two values of attribute i, d_{im} and d_{in} are dissimilar. If the distributions of values are similar, then $v^j_{d_{im},d_{in}}$ converges on 0, and this means that d_{im} and d_{in} are close with respect to an attribute j. If they are dissimilar, then $v^j_{d_{im},d_{in}}$ converges on 1. Now, the value difference between two values in an attribute i with respect to the other attributes are drawn by summation of each VD as follows. For k-dimensional objects,

$$v_{d_{im},d_{in}} = \sum_{j=1, j\neq i}^{k} \frac{v^j_{d_{im},d_{in}}}{\delta} \ . \tag{4}$$

The dissimilarity of values is represented by the summation of each value difference with respect to the other attributes and it is affected by the number of attributes. The parameter δ is a normalization factor to eliminate this dimensional influence and can be selected in the range from 1 to k. If δ is 1, the maximum of dissimilarity converges on k and it could reflect intrinsic distance at most, whereas it converges on 1 for δ as k.

For example, VDs of "buff" and "cinnamon", "buff" and "gray" with respect to an attribute "odor" are 0.438 and 0.047 respectively. VDs of them with a normalization factor δ as 1 for all attributes are 6.727 and 3.888 respectively. The mushroom data has small number of values missed and the dimensional influence is not conspicuous. These VDs mean that the value "buff" is about twice more dissimilar with the value "cinnamon" than with the value "gray".

So far we derived the dissimilarity of values in an attribute from the distribution of other attributes values based on the hypothesis mentioned formerly. Now the dissimilarity of objects can be measured by the value difference in each attribute.

Definition 4. *There are two k-dimensional objects $x = (x_1, x_2, \ldots, x_k)$ and $y = (y_1, y_2, \ldots, y_k)$ and the value difference v_{x_i, y_i} between two attribute domain values, x_i and y_i. The Domain Value Dissimilarity (DVD) function to measure distance between x and y is defined as follows.*

$$f(x, y) = \sum_{i=1}^{k} v_{x_i, y_i} \; . \tag{5}$$

Now, the difference of any objects in Mushroom dataset can be measures using DVD. For example, let two data objects, x and y be $x =$(x, s, n, t, p, f, c, n, k, e, e, s, s, w, w, p, w, o, p, k, s, u) and $y =$(x, s, y, t, a, f, c, b, k, e, c, s, s, w, w, p, w, o, p, n, n, g) selected from dataset. The difference of x and y is measured as $(0 + 0 + 5.676 + 0 + 3.498 + 0 + 0 + 4.655 + 0 + 0 + 6.283 + 0 + 0 + 0 + 0 + 0 + 0 + 0 + 0 + 0.010 + 1.970 + 2.358) = 24.45$ by DVD with $\delta = 1$.

DVD is a distance measure using VD and can reflect intrinsic dissimilarity among values in attribute domains and quantify how dissimilar an categorical object is to the other one effectively. DVD can be applied to various application based on dissimilarity such as clustering, k-nearest neighor finding, outlier detection. In this paper, clustering is our main application of DVD because clustering is a representative unsupervised grouping process and our measuring process does not assume any supervised knowledge too.

4   Clustering with Domain Value Dissimilarity

In this section, we describe the clustering process using DVD. We focused on devising dissimilarity measure, which can be applied to any of partitioned and hierarchical clustering algorithms, instead of developing new algorithm. So DVD can be applied to general clustering process as shown in Fig. 3.

After constructing GOVs for each values appearing in dataset, VDs of them are figured out. In main clustering process, DVD is employed as distance measure for objects.

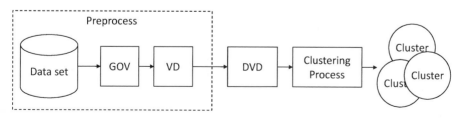

Fig. 3. Overview of clustering steps using DVD

GOV is constructed from co-occurrence information of domain values of attributes. Co-occurrence information for k-dimensional space is recorded into $_kC_2$ matrices during scanning a whole dataset once, for one matrix is for two attributes respectively. Using GOV information for each attribute, VDs of all domain values are figured out and is stored in VD matrices. There are generated k VD matrices for dataset k-dimensional space. Each VD matrix contains the VD of all domain values in a specific attribute.

Constructions of GOVs and VD matrices are conducted during preprocessing time. After that, DVD of objects is calculated based on VD matrices when dissimilarity measurement is required in any clustering algorithm.

5   Experimental Evaluation

We conducted experiments designed to show the effectiveness of DVD in terms of accuracy. DVD is a dissimilarity measure. Its effectiveness is evaluated in cooperated with traditional prototype-based algorithms, partitioning and hierarchical clustering methods. We compare DVD measure with Jaccard Coefficient which is a general measure for categorical attributed value.

For partitioning method, k-modes [12] is adequate to data with categorical attribute rather than k-means because k-means uses the average or mean values of all data objects in a same group for constructing a prototype of the group, i.e. cluster. The average value is, however, irrelevant to categorical value. In k-modes, the prototype of a cluster consists of the most frequent values in the cluster and this is more suited to process data with categorical values.

Our approach was evaluated on synthetic and real datasets. Firstly, we evaluated our approach being applied to hierarchical and partitioning clustering methods on a synthetic dataset. Secondly, we evaluated the performance of our approach on Mushroom dataset in Machine Learning Repository and analyzed the result. Finally, we compared the results of ours with those of the previous study on Wisconsin breast cancer dataset and Letter recognition dataset[8].

A. *Synthetic dataset*: We generated a synthetic dataset with 10,000 objects of 10-dimensional space $S = D_1 \times D_2 \times \cdots \times D_{10}$. Each attribute of data has one of 9 relative values (i.e., $|D_i| = 9$ for $1 \le i \le 10$) and the whole dataset have been grouped into 3 clusters as shown in Table 1. To guarantee the effect of relative importance among attributes, the values of half attributes $(D_1, D_2, \ldots, D_5)$ are randomly selected in each attribute domain respectively. Attributes with values selected randomly are independent on cluster. The values of the other half attributes $(D_6, D_7, \ldots, D_{10})$ are selected from designated value sets with size of 3 for each cluster in order to preserve inter choice relationship depending on cluster among values selected. In practice, the value of attribute i ($6 \le i \le 10$) is selected in the subset of each attribute domain: $\{d_{i1}, d_{i2}, d_{i3}\}, \{d_{i4}, d_{i5}, d_{i6}\}, \{d_{i7}, d_{i8}, d_{i9}\}$ for Cluster A, B, C respectively.

From the result of k-modes clustering algorithm with our approach and the traditional measure on the synthetic dataset, the confusion matrix is constructed

---

[8] http://www.sgi.com/tech/mlc/db/

Table 1. Synthetic dataset with size of 10,000

Cluster	Dim size	Dataset size
A	10	3329
B	10	3256
C	10	3415

Table 2. Confusion matrix for k-modes clustering on synthetic dataset

	DVD			Jaccard		
	1	2	3	1	2	3
A	3329	0	0	1897	1904	665
B	0	3256	0	880	58	2658
C	0	0	3415	552	1294	92

Table 3. Confusion matrix for hierarchical clustering on synthetic dataset

	DVD			Jaccard		
	1	2	3	1	2	3
A	3329	0	0	111	51	3362
B	0	3256	0	36	3160	0
C	0	0	3415	3182	45	53

as shown in Table 2. The column of confusion matrix is the original cluster of dataset and the row is the cluster acquired from experiment. In confusion matrix, the ideal result is that only one cell has non-zero value and all the others have zero value. The method using Jaccard Coefficient produced inaccurate results even if we select the best of several trial, but the clustering with DVD shows more accurate result.

This result can be also found from hierarchical clustering algorithm in Table 3. Jaccard Coefficient got about 3% false alarm, but DVD result in the ideal result. The result explains that DVD metric can reflect intrinsic similarity and dissimilarity between categorical values properly. Considering VD of attributes, appropriateness of DVD can be forseen easily.

Table 4 and Table 5 are VD matrices of the 5th attribute and the 6th attribute of the synthetic dataset respectively. The 5th attribute is one of attributes whose values are distributed randomly whereas some values of the 6th attribute are correlated with on another. The VDs of all values in the 5th attribute shown in Table 4 are almost same and this means these values of the attribute are meaningless to distinguish objects. The VDs of values in the 6th attribute shown in Table 5 are remarkable difference with one another. There could be, however, found three pairs of 3 close values: Value No 1, 2, 3, No 4, 5, 6, and No 7, 8, 9 respectively. The pairs of close values like this could be also found in the

7th, 8th, 9th, and 10th attributes. Their distribution was designed in dataset generation considering cluster in advance. VDs of correlated values are small but those of the others are large. The values with large VDs, (i.e., not correlated) should dominate to distinguish groups of objects in our DVD. Our method could measures the difference of values based on their correlation properly and cluster the dataset successfully in the result.

Table 4. Value Difference Matrix of the 5th attribute of the sysnthetic dataset with $\delta = 1$

Value No	1	2	3	4	5	6	7	8	9
1	0.00	0.07	0.06	0.08	0.06	0.05	0.06	0.06	0.06
2	0.07	0.00	0.06	0.06	0.05	0.06	0.05	0.07	0.07
3	0.06	0.06	0.00	0.07	0.06	0.05	0.06	0.06	0.06
4	0.08	0.06	0.07	0.00	0.09	0.05	0.07	0.09	0.06
5	0.06	0.05	0.06	0.09	0.00	0.07	0.08	0.06	0.07
6	0.05	0.06	0.05	0.05	0.07	0.00	0.07	0.07	0.06
7	0.06	0.05	0.06	0.07	0.08	0.07	0.00	0.07	0.04
8	0.06	0.07	0.06	0.09	0.06	0.07	0.07	0.00	0.06
9	0.06	0.07	0.06	0.06	0.07	0.06	0.04	0.06	0.00

Table 5. Value Difference Matrix of the 6th attribute of the sysnthetic dataset with $\delta = 1$

Value No	1	2	3	4	5	6	7	8	9
1	0.00	0.04	0.04	4.04	4.05	4.03	4.03	4.04	4.03
2	0.04	0.00	0.04	4.03	4.04	4.03	4.04	4.03	4.04
3	0.04	0.04	0.00	4.04	4.03	4.03	4.04	4.04	4.04
4	4.04	4.03	4.04	0.00	0.04	0.04	4.04	4.04	4.04
5	4.05	4.04	4.03	0.04	0.00	0.04	4.04	4.05	4.04
6	4.03	4.03	4.03	0.04	0.04	0.00	4.04	4.04	4.02
7	4.03	4.04	4.04	4.04	4.04	4.04	0.00	0.04	0.04
8	4.04	4.03	4.04	4.04	4.05	4.04	0.04	0.00	0.05
9	4.03	4.04	4.04	4.04	4.04	4.02	0.04	0.05	0.00

B. *Mushroom dataset*: Each object of dataset is a description of gilled mushrooms in the Agarigus and Lepiota Family. This dataset has 22 attributes of physical characteristics (e.g., cap-shape, cap-surface, cap-color, bruises, odor) and all values of them are categorical, for example, the values of "cap-shape" attribute are *bell, conical, convex, flat, knobbed, and sunken*. This dataset is composed of 8,124 edible and poisonous mushrooms corresponding to 23 species and The number of edible and poisonous objects are 4208 and 3916 respectively. The mushroom data has two groups, edible and poisonous on a large scale but, furthermore, each groups can be divided into smaller several groups due to species. We clustered the mushroom data into 20 groups as ROCK did [4]. The method with Jaccard Coefficient has about 5.5% (447 of 8124) false alarm as shown in

Table 6. Clustering Result on Mushroom Data for Traditional Similarity Measure

Hierarchical Algorithm based on Jaccard Coefficient					
Cluster No	Eidible	Poisonous	Cluster No	Edible	Poisonous
1	1728	4	11	705	256
2	1	1296	12	0	295
3	50	288	13	192	0
4	196	0	14	192	0
5	0	623	15	32	72
6	0	595	16	48	0
7	0	211	17	48	8
8	376	0	18	0	68
9	528	0	19	16	0
10	96	192	20	0	8

Table 7. Clustering Result on Mushroom Data for DVD Measure

Hierarchical Algorithm based on DVD					
Cluster No	Eidible	Poisonous	Cluster No	Edible	Poisonous
1	0	1296	11	0	192
2	1732	0	12	192	0
3	192	0	13	192	0
4	768	0	14	0	72
5	288	0	15	96	0
6	0	288	16	48	0
7	512	0	17	48	0
8	0	1728	18	39	0
9	96	8	19	0	36
10	0	256	20	5	40

Table 6 and one with DVD measure has only about 0.2% (13 of 8124) false alarm as shown in Table 7. This is superior to ROCK.

C. *Wisconsin breast cancer dataset*: The dataset consists of 699 objects and each object has 9 attributes. The attributes have integer values but they are neither continuous nor ordered. So they can be handled as categorical values. It has two groups, *Benign* (458 objects) and *Malignant* (241 objects). The results for k-modes with our approach and AD07 [8] is as shown in Table 8. Though AD07 groups 538 objects (77%) in correct clusters, our approach with DVD measures does 634 objects (90.7%) in correct cluster.

D. *Letter recognition dataset*: This dataset consists of 20000 objects and each object represents one of the 26 capital letters in the English alphabet with 16 attributes. As the study of [8], we take data objects with similar looking letters, E and F from the dataset. The extracted subset of data consists of 1543 objects: 768 objects of E and 775 objects of F. In the result, our approach shows

Table 8. Confusion matrix for k-modes clustering on Wisconsin breast cancer dataset

	DVD		AD07	
	Benign	*Malignant*	*Benign*	*Malignant*
1	7	183	130	210
2	451	58	328	31

Table 9. Confusion matrix for k-modes clustering on Letter recognition dataset (E and F alphabets)

	DVD		AD07	
	E	*F*	*E*	*F*
1	677	71	708	203
2	91	704	60	572

Table 10. Value Difference Matrix of "cap-color" attribute of the mushroom dataset with $\delta = 1$

Value	b	c	e	g	n	p	r	u	w	y
b	0.000	6.727	5.664	3.888	4.501	3.216	7.955	7.955	3.332	5.618
c	6.727	0.000	5.273	5.756	4.574	4.562	6.938	6.938	6.117	6.988
e	5.664	5.273	0.000	3.846	0.522	4.401	6.615	6.615	5.504	6.376
g	3.888	5.756	3.846	0.000	2.898	4.038	6.129	6.129	2.942	1.599
n	4.501	4.574	0.522	2.898	0.000	3.314	5.972	5.972	3.330	5.676
p	3.216	4.562	4.401	4.038	3.314	0.000	6.708	6.708	3.005	6.145
r	7.955	6.938	6.615	6.129	5.972	6.708	0.000	0.000	7.501	7.249
u	7.955	6.938	6.615	6.129	5.972	6.708	0.000	0.000	7.501	7.249
w	3.332	6.117	5.504	2.942	3.330	3.005	7.501	7.501	0.000	4.999
y	5.618	6.988	6.376	1.599	5.676	6.145	7.249	7.249	4.999	0.000

considerable improvement with 89.5% correct grouping result while AD07 [8] gets 83% correct answer as shown in Table 9. The result of traditional measure (Jaccard Coefficient) on this dataset was worse than that of AD07.

The result of real data is caused that DVD can discriminate the dissimilarity of value in attributes well. Table 10 is VD matrix of "cap-color" attribute of the mushroom dataset. In the VD matrix, the dissimilarities of domain values in the attribute are not identical. This matrix represents hidden correlation information of values. Especially the dissimilarity of the value r, "green" and the value u, "purple" is 0 in our VDM. On examination of the mushroom dataset, objects whose cap-color is "green" are identical to those whose cap-color is "purple". Though previous traditional dissimilarity and distance measures should considier these values distinct, our method discriminates the correlation of two values correctly and considers them identical.

Our approach has some drawbacks of performance. To establish GOV and VD, extra scanning of dataset and memory to store that information are required. For k-dimensional data space, information, C_k^2 matrices with size of $|D_i| \times |D_j|$, where $1 \le i \le k$, $1 \le j \le k$ and $i \ne j$, is required to construct GOV and k matrices with size of $|D_i| \times |D_i|$ where $1 \le i \le k$ in needed for VD information used by DVD measure.

6 Concluding Remarks

Categorical attribute has been regarded as binary and it has been assumed that the difference of domain values in the attribute is identical. In real application, however, there exists the intrinsic difference for dissimilarity of categorical values and this difference information can enhance the clustering method.

We find out the dissimilarity of values in an attribute from the value distributions of other attributes in respective groups of data objects containing the values. To figure out the value distribution in a data group, we construct GOV. GOV of a specific value in an attribute represents distribution of values in the other attribute in the group. So VD of two values can be derived from dissimilarity of two GOVs for values respectively.

DVD measure is a distance measure of categorical values based on VD information and can be applied to any clustering algorithms adopting the concept of dissimilarity of objects. The experiments with synthetic and real data have shown that our approach can effectively discover the level of dissimilarity among categorical values and DVD measure based on can effectively quantify the dissimilarity among objects with those values.

Our method requires more computational time and memory space for extra information. Since information of GOV and VD is constructed during preprocessing of a whole clustering process, the time and space required to generate VD are relatively smaller than those needed by clustering algorithms themselves. However, the cost of our DVD measure could be expensive considerable when the size of dataset and the number of attributes increase significantly. Some summarization and approximation techniques of whole dataset information like sampling can be adopted to reduce time consuming. Our VD matrix is symmetric and sparse, so some compression methods using this characteristics could be applied to alleviate space requirement.

References

1. Myatt, G.J.: Making Sense of Data: A Practical Guide to Exploratory Data Analysis and Data Mining. John Wiley & Sons, Inc., Chichester (2007)
2. Jain, A.K., Dubes, R.C.: Algorithms for Clustering Data. Prentice-Hall, Englewood Cliffs (1988)
3. Ganti, V., Gehrke, J., Ramakrishnan, R.: Cactus-clustering categorical data using summaries. In: Proc. of ACM SIGKDD, pp. 73–83 (1999)
4. Guha, S., Rastogi, R., Shim, K.: Rock: A robust clustering algorithm for categorical attributes. In: Information Systems, pp. 512–521 (1999)

5. Zhang, Y., Fu, A.W.C., Cai, C.H., Heng, P.A.: Clustering categorical data. In: ICDE, p. 305 (2000)
6. Tan, P.N., Steinbach, M., Kumar, V.: Introduction to Data Mining. Addison-Wesley, Reading (2005)
7. Cost, S., Salzberg, S.: A weighted nearest neighbor algorithm for learning with symbolic features. Machine Learning 10, 57–78 (1993)
8. Ahmad, A., Dey, L.: A method to compute distance between two categorical values of same attribute in unsupervised learning for categorical data set. Pattern Recognition Letters 28(1), 110–118 (2007)
9. Barbará, D., Li, Y., Couto, J.: COOLCAT: an entropy-based algorithm for categorical clustering. In: Kalpakis, K., Goharian, N., Grossmann, D. (eds.) Proceedings of the Eleventh International Conference on Information and Knowledge Management (CIKM 2002), November 4–9, pp. 582–589. ACM Press, New York (2002)
10. Andritsos, P., Tsaparas, P., Miller, R.J., Sevcik, K.C.: LIMBO: Scalable clustering of categorical data. In: Bertino, E., Christodoulakis, S., Plexousakis, D., Christophides, V., Koubarakis, M., Böhm, K., Ferrari, E. (eds.) EDBT 2004. LNCS, vol. 2992, pp. 123–146. Springer, Heidelberg (2004)
11. Boriah, S., Chandola, V., Kumar, V.: Similarity measures for categorical data: A comparative evaluation. In: SDM, pp. 243–254. SIAM, Philadelphia (2008)
12. Huang, Z.: A fast clustering algorithm to cluster very large categorical data sets in data mining. In: Research Issues on Data Mining and Knowledge Discovery, pp. 1–8 (1997)

The Normalized Compression Distance as a Distance Measure in Entity Identification

Sebastian Klenk, Dennis Thom, and Gunther Heidemann

Stuttgart University
Intelligent Systems Group
Universitätsstrasse 38, 70569 Stuttgart, Germany
ais@vis.uni-stuttgart.de

Abstract. The identification of identical entities accross heterogeneous data sources still involves a large amount of manual processing. This is mainly due to the fact that different sources use different data representations in varying semantic contexts. Up to now entity identification requires either the – often manual – unification of different representations, or alternatively the effort of programming tools with specialized interfaces for each representation type. However, for large and sparse databases, which are common e.g. for medical data, the manual approach becomes infeasible.

We have developed a widely applicable compression based approach that does not rely on structural or semantical unity. The results we have obtained are promising both in recognition precision and performance.

1   Introduction

The integration of different data sources is an ever increasing obstacle in the process of knowledge discovery. The continuous employment of department wide databases and specialized information systems are fostering the scattering of data. At the same time, the increasing computing power makes a more holistic view on data possible and increases the need for data integration.

Health care is an important field of application which suffers from the scattering of data. In hospitals it is common practice to store patient information in distributed and independent database systems. The information system of the hospital administration stores financial and accounting specific information, the laboratory data is stored in lab-databases while radiology employs its own problem specific system. But for data mining and statistical analysis, a "holistic" view on patient data for is required, so all relevant information must be gathered and integrated for the different data sources. Therefore records or entities from one data source must be linked to the entities of another source. This task is easy as long as all records share a global identifier (for example a patient identification number), but this is not often the case. As a rule, there is no unique way of identification. Alternative ways to identify identical entities, e.g. a combination of name, birth date and further information, are usually not present in all of the representations and/or are erraneous. So for a successful approach

P. Perner (Ed.): ICDM 2009, LNAI 5633, pp. 325–337, 2009.

to entity identification we need a method which is fault tolerant and can cope with missing or mixed up information. Further, because of the high number of records in current databases, the method should be computationally inexpensive to allow for high throughput.

We will present an entity identification method based on the *Normalized Compression Distance* (NCD) similarity measure introduced by Cilibrasi and Vitànyi [9] which yields not only good performance but has significant advantages (high robustness to noise and structural heterogeneity) compared to currently employed methods.

1.1   Entity Matching

The basic procedure when matching identical entities in different data sources is this: First, identify related fields. This step is usually carried out manually. Then calculate for each pair of related fields a measure of similarity and merge the field specific similarities to a general measure that describes the degree of similarity between the two data entities. Within this process several data structural and lexical issues can occur which make the matching difficult. Problems range from data entrance errors or spelling ambiguities to structural differences, where in one data source a data element is encoded in a single field whereas the other source employs several fields. For example, an address can be stored in a single large text field or it may be spread over separate fields for street, city, country, etc. More formally, these problems can be grouped in three categories: *lexical heterogeneity, structural heterogeneity* and the lack of *data quality*.

Current research addresses these problems usually independent of each other. First the structure of the data sets is detected and unified, then identical fields are matched. Already in 1969, Fellegi and Sunter described the need to identify corresponding "blocks" and work on these [11]. The overview given by Winkler [23] proposes a related approach, Elmagarmid et. al. [10] focus solely on lexical heterogeneity, similar to Hernández and Stolfo [16]. Kimball states that 70% of the work on a data warehouse project goes into the ETL process [17]. Much of this work is due to the manual part of entity identification. By contrast, our approach based on the Normalized Compression Distance does not require any manual processing, in particular, the tedious work of field identification can safely be skipped.

For the actual matching of records, most current approaches are either *learning based* [22,26,25,27,4] or *rule based* [16,10,20]. Both methods suffer from different kinds of drawbacks. The learning approach requires training data which usually must be prepared manually under great effort. When developing or selecting training data it is of great importance to cover as many true-positive cases as possible, because a classification method that is biased towards a certain decision would be of little help. Interestingly, a much greater challenge seems to be the selection of good "non-matching" samples. Sarawagi and Bhamidipaty circumvent this problem by their active learning procedure that actively generates the training set by presenting samples to the user [22]. Distance based methods in general require no or little training (in case of weighting issues). The threshold selection, which we will discuss in more detail later, can be done with the given data. Rule based methods require a strong knowledge of the structure of the data and are not as robust to noise as learning based methods [10]. The method we propose, which is based on the Normalized Compression Distance, is robust to noise as was shown by Cebrian et. al. [6], requires no previous learning procedures and is well suited as a distance measure for clustering, classification and similarity matching tasks [15,9,2]. The broad range of classification subjects that are covered by almost identical procedures also shows that there is no need for the inclusion of domain specific knowledge which makes this procedure a good candidate for areas where only little is known about the domain itself.

2   Similarity-Measurement by Compression

Several concepts for the measurement of string- or file-similarity by means of compression algorithms have been proposed so far. The one with the most profound theoretical fundament is however the Normalized Compression Distance, formulated by Rudi Cilibrasi and Paul Vitànyi [9].

Although there are many known concepts and methods, that give an operationalization of the resemblance of two given data pieces with specific content, such as image-, text- or audio-files, most of them are highly specialized, complex and work on high-level aspects of the data. The idea behind the compression-distance is however, to measure closeness of any given files or strings unregarded of their specific content or structure by means of fast and simple algorithms, that work on a very low level. Even though it first seems arbitrary to attend to compression-algorithms in means of computing a similarity distance, the astonishing results, that already have been made with the approach, encourage this choice [15,9,2,5,1]. Furthermore there are several reasons to assume, that compression takes a vital part in the challenge of generalized pattern recognition.

2.1   Information Distance

We define the Kolmogorov complexity $K(x)$ of a string $x \in \Sigma^*$ as the length of a shortest Turing machine program, that, given the empty string ϵ as input, generates x on its output tape and then halts. Even though the function, that maps strings x to their complexity $K(x)$ is evidently not computable. The Kolmogorov complexity can be considered as an idealized measurement of the informational quantity given in a string (or file). Based on this proposal the informational distance of two strings could also be measured by consulting the concept. In doing so, we are talking about Turing machines, that are given the information of one string to resemble the other one. The less different the two strings are, the less complex this task should be and therefore the less the size of the smallest Turing machine, that does the job. To formalize this thought, we define the Normalized Information Distance [18] of two strings $x, y \in \Sigma^*$ as

$$NID(x,y) = \frac{\max \{K(x|y), K(y|x)\}}{\max \{K(x), K(y)\}}$$

where $K(x|y)$ is the conditional Kolmogorov complexity, i.e. the size of a smallest Turing machine, that, given y as input, generates x on its output tape and then halts. The normalization term $\max\{K(x), K(y)\}$ is included to generalize the approach on parameters of variable size and informational quantity. Based on the assumption, that always $K(x|y) \leq K(x)$ the NID should always take values in $[0,1]$, with $NID(x,x) \approx 0$ and $NID(x,\epsilon) \approx 1$. It can be proven, that the NID satisfies the metric (in-)equalities. Besides, it has been proven in [18], that the NID minorizes all normalized functions in a wide class of relevant distance-functions, meaning, that if two strings are close according to an aspect measured by one of these functions, they are also close according to the NID.

2.2 Compression Distance

Even if we want to accept the NID as a generalized measurement for informational distance, there remains the problem of non-computability of $K(x)$. One obvious real world approximation to the NID is to see the Turing machines as a maximal possible compression of the string x. Therefore the size of the output $C(x)$, that any real world compression algorithm, like Lempel-Ziv or Huffman, produces on input x could be taken to approximate $K(x)$ in the formula. Furthermore $\max\{K(x|y), K(y|x)\}$ could be approximated by $C(xy) - \min\{C(x), C(y)\}$, where xy is the concatenation of x and y. This way it can be seen how good the compression algorithm can use the information given in one string to better compress the other one and vice versa. By subtraction of $\min\{C(x), C(y)\}$ the term approximates the remaining compressed size of the more complex string if the information in the other one was used. Therefore the term should be smaller, the more similar the two strings are. Considered together, we define the Normalized Compression Distance as

$$NCD(x,y) = \frac{C(xy) - \min\{C(x), C(y)\}}{\max\{C(x), C(y)\}}$$

Obviously there is no way to prove, that this function approximates, what would be seen as an intuitive distance of file-contents, like similarity of pictures or tunes, i.e. giving an absolute and adequate universal and parameter-free similarity metric. But, it has been proven in [9], that the NCD minorizes relevant distance functions up to an additive term, vanishing with the quality of the approximation of $K(x|y)$ by the chosen compressor. Besides, as already mentioned, the experimental results, that have been observed with the method, encourage the approach.

3 ETL and Duplicate Detection

Even despite it's great age, entity identification, especially in the context of data integration and the Extract Transform Load (ETL) process, still is a very active field of research. This can be seen by the great number of recent publications [25,27,13,7,24] but also by the books that cover this subject or address related issues [14,17]. In this section we will present the basic procedures that occur during the ETL process and how the use of the Normalized Compression Distance improves the entire process.

3.1 Data Integration

When it comes to integrating data from different sources there are a number of difficulties that can arise. Biomedical data is a particularly good example. Here structural and lexical elements as well as data quality (which can be attributed to lexical heterogeneity) differ to a large degree from one data source to an other. Practically speaking this means that similar data is neither stored in a common structure (i.e. in some sources the data is spread over multiple fields whereas in others it is stored in a single one) nor is it encoded with similar values. These differences can range from different character sets, for example plain ASCII as opposed to UTF-8 where special characters like *ä* or the like will be coded as *ae*, complex abbreviations where *5th Street, New York* could be written as *5th St NY*. Detecting these inter-field similarities is usually a computationally expensive task and as Christen [7] points out the bottleneck of a record linkage system. Commonly used techniques like string matching algorithms require usually $O(|\sigma_1| \cdot |\sigma_2|)$ comparisons for two strings $\sigma_{\{1|2\}}$ [10]. Learning based methods can improve this number but involve training which is in itself an expensive task.

Besides record comparison being expensive, the number of necessary calculation for the whole data set can also become prohibitively large. For two sets A and B an exhaustive search for similar records would require $|A| \cdot |B|$ comparisons which would amount to 10^{10} for $|A| = |B| = 100.000$. One approach to reduce this number is *blocking*. Thereby the data sets will be grouped in similar blocks depending of the value of certain fields. For example the zip-code of the address field could be used to assign records into blocks with identical value. Now the comparison can take place only in blocks which share the same zip code. This of course requires that the field used for blocking is free of errors. A more advanced version of this method is canopies based clustering where very simple distance measures are used to efficiently calculate a number of overlapping clusters that each, in a second step, will be clustered with the detailed and expensive distance measure. This method is, opposed to the blocking scheme, tolerant to noise as the clusters overlap [19].

3.2 Procedure

The comparison of two strings based on the Normalized Compression Distance can be implemented as a linear algorithm [21] which improves significantly on the above mentioned values. Even though it is not sufficient to come around blocking or clustering it can increase the performance of a system significantly. Here we will describe an entity identification procedure that employs NCD as a distance measure.

The standard procedure as described by Christen and Goiser [8] and presented in figure 1 places the cleaning process before blocking and comparison.

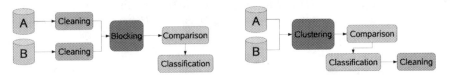

Fig. 1. The standard entity matching procedure where the cleaning comes before the matching

Fig. 2. The NCD based entity matching procedure where the matching can be done one the noisy data and the cleaning can use the extra information of entity similarity

This is necessary because most string matching and dissimilarity measures are highly domain specific and therefor can only be employed to certain fields. These need to be known in advance and, in case of structural heterogeneity, joined or split to fit the other data sources. In the light of data cleaning this has several drawbacks. First the knowledge about similar records, and thereby about identical values, can not be used during cleaning. Second a preceding data scrubbing might pose the risk of worsening the identification procedure by introducing new, and possibly more severe, errors.

The NCD based approach we propose follows the procedure presented in figure 2. Here clustering, as the first step, is required for performance reasons on large databases. The second step is already the record comparison. Therefor, for two records ρ_i and ρ_j, all fields (or a relevant subset depending on the size of the record) of each record are concatenated to form two large string σ_i and σ_j. As we will see in section 4 the order of the fields in the string is largely irrelevant. If for both records only a subset of fields is used, the two subsets should have some overlap. The two resulting strings are now compared by the NCD to get the degree $NCD(\sigma_i, \sigma_j) = d_{ij}$ of similarity between them. Now the knowledge gained about the similarity of two strings can be used to improve the cleaning of the fields. If one knows that fields contain similar values it is much easier to correct possible mistakes than if one is left to guess whether the values differ on purpose or not.

To decide whether two strings are similar enough to represent the same entity is usually being done by comparison with a given threshold t. Is the value of the similarity measure $NCD(\sigma_i, \sigma_j) < t$ is less than the threshold, both strings are considered matching. The difficult part is how to chose a t which is small enough to ignore similar, but not identical, entities but also large enough to not miss actual true hits. In terms of information retrieval this can be described as to find an equilibrium between recall and precision [3].

Of great help in this situation is that identical entities usually are significantly closer to each other than to all others. This means that the actual neighborhood of an entity – those other entities that are close to it – is sparse. Statistically this can be described as distances that fall below a significance or confidence level [12].

Once one knows the statistical properties (mean μ and variance ν) of the distances it is easy to calculate a good threshold

$$t = \mu - c\sqrt{\nu}$$

given the appropriate value $c = \Phi^{-1}(\alpha)$ of the distribution-quantile function where α specifies the confidence region. Figure 3 shows the distances of one entity to all matching candidates, the mean value and the confidence regions. There also the spares neighborhood (the small number of deep spikes) can be seen very good.

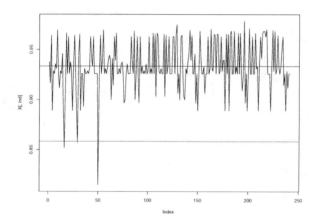

Fig. 3. The distances with mean value and confidence regions for a large noise level of $\alpha = 0.6$

4 Application

The main intention when developing this method was a fast and easy algorithm which could assist in finding duplicates and identical entities in unstructured, noisy and heterogeneous data. One of the central applications we had in mind was the integration of several large medical databases stored at different locations. These contain patient data stored in various formats, ranging from Microsoft Excel spread-sheets to relational and object oriented databases. Most of these systems don't share a common identifier and the content and structure of the data varies significantly.

4.1 Experiments

To get an impression of the usefulness of the Normalized Compression Distance for this task we prepared two test cases. One which allowed us to vary parameters like record structure or noise level in a controlled environment and a second one which represents the actual task at hand, namely the detection of entities across different tables.

The first test case consists of 241 employee records from the computer science department at the Stuttgart University. The records contains fields like name, room, telephone number and institute. The data is clearly structured and free from syntactic ambiguities. Because of the simple structure it is easy to see the influence of changes in structure like a flip of two fields or how noisy data changes the result set. We will therefor use this data set to manually add noise and test the influence thereof.

The second test consisted of two tables. One taken from a tumor center of a Stuttgart hospital (slightly more than 6000 entities) and the other one from the local cancer registry (about 600 entities). Both represent data of practicing physicians like name, address or further contact details. They are very heterogenous in structure and don't share a common identifier. Because data was inserted separately most fields also don't share common representation schemes such that telephone numbers or street addresses are represented differently even within the same table.

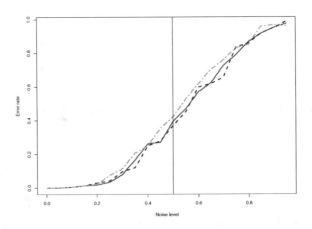

Fig. 4. The evolvement of the error-rate with increasing noise

Employee Data. In our tests we took the employee data base as reference set and compared it to the same data which was also augmented with various changes in structure and content. First we began to increase the noise level of the data in comparison. This means that, given a noise level α we draw for each letter a random number $r \in [0, 1]$ and checked whether it was smaller than α. If this was the case we replaced the letter by a randomly chosen letter. For a noise level of $\alpha = 0.5$ each letter would be replaced with a $50 - 50$ chance by a random letter which means that only about half of the string matches the actual string. To see the influence of this we compared for each noise level if the best match actually is the identical string. The results of this first evaluation are plotted in figure 4 (the solid line). Because we are not only interested in the best match but also in all matches we calculated for each noise level the recall R and precision P

$$R = \frac{r_m}{r_r} \quad \text{and} \quad P = \frac{r_m}{m}$$

where m is the number of matched entities, r_m the number of relevant matched entities and r_r all relevant entities. Figure 5 shows how precision and recall decrease with increasing noise.

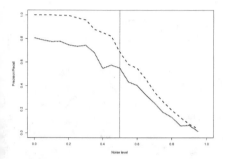 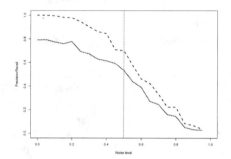

Fig. 5. The evolvement of precision and recall with increasing noise

Fig. 6. The evolvement of precision and recall with increasing noise for a matching string with field switches

One major advantage of the Normalized Compression Distance is the resistance to structural changes. To see how the NCD masters the task of identity identification we started flipping and changing field position within the matching string to see how error, precision and recall change by increasing structural heterogeneity. So we repeated the test scenario described above but now we switched the position of fields in the matching string, the reference data was left unchanged. Figure 6 shows the influence of two field switches on precision and recall. In 4 also the error levels for the structural changes are plotted. The solid line represents the error for no structural change, the dashed line involves one field flip and the double dashed line shows the error-rate for two field changes. It can be seen that there is only a very small increase in the error rate which is quite interesting if one considers the large change such a field flip poses for the matching string.

Physician Data. The second data set consists of contact information from physicians in the area of Stuttgart. The two tables, one taken from a tumor center the other one from a cancer registry, represent two non fully overlapping sets. For evaluation we prepared one test set and one reference set. The test set represents the first 100 elements out of the intersection of both sets, taken from the cancer registry. The reference set is the unaltered data from the tumor center. The task was then to find the corresponding element in the reference set that belonged to the data in the test set. This set up was chosen such that it is possible to count the correctly matching entities without having to handle not existing entities (these would come from the difference of the two sets) and above all to still be able to interpret the results by visual inspection.

The structure of the data tables is highly heterogeneous. The cancer registry table consists of the following fields in the exact order:

```
Code, CID, Salutation, Name, Field of Work, Street, Town, Status,
Family Physician
```

The fields of the tumor center table are:

```
TID, Date_edit, User_edit, ID_SAP, Zip, Country, Town, Title,
Street, Name, Salutation, GivenName, Tel, Fax, ID_int, ...
```

As can be seen the two tables not only contain a different ordering, there are also fields that share similar content. The field `Name` from the cancer registry contains the fields `Title`, `Name` and `GivenName` form the tumor center data. To make things worse these fields don't have to be in this order or can contain further information. There are for example physician that have a foreign title which augments the, in Germany, commonly used `Dr. med.` or the title is positioned after the given name (`Mr. XXXX YYYY (Dr. med.)`). Besides all that there are also differences like short names or different separations. Examples are:

- `Dres. med. Maier/Müller/Schultze` vs. `D. Schultze & C. Müller`
- `CA PD Dr. med. Andrea Müller` vs. `Prof. Andrea Mueller`
- `Dr.-Medic/IM Temeschburg Erich-Maria Schultze`   vs.   `Herr Erich-Maria Schultze`

To handle difficulties like these in a classical setting leads to a large deal of manual work. With our method it is possible to reduce the degree of human involvement to a minimum. The Idea here is to let the NCD calculate the differences and estimate the best fit (see Figures 7). If this is a miss match still manual labor is necessary but in majority of cases it is not.

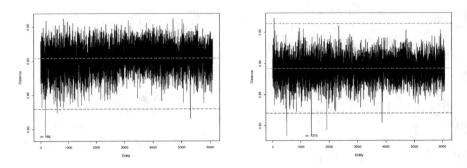

Fig. 7. The distances of the reference data to a select record in the test data set

In our setting we tried to match the first 100 entities of the test set that existed also in the reference set. There was no preprocessing of any kind. We have only compared the raw data coming from the data tables. Our results showed more that 80% of the entities were correctly identified. Compared to the number of preprocessing steps needed to acchive similar matching results this is an impressively high number.

5   Conclusions

In this paper we have introduced and evaluated a procedure to match entities with the Normalized Compression Distance. We have presented the rates of error, precision and recall for varying noise and structural heterogeneity levels. The results show that the NCD is capable of correctly identifying almost all of the identical entities, even when confronted with a large level of noise.

We also have tested the NCD based procedure on real world data with a very high level of structural and syntactical heterogeneity. Here the method allows for a significant degree of reduction in manual work when considered with other commonly used methods.

Our tests show that the use of the NCD is especially interesting if one considers its resistance to structural and syntactical changes and its linear computational complexity. This can be of great help in situations where large and unstructured data sets, as is the common case for biomedical or genome related data, must be identified and matched. Especially when it comes to first processing steps the proposed procedure allows for a first review of the data without much hands on labor.

As for further work we are currently working on an integrative software that will be applied to such task and we are positive that the obtained results will ease the work of integration unknown and heterogenous data sources.

References

1. Alfonseca, M., Cebrián, M., Ortega, A.: Testing genetic algorithm recombination strategies and the normalized compression distance for computer-generated music. In: AIKED 2006: Proceedings of the 5th WSEAS International Conference on Artificial Intelligence, Knowledge Engineering and Data Bases, Stevens Point, Wisconsin, USA, pp. 53–58. World Scientific and Engineering Academy and Society (WSEAS) (2006)
2. Amitay, E., Yogev, S., Yom-Tov, E.: Serial sharers: Detecting split identities of web authors. In: Stein, B., Koppel, M., Stamatatos, E. (eds.) PAN. CEUR Workshop Proceedings, vol. 276, CEUR-WS.org (2007)
3. Baeza-Yates, R., Ribeiro-Neto, B.: Modern Information Retrieval. Addison-Wesley, Reading (1999)
4. Bilenko, M., Mooney, R.J.: Adaptive duplicate detection using learnable string similarity measures. In: KDD 2003: Proceedings of the ninth ACM SIGKDD international conference on Knowledge discovery and data mining, pp. 39–48. ACM, New York (2003)

5. Casey, M.A., Veltkamp, R., Goto, M., Leman, M., Rhodes, C., Slaney, M.: Content-based music information retrieval: Current directions and future challenges. Proceedings of the IEEE 96(4), 668–696 (2008)
6. Cebrian, M., Alfonseca, M., Ortega, A.: The normalized compression distance is resistant to noise. IEEE Transactions on Information Theory 53(5), 1895–1900 (2007)
7. Christen, P.: A two-step classification approach to unsupervised record linkage. In: AusDM 2007: Proceedings of the sixth Australasian conference on Data mining and analytics, Darlinghurst, Australia, pp. 111–119. Australian Computer Society, Inc. (2007)
8. Christen, P., Goiser, K.: Quality and complexity measures for data linkage and deduplication. In: Guillet, F., Hamilton, H.J. (eds.) Quality Measures in Data Mining. Studies in Computational Intelligence, vol. 43, pp. 127–151. Springer, Heidelberg (2007)
9. Cilibrasi, R., Vitanyi, P.: Clustering by compression. IEEE Transactions on Information Theory 51(4) (2005)
10. Elmagarmid, A.K., Ipeirotis, P.G., Verykios, V.S.: Duplicate record detection: A survey. IEEE Transactions on Knowledge and Data Engineering 19(1), 1–16 (2007)
11. Fellegi, I.P., Sunter, A.B.: A theory for record linkage. Journal of the American Statistical Association 64(328), 1183–1210 (1969)
12. Feller, W.: An introduction to probability theory and its applications, vol. 1. Wiley, Chichester (1950)
13. Goiser, K., Christen, P.: Towards automated record linkage. In: AusDM 2006: Proceedings of the fifth Australasian conference on Data mining and analytics, Darlinghurst, Australia, pp. 23–31. Australian Computer Society, Inc. (2006)
14. Han, J., Kamber, M.: Data mining. Morgan Kaufmann, San Francisco (2001)
15. Heidemann, G., Ritter, H.: On the Contribution of Compression to Visual Pattern Recognition. In: Proc. 3rd Int'l Conf. on Comp. Vision Theory and Applications, Funchal, Madeira - Portugal, vol. 2, pp. 83–89 (2008)
16. Hernández, M.A., Stolfo, S.J.: Real-world data is dirty: Data cleansing and the merge/purge problem. Data Min. Knowl. Discov. 2(1), 9–37 (1998)
17. Kimball, R., Caserta, J.: The Data Warehouse ETL Toolkit: Practical Techniques for Extracting, Cleanin. John Wiley & Sons, Chichester (2004)
18. Li, M., Chen, X., Li, X., Ma, B., Vitanyi, P.: The similarity metric (2001)
19. McCallum, A., Nigam, K., Ungar, L.H.: Efficient clustering of high-dimensional data sets with application to reference matching. In: KDD 2000: Proceedings of the sixth ACM SIGKDD international conference on Knowledge discovery and data mining, pp. 169–178. ACM Press, New York (2000)
20. Navarro, G.: A guided tour to approximate string matching. ACM Comput. Surv. 33(1), 31–88 (2001)
21. Rodeh, M., Pratt, V.R., Even, S.: Linear algorithm for data compression via string matching. J. ACM 28(1), 16–24 (1981)
22. Sarawagi, S., Bhamidipaty, A.: Interactive deduplication using active learning. In: KDD 2002: Proceedings of the eighth ACM SIGKDD international conference on Knowledge discovery and data mining, pp. 269–278. ACM Press, New York (2002)
23. Winkler, W.E.: Overview of record linkage and current research directions. Technical Report RRS2006/02, US Bureau of the Census (2006)
24. Yan, S., Lee, D., Kan, M.-Y., Giles, L.C.: Adaptive sorted neighborhood methods for efficient record linkage. In: JCDL 2007: Proceedings of the 7th ACM/IEEE-CS joint conference on Digital libraries, pp. 185–194. ACM Press, New York (2007)

25. Zhao, H.: Semantic matching across heterogeneous data sources. Commun. ACM 50(1), 45–50 (2007)
26. Zhao, H., Ram, S.: Entity identification for heterogeneous database integration: a multiple classifier system approach and empirical evaluation. Inf. Syst. 30(2), 119–132 (2005)
27. Zhao, H., Ram, S.: Entity matching across heterogeneous data sources: An approach based on constrained cascade generalization. Data & Knowledge Engineering (corrected proof) (in press, 2008) (available online May 4)

Attribute Constrained Rules for Partially Labeled Sequence Completion

Chad A. Williams[1,*], Peter C. Nelson[1], and Abolfazl (Kouros) Mohammadian[2]

[1] Dept. of Computer Science
University of Illinois at Chicago
851 South Morgan Street
Chicago, IL 60607-7053
cwilliam@cs.uic.edu, nelson@cs.uic.edu
[2] Dept. of Civil and Materials Engineering
University of Illinois at Chicago
842 West Taylor Street
Chicago, IL 60607-7023
kouros@uic.edu

Abstract. Sequential pattern and rule mining have been the focus of much research, however predicting missing sets of elements within a sequence remains a challenge. Recent work in survey design suggests that if these missing elements can be inferred with a higher degree of certainty, it could greatly reduce the time burden on survey participants. To address this problem and the more general problem of missing sensor data, we introduce a new form of constrained sequential rules that use attribute presence to better capture rule confidence in sequences with missing data than previous constraint based techniques. Specifically we examine the problem of given a partially labeled sequence of sets, how well can the missing attributes be inferred. Our study shows this technique significantly improves prediction robustness when even large amounts of data are missing compared to traditional techniques.

Keywords: frequent sequence mining, constrained rules, missing data.

1 Introduction

Frequent pattern mining of sequences has been a prominent research theme since its introduction by Agrawal and Srikant [1], yet how to effectively use pattern-based mining for classification and prediction remains a challenge [2]. The problem examined in this work is given a sequence of sets of attribute values where one to all fields within a set can be missing, populate the missing attribute values in the sequence. We refer to this problem as *partially labeled sequence completion*. Two common versions of this problem are:

---

* This research was supported in part by the National Science Foundation IGERT program under Grant DGE-0549489.

P. Perner (Ed.): ICDM 2009, LNAI 5633, pp. 338–352, 2009.

1. given the prior sequence, complete the missing elements of the current/next set; and
2. given a target set anywhere in a sequence complete the missing attribute values using the sets both before and after the target set

Many studies have focused on the next step prediction form, particularly for web applications such as pre-fetching and personalization [3,4]. In this study, we examine the more general form of the problem, since it also addresses the growing number of applications that would benefit from inferring additional information about an event given both the events before and after the event of interest. An example of this would be a group of mobile sensors that are periodically collected, where any number of the readings may be missing from any particular time step [5]. Other work in survey design suggests that if missing elements within a sequence can be inferred with a higher degree of certainty, it would greatly reduce the time burden on survey participants [6,7].

This work presents a new form of constrained sequential rules to address the more general form of the problem, which can also be applied for next step set prediction. Attribute constrained rules (ACR) are based on traditional sequential rules that can be derived from frequent sequential pattern mining, however extensions are made to better address attribute labeled sequences with missing value data. This problem of partially labeled sequence completion is formally stated below, followed by an overview of related work and an illustrative example of the value of ACR rules and how they are mined. Our study then shows this technique significantly improves prediction robustness for even large numbers of missing values compared to traditional sequential rules using a publicly available travel survey data set.

2   Partially Labeled Sequence Completion

Algorithms for rule mining of sequential patterns have been a major source of interest since they were first introduced in [8]. Much of prior work has focused on mining sequential rules for predicting future patterns, however being able to infer missing information within an observed sequence of sets of attribute values also has many useful applications that have largely been overlooked. One example of this is multi-day travel survey design, where being able to reliably infer missing values from the surrounding data set would allow respondent burden to be greatly reduced if only a portion of the data points needed to be collected regularly rather than the full set of questions [7]. To address problems such as these, we examine the more general problem of missing information within a sequence, however the technique also applies to traditional prediction as well. Specifically the problem we address is given a sequence where there are a known set of attributes that describe an event within the sequence, infer any missing values of the attributes for a target set. In this section, we formalize this constrained sequence problem and introduce a technique for mining and applying rules specifically for this task.

2.1 Problem Statement

For the problem of partially labeled sequence completion, let

$$H = \{H_1, H_2, \cdots, H_n\}$$

be a database of sequences, and let:

$$H_i = < S_1, S_2, \cdots, S_n >$$

be the sequence of sets of observations in a sequence i; where each observation set S_j is composed of 1 to m attributes $\{a_{j,1}, a_{j,2}, \cdots, a_{j,m}\}$. Each attribute $a_{j,k}$ has a discrete set of values for the kth position that is shared across all observation sets S. Intuitively the sequence H_i can be thought of as a series of recordings by a survey instrument or sensor with a fixed set of discrete measures (the attributes), where at each event j all measurements are relevant, but only a portion of these measures may actually be recorded in the set S_j. Given a sequence H_{target} of length l and a target set S_t to be completed where $1 \leq t \leq l$ and between 1 to m arbitrary attributes are missing values. Determine the values of all missing attributes $\{a_{t,1}, a_{t,2}, \cdots, a_{t,m}\}$ in S_t. Thus our goal is to use the surrounding sequence information in H_{target} to populate any missing values to complete the set S_t.

3 Background and Motivation

3.1 Related Work

One of the common limitations of the majority of association mining algorithms is they do not take advantage of variable information or variable presence which becomes particularly important in sequences with missing values. For these types of sequences, which we refer to as partially labeled sequences, if we consider the sets being observed as a set of possible attribute assignments from a portion of the set of attributes (such as instrument output) we are observing, the problem of predicting future sets can become far more well defined in terms of the possible values given a set of possible attributes with either known or inferred constraints.

While techniques have been introduced for mining sequential patterns given regular expression constraints [9,10], the expression constraints in these works are best suited for matching a value pattern. For example, while an expression can be defined to match any sequence of values that can be described by a regular expression, the language does not provide for a more sophisticated notion of attribute value restrictions. While some aspects of this type of functionality could be encoded to restrict attribute values to a particular value such as the regular expression constraint: $\{$"$a_1 = 1$",("$a_2 = 2$"$|$"$a_2 = 3$")$\}$, this type of encoding is insufficient for constraining the general presence of an attribute if all values are not known ahead of time.

Other work such as [11] has sought to identify the specific types of constraints needed for sequential pattern mining beyond those that can be expressed with regular expressions. Their work introduces the concept of *item constraints*, which are intended to allow the presence (or absence) of a particular individual or group of items to be specified in the mining process. Given a particular query described in terms of constraints the algorithm they introduced, PrefixGrowth, finds patterns matching the constraints through a prefix extension method. While this algorithm is effective for finding patterns matching a particular query, it does not address being able to identify the set of possible constraint based rules for completing the values of a pattern in general.

3.2   Attribute Constrained Rule Mining

To address rules of this form we extend an idea from the text mining community called *label sequential rules* (LSR) [12,13]. Originally introduced for analyzing opinions in text reviews, this rule form was proposed for identifying common sentence forms or templates where a type of word of interest, termed a label, would likely appear. These rules form a more constrained matching pattern through wild cards producing rules of the form:

$$\langle \{1\} \{3, *, *\} \{6\} \rangle \rightarrow \langle \{1\} \{3, 4, 7\} \{6\} \rangle$$

where confidence of the rule would be defined with respect to the likelihood of the right hand side (RHS) given all sequences that contain the wild card constrained pattern. Thus, if we are only interested in rules that address completing two items in the middle set, these constraints allow a more meaningful measure of rule confidence since the likelihood is only measured in relation to patterns that match the LSR template.

For the task of partially labeled sequence completion, we propose a similar idea for identifying templates of sequential patterns of attribute values which we refer to as *attribute constrained rules* (ACR). Whereas with LSR the confidence of rules specify how likely a generalization is about elements within a pattern, with ACR the rule's confidence specifies the likelihood of a specific attribute value or combination of attribute values given a surrounding pattern.

Illustrative Example. In this section, we provide a set of illustrative examples of the benefit of constrained rules such as ACR and LSR. Throughout these examples refer to Table 1 as the sequence database.

Below we use the standard definitions of *support* and *confidence* defined as:

Definition 1. *The **support** of the sequential rule $X \rightarrow Y$ is the fraction of sequences in the database that contain Y.*

Table 1. Example sequence database

H_1	$< \{a_1\}\{a_2, b_2\}\{b_1\} >$
H_2	$< \{a_1\}\{a_2, b_2\}\{a_2, b_1\} >$
H_3	$< \{a_1\}\{b_2, c_2\}\{a_2\}\{b_1\} >$
H_4	$< \{a_1\}\{a_2, c_1\}\{b_1\} >$

Definition 2. *The **confidence** of a sequential rule $X \rightarrow Y$ is the fraction of sequences in the database that contain X that also contain Y.*

For an example of how constrained rules can better represent the applicable confidence, consider the following scenario: $H_{target} = \langle \{a_1, b_1\} \{a_2, \ ?\} \{b_1\} \rangle$, where $S_2 = \{a_2, \ b{:}?\}$ is the target set. The following traditional sequence associative rule would be applicable:

$$\langle \{a_1\} \{a_2\} \{b_1\} \rangle \rightarrow \langle \{a_1\} \{a_2, b_2\} \{b_1\} \rangle$$
$$[\text{sup} = 2/4, \text{conf} = 2/4]$$

Which can be interpreted as S_2 can be completed $\{a_2, b_2\}$ with a confidence of $2/4$. By contrast a label constrained version of the same rule:

$$\langle \{a_1\} \{a_2, *\} \{b_1\} \rangle \rightarrow \langle \{a_1\} \{a_2, b_2\} \{b_1\} \rangle$$
$$[\text{sup} = 2/4, \textbf{conf} = \mathbf{2/3}]$$

Where the notation $\{a_2, *\}$ indicates a set containing value a_2 for attribute a and a second value within the same set. As this example shows, by further constraining the attribute and location of pattern extension with LSR constraints, the confidence of the pattern is raised to $2/3$ or roughly 67%. With ACR this idea is extended to constrain pattern matches to particular attribute values of interest. In our example, since we are specifically interested in the value of attribute b, the ACR version of the same rule would be:

$$\langle \{a_1\} \{a_2, \mathbf{b{:}*}\} \{b_1\} \rangle \rightarrow \langle \{a_1\} \{a_2, \mathbf{b_2}\} \{b_1\} \rangle$$
$$[\text{sup} = 2/4, \textbf{conf} = \mathbf{2/2}]$$

which is able to further clarify the confidence in populating attribute b, since it is able to discount sequence H_4 as it does not match the attribute constrained pattern. This advantage in accurately evaluating the value of the constrained sequence rule is the reason we examine ACR for partially labeled sequence completion.

The left hand side of Figure 1 shows the frequent sequence graph using a minimum support of 40% for Table 1 along with the support counts for each frequent sequence. All frequent ACR antecedents can be easily identified from the frequent item sets by expanding the remaining possibilities. For example the following antecedents can be generated from the frequent item set $\langle \{a_1\} \{b_2\} \{b_1\} \rangle$:

$$\langle \{a_1\} \{b_2\} \{b_1\} \rangle \Rightarrow \begin{bmatrix} \langle \{a : *\} \{b_2\} \{b_1\} \rangle, \langle \{a : *\} \{b_2\} \{b_1 : *\} \rangle, \\ \langle \{a_1\} \{b : *\} \{b_1\} \rangle, \langle \{a_1\} \{b : *\} \{b : *\} \rangle, \\ \langle \{a_1\} \{b_2\} \{b : *\} \rangle, \langle \{a : *\} \{b : *\} \{b_1\} \rangle, \\ \langle \{a : *\} \{b : *\} \{b : *\} \rangle \end{bmatrix}$$

As this example shows, due to the combinatorial growth of the attribute constraint sets this problem quickly becomes impractical for datasets with a large number of attribute values or lengthy sequences if all completion rules are considered. For example with even this small example, the 17 frequent sequences have over 40 potential ACR antecedents. For the problem as stated in Section 2.1 there are some properties that can be taken advantage of to reduce the number of ACR antecedents. Specifically, one of the key features of the problem we address is that every observation set S_j is composed of the same m attributes $\{a_{j,1}, a_{j,2}, \cdots, a_{j,m}\}$, and only one target set is examined at a time. The implication of this fact is the property that for any set S_i, there is a value (whether known or not) for every attribute $a_{j,i}$. This property means that while the number of possible antecedents may grow quickly, the only ones that need to be kept are those with all constraints within a single set within the sequence. Once all possible ACR antecedents for the frequent pattern sets have been enumerated, the support for all the patterns can be updated with a single pass of the data set. By adding this subset and associated support to the frequent sequence graph as shown in the right hand side of Figure 1, ACR predictions and completion can be quickly determined using just the information in the extended frequent sequence graph. Note that while not shown in the figure, links would also exist between the ACR antecedents and the frequent sequence completions in the graph.

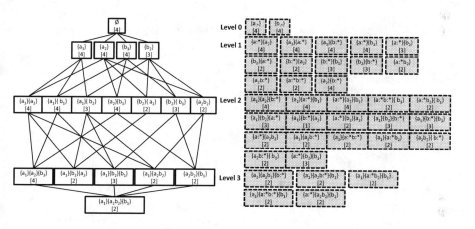

Fig. 1. ACR Frequent sequence graph

4   Experiments

4.1   Evaluation Metrics

For measuring prediction performance, we use the information retrieval metrics of precision and recall [14]. The basic definition of recall and precision can be written as:

$$\text{precision} = \frac{\#\text{ true positives}}{(\#\text{ true positives} + \#\text{ false positives})}$$

$$\text{recall} = \frac{\#\text{ true positives}}{(\#\text{ true positives} + \#\text{ false negatives})}$$

For the purpose of this study, since we are primarily interested in the correctness of the attribute value (if the attribute appeared at all). Thus # *true positives* is the number of attribute values predicted correctly; # *false positives* is the number of attribute values incorrectly predicted where the attribute did appear in the time step, and # *false negatives* is the number of attributes values which no prediction was made, but some value for the attribute appeared in the time step. Since these two measures are often associated with a tradeoff of one for the other, we also examine a combined metric the F-measure [15] which can be calculated as:

$$\text{F-measure} = \frac{(2 \cdot \text{precision} \cdot \text{recall})}{\text{precision} + \text{recall}}$$

We use this metric to compare the balanced performance of the algorithms.

4.2   Experimental Setup

Data. To evaluate the proposed ACR technique, we chose the 2001 Atlanta Household Travel Survey for several reasons. This dataset contains a large number of sequences of sets that are known to have a strong relationship between the entire set of attributes at each step and their ordering, making it well suited for sequential set prediction. Second, the type of data collected in this survey is very similar to one of the proposed applications of this type of partially labeled sequence learning, reducing survey participant burden [6,7]. Demonstrating that a significant portion of this type of data can be removed (i.e. a number of survey questions reduced) while limiting the impact on predictions is a significant step in showing the feasibility of this type of application. Finally, this data set represents one of the larger publically available data sets of this type, making the results of this study open to competitive comparisons by other work in this area.

The 2001 Atlanta Household Travel Survey was conducted from April 2001 through April 2002 on behalf of the Atlanta Regional Commission (ARC) [16]. The data consists of travel activity information for 21,323 persons from 8,069 households and 126,127 places visited during the 48-hour travel period. This survey focused on observed activity type, timing, and travel associated with each person's activity schedule captured over a 2 day period. The survey captured a wide range of in-home and out-of-home activity types which were broken down by a high-level classification. The survey captures over 250 attributes relating to the travel, activity, and demographic characteristics of the individual for each activity sequence that was recorded. The data is structured such that each event corresponds to an activity in the person's schedule with the set of attributes corresponding to the characteristics of the activity and travel associated with getting to the activity.

In the experiments below, we focus on a subset of 6 of these attributes: activity type, mode of transportation, arrival time, departure time, activity duration, and traveler age. These attributes were selected as they represent a mix of information about the type of activity, the travel, relative time the activity took place, activity duration, and features of the person involved that have been shown to be highly related both intra-event and inter-event in predicting traveler activity patterns [17,18]. Thus, the dataset can be thought of as a database of sequences of events with the set of attribute values at each event being highly related. For the subset of the data set we use, there are 49,695 sets of activity information, with an average sequence length of just over 7.4 sets. Additional information about the data set can be found in Appendix A.

Methods. In the results below, we present the results of both the ACR technique, introduced in this work, and traditional sequence rules for a comparative baseline. As both rule-based techniques are frequent pattern based, which is deterministic for a given minimum support threshold, in all experiments below the same set of frequent patterns were used for both the traditional sequential mining and the ACR mining to ensure a fair comparison. In all experiments, both sets of rules were mined using the same minimum confidence threshold, and only rules that produced at least one target item in the target pattern were considered.

To generate predictions given a set of many potentially applicable rules, a ranking prediction scheme was utilized. Specifically, the rules were ranked in order by confidence, number of target productions, support, antecedent length, and finally a string based comparison to ensure repeatability if all other factors were equal. The productions of the top ranked rule were then applied to the target set, the remaining matching rules were then re-ranked as the number of target productions may have dropped due to the previous rule's productions. The rule with the highest rank of those remaining was then applied and this process continued until either a prediction had been made for all target items or no rules remained.

As described in Section 2.1, the problem of partially labeled set completion involves taking a sequence and trying to fill in or predict items within a single target set within a sequence. Since the problem of partially labeled set completion can take the form of predicting anywhere from a single item in a target set to all items in the target set, the results below reflect the average of all possible combinations of the target pattern in all possible positions for the target set. Where target pattern means: the set of attribute values in the target set that are being evaluated. Thus in the experiments below, for the target set any attribute value that is not specifically of interest as specified by the target pattern retains its original value for determining matching rules. For example if the target pattern included attributes a and c ($S_T = \{a_T c_T\}$). In testing the sequence:

$$\langle \{a_1 b_2 c_1\} \{a_2 b_2 c_2\} \{a_1 b_1 c_2\} \rangle$$

If the target set was S_2 for the sequence, the test sequence would thus be:

$$H_{target} = \langle \{a_1 b_2 c_1\} \{a_T b_2 c_T\} \{a_1 b_1 c_2\} \rangle$$

In the base experimental data set described above, no attribute values were missing. The missing data scenarios were created by randomly removing the specified percentage of values from both the training and test sets for any attribute appearing in the target pattern. All experiments below were run using a minimum support threshold of 60% for frequent patterns and a minimum rule confidence threshold of 80%. To ensure the significance of our results, all results shown are the average of a 10 times cross-folding methodology.

4.3   Experimental Evaluation

ACR vs. Sequential Rule Comparison. In the first set of experiments, we examine the impact of missing data on each rule mining technique. Figure 2 portrays the recall of the missing attribute values. In the figure, *ACR-Avg* and *SEQ-Avg* represent the ACR results and the traditional sequential rules results respectively averaged over all possible number of target items in the target pattern. As these results show, the ACR technique produces both a higher overall recall as well as less degradation as values are removed and the data becomes sparser.

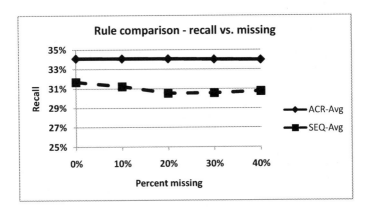

Fig. 2. Comparison of recall for ACR and traditional sequential rules as the percent of missing data increases

Since higher recall can indicate an increase in the quantity of predictions at the cost of accuracy, it is important to consider the precision as well. As Figure 3 shows, the boost in recall from ACR comes at a tradeoff of less than a .3% drop in precision. While this small drop is statistically significant, in practice the benefit of the additional 3-3.5% of attribute values with good predictions (recall) is likely to outweigh this small drop in precision for most applications.

The combined performance, as evaluated using the F-measure, is shown in Figure 4. As these results demonstrate, the ACR technique results in far better combined predictive robustness compared to traditional rules as the amount of missing data increases.

Number of Target Items Comparison. In the next set of experiments, a more in depth look is taken on the impact of the number of items trying to be predicted in the target pattern. In these results, *ACR-X* and *SEQ-X*

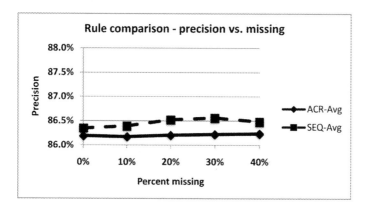

Fig. 3. Comparison of precision for ACR and traditional sequential rules as the percent of missing data increases

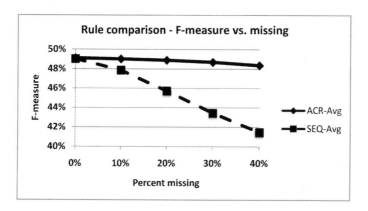

Fig. 4. Comparison of F-measure for ACR and traditional sequential rules as the percent of missing data increases

represent the ACR results and the traditional sequential rules results respectively averaged over all possible target patterns with X number of target items in the target pattern. Thus, $ACR\text{-}3$ would indicate the average performance of the ACR technique averaged across all possible target patterns with exactly 3 target items. The reason this is of interest, as Figure 5 shows, is that the number of items that are trying to be predicted can have a significant impact on how missing data affects prediction performance. As the results demonstrate for traditional sequential rules, while a small amount of missing data (10%) has a greater impact when predicting fewer items; as the amount of missing data increases (30%) this relationship is reveresed.

As Figure 6 shows, with traditional sequential rules, as the number of values needing to be predicted increases, it becomes increasingly harder to recall all

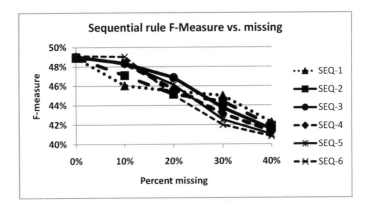

Fig. 5. Comparison of differences in predictive performance for traditional sequential rules for different target set sizes as the percent of missing data increases

of the target values and becomes even harder as the amount of missing data increases. Comparing this to the rules produced by the ACR technique shows that while this is still the case; the attribute constraints significantly increase the recall for this scenario even without missing values. Intuitively this is because in a less restricted target set, the attribute constraints better capture the likelihood of the predicted attributes all occurring simultaneously in the sequence than the unconstrained form of the rule.

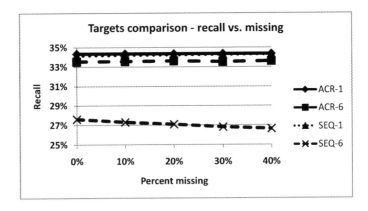

Fig. 6. Comparison of recall for ACR and traditional sequential rules for different target set sizes as the percent of missing data increases

A look at the precision for this same scenario, Figure 7, shows perhaps an intuitively unexpected result that the precision when predicting a full set is actually slightly higher than when predicting a single target item and furthermore increases slightly with some missing data. The reason for the higher precision with more target items is due largely to a smaller percentage of attribute values actually being predicted (as reflected in the recall) and in this case, is likely in part due to a feature of the data set such that some attribute values are easier to predict than others (not shown in this work). Likewise the small elevation in precision associated with a percentage of the elements being removed likely reflects the benefit of randomly reducing the appearance of some of the less frequent items which may have some similarity to noise when considering full set prediction.

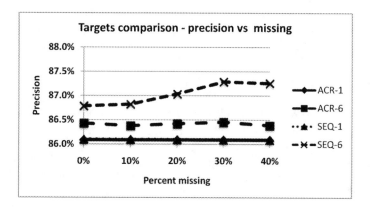

Fig. 7. Comparison of precision for ACR and traditional sequential rules for different target set sizes as the percent of missing data increases

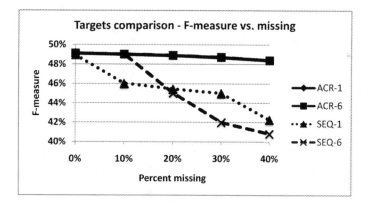

Fig. 8. Comparison of F-measure for ACR and traditional sequential rules for different target set sizes as the percent of missing data increases

Finally, the F-measure comparison of combined prediction performance is shown in Figure 8. As the results show, with traditional sequential rules while a small amount of data can be removed (less than 10% for this data set) with limited impact on full set prediction; as the amount of missing data increases beyond this the performance quickly degrades. Single target prediction displays a slightly different trend being much more affected by even a slight increase in the amount of missing data, but being slightly more resilient than full set prediction as the amount of missing data increases beyond the initial amount.

This general pattern for traditional sequential rules show that the fewer the number of target items, the more significant any increase in missing data becomes; but also the less effected by subsequent increases in missing data is further illustrated in Figure 5. In contrast, the ACR technique proves much more resilient in either scenario as the amount of missing data increases, demonstrating nearly identical balance in predictive performance in both scenarios as the amount of missing data increases. This same nearly identical F-measure trend was observed for all target set sizes with the ACR technique (not shown).

5   Discussion and Future Work

In this work we introduced attribute constrained rules, a technique for mining and better estimating rule performance for partially labeled sequence completion. As our results demonstrate, this technique shows significant promise for accurately predicting sets of attributes within a sequence with missing data compared to a traditional sequential rule-based approach. In the context of survey applications aimed at reducing the time burden on participants such as those described in [6,7]; this represents a significant time savings opportunity. Specifically, the implications of the results presented in this work are that rather than needing to ask the full set of questions for each event as is currently done; a participant could be asked a much smaller portion of the questions with minimal impact on the benefits gained by pre-populating their likely responses. In the context of other applications such as mobile sensor data, this might represent a chance to reduce costly communication without reducing the ability to reliably predict missing values.

While the technique introduced is well suited for completing multiple attributes within a set of a sequence, a heuristic take on this technique may be necessary if it were to be used for predicting multiple sets simultaneously due to the combinatorial growth of possible ACR antecedents. In future work, we intend to explore ways this type of approach can be adapted to handle streaming data. Finally, a study is underway to confirm the benefits of this technique in practice for interactive survey applications such as those described above.

References

1. Agrawal, R., Srikant, R.: Mining sequential patterns. In: Yu, P.S., Chen, A.S.P. (eds.) Eleventh International Conference on Data Engineering, Taipei, Taiwan, pp. 3–14. IEEE Computer Society Press, Los Alamitos (1995)
2. Han, J., Cheng, H., Xin, D., Yan, X.: Frequent pattern mining: current status and future directions. Data Mining and Knowledge Discovery 15(1), 55–86 (2007)
3. Yang, Q., Zhang, H.H., Li, T.: Mining web logs for prediction models in www caching and prefetching. In: KDD '01: Proceedings of the seventh ACM SIGKDD international conference on Knowledge discovery and data mining, pp. 473–478. ACM, New York (2001)

4. Mobasher, B., Dai, H., Luo, T., Nakagawa, M.: Using sequential and non-sequential patterns in predictive web usage mining tasks. In: ICDM 2002: Proceedings of the 2002 IEEE International Conference on Data Mining, Washington, DC, USA, p. 669. IEEE Computer Society Press, Los Alamitos (2002)

5. North, R., Richards, M., Cohen, J., Hoose, N., Hassard, J., Polak, J.: A mobile environmental sensing system to manage transportation and urban air quality. In: IEEE International Symposium on Circuits and Systems, 2008. ISCAS 2008, May 2008, pp. 1994–1997 (2008)

6. Marca, J.E., Rindt, C.R., McNally, M.G.: Collecting activity data from gps readings. Technical Report Paper UCI-ITS-AS-WP-02-3, Institute of Transportation Studies, Center for Activity Systems Analysis, University of California, Irvine (July 2002)

7. Auld, J., Williams, C.A., Mohammadian, A., Nelson, P.C.: An automated GPS-based prompted recall survey with learning algorithms. Transportation Letters: The International Journal of Transportation Research 1(1), 59–79 (2009)

8. Agrawal, R., Imielinski, T., Swami, A.: Mining association rules between sets of items in large databases. SIGMOD Rec. 22(2), 207–216 (1993)

9. Garofalakis, M.N., Rastogi, R., Shim, K.: Spirit: Sequential pattern mining with regular expression constraints. In: VLDB 1999: Proceedings of the 25th International Conference on Very Large Data Bases, pp. 223–234. Morgan Kaufmann Publishers Inc., San Francisco (1999)

10. Garofalakis, M., Rastogi, R., Shim, K.: Mining sequential patterns with regular expression constraints. IEEE Transactions on Knowledge and Data Engineering 14(3), 530–552 (2002)

11. Pei, J., Han, J., Wang, W.: Mining sequential patterns with constraints in large databases. In: CIKM 2002: Proceedings of the eleventh international conference on Information and knowledge management, pp. 18–25. ACM, New York (2002)

12. Liu, B., Hu, M., Cheng, J.: Opinion observer: analyzing and comparing opinions on the web. In: WWW 2005: Proceedings of the 14th international conference on World Wide Web, pp. 342–351. ACM Press, New York (2005)

13. Liu, B.: Web data mining: exploring hyperlinks, contents, and usage data. In: Data-Centric Systems and Applications. Springer, Heidelberg (2007)

14. Cleverdon, C.: Evaluation of tests of information retrieval systems. Journal of Documentation 26, 55–67 (1970)

15. van Rijsbergen, C.: Information Retrieval. Butterworth, London (1979)

16. NuStats: 2001 atlanta household travel survey: Final report. Technical report, Atlanta Regional Commision (April 2003)

17. Timmermans, H. (ed.): Progress in Activity-Based Analysis. Elsevier, Oxford (2005)

18. Ettema, D., Schwanen, T., Timmermans, H.: The effect of location, mobility and socio-demographic factors on task and time allocation of households. Transportation: Planning, Policy, Research, Practice 34(1) (2007)

A   Appendix: Data Set Information

The data selected for use in the experiments from the 2001 Atlanta Household Travel Survey was all events from the survey that didn't contain any missing values for the attributes: activity, mode of trasportation, arrival time, departure time, duration, and age. This filtering was done to allow the experiments with the amount of missing data conducted in this work to be performed in a more controlled manner. These attributes were broken down into the following discrete values:

- **Activity** - 29 values: 'Other at home activities', 'Working', 'Medical or dental', 'Getting ready', 'Major shopping', 'Personal', 'Watching children', 'Pick up something or drop off something', 'Worship or religious meeting', 'Visit friends or relatives', 'Household work or outdoors work', 'Entertainment', 'Outdoor recreation', 'Fitness or exercising', 'Rest or relax', 'Waiting for transportation', 'No other activities', 'Personal business', 'Other', 'Eating', 'Volunteer work', 'Work related from home', 'Community meetings', 'ATM, banking, post office, bill payment', 'Sleep', 'Work related business', 'School', 'Drop off or pickup someone', 'Incidental shopping'
- **Mode of transportation** - 12 values: 'Walk', 'Auto or van or truck - passenger', 'Other', 'Airplane', 'Intercity train (Amtrak)', 'Transit - MARTA bus', 'Dial a ride or paratransit', 'Intercity bus (greyhound, trailways)', 'School bus', 'Taxi, shuttle bus or limousine', 'Auto or van or truck - driver', 'Motorcyle or moped', 'Bicycle', 'Transit - CCT bus', 'Heavy rail - MARTA'
- **Arrival time** - 8 values: '3am-8am', '8am-10am', '10am-12pm', '12pm-2pm', '2pm-4pm', '4pm-6pm', '6pm-8pm', '8pm-3am'
- **Departure time** - 8 values: '3am-8am', '8am-10am', '10am-12pm', '12pm-2pm', '2pm-4pm', '4pm-6pm', '6pm-8pm', '8pm-3am'
- **Duration** - 7 values: '10 minutes or less', '10-30 minutes', '30-60 minutes', '1-2 hours', '2-4 hours', '4-8 hours', 'Greater than 8 hours'
- **Age** - 9 values: '10 years old or less', '10-20 years old', '20-30 years old', '30-40 years old', '40-50 years old' '50-60 years old', '60-70 years old', '70-80 years old', 'greater than 80 years old'

Mining Determining Sets for Partially Defined Functions

Dan A. Simovici, Dan Pletea, and Rosanne Vetro

Univ. of Massachusetts Boston, Dept. of Comp. Science, 100 Morrissey Blvd. Boston,
Massachusetts 02125 USA
{dsim,dpletea,rvetro}@cs.umb.edu

Abstract. This paper describes an algorithm that determines the minimal sets of variables that determine the values of a discrete partial function. The Apriori-like algorithm is based on the dual hereditary property of determining sets. Experimental results are provided that demonstrate the efficiency of the algorithm for functions with up to 24 variables. The dependency of the number of minimal determining sets on the size of the specification of the partial function is also examined.

1 Introduction

Partially defined finite functions are studied by both mathematicians and engineers due to their many technical applications, particularly in designing switching circuitry. They model such diverse circuits as logical programmable arrays, or content addressable memory. The performance of such circuits (including wiring complexity, power dissipation, etc.) is heavily influenced by the number of arguments on which the function implemented by the circuit depends effectively.

The goal of this paper is to present an algorithm to generate various sets of input variables on which a partial function depends using an approach inspired by Apriori, a well-known data mining algorithm developed for determining frequent item sets in transactional databases [1,2,3].

The proposed algorithm is based on the fundamental observation that and superset of a determining set for a partially defined function f is itself a determining set for f. We use this dual heredity property of determining sets to formulate an Apriori-like algorithm that computes the determining sets by traversing the lattice of subsets of the set of variables.

This problem has been addressed in [4] using an algebraic minimization algorithm that applies to functions that depend on small number of variables. Our approach is distinct and involves techniques inspired by data mining. Additionally, it has the advantage of not being linked to any value of the input or output radix of the partial function f.

The rest of the paper is organized as follows. In Section 2 we introduce the notion of determining set for a partial function and examine a few properties of these sets that play a role in our algorithm. This algorithm is presented in Section 3. Section 4 discusses experimental work related to the algorithm. In the last section (Section 5) we present conclusions and future work.

P. Perner (Ed.): ICDM 2009, LNAI 5633, pp. 353–360, 2009.
© Springer-Verlag Berlin Heidelberg 2009

2 Determining Sets for Partially Defined Functions

We denote the finite set $\{0, 1, \ldots, n - 1\}$ by $\mathbf{n}$. The partial functions that we study have as domain a subset of the finite set $\mathbf{r}^n$ and as range a subset of the finite set $\mathbf{p}$ for some positive natural numbers r and p, referred to as the *input radix* and the *output radix* of the function, respectively. The set of all such partial functions is denoted by $\mathsf{PF}(\mathbf{r}^n, \mathbf{p})$. If $f \in \mathsf{PF}(\mathbf{r}^n, \mathbf{p})$ we denote by $\mathrm{Dom}(f)$ the set of all n-tuples $(a_1, \ldots, a_n)$ in $\mathbf{r}^n$ for which $f(a_1, \ldots, a_n)$ is defined.

A partial function $f \in \mathsf{PF}(\mathbf{r}^n, \mathbf{p})$ is specified as a table T_f having columns labelled by the argument variables $x_1, \ldots, x_n$ and by the output variable y. If $f(a_1, \ldots, a_n) = b$ we have in the table T_f the $(n + 1)$-tuple $t = (a_1, \ldots, a_n, b)$. For example, in Table 1 we show a partial function defined on all triplets in $\mathbf{3}^3$ that contain at least two non-zero elements, and ranging in the set $\mathbf{4}$: The number of rows of the table that represents

Table 1. Tabular Representation of a Partial Function

x_1	x_2	x_3	y
0	1	1	0
0	1	2	1
0	2	1	2
0	2	2	2
1	0	1	3
1	0	2	3
2	0	1	3
2	0	2	3
1	1	0	2
1	2	0	2
2	1	0	1
2	2	0	0

a partial function defined on $\mathbf{r}^n$ can range between 0 and r^n. Usually, the number of rows of such a function is smaller than r^n and, often this number is much smaller. Tuples $(a_1, \ldots, a_n)$ that do not belong to the definition domain of f are considered as "don't care" tuples, that is, as input sequences that are unlikely to occur as inputs of the functions, or the output of the function for such inputs is indifferent to the designer.

For a tuple t in T_f and a set of variables $U \subseteq \{x_1, \ldots, x_n, y\}$ we denote by $t[U]$ the *projection* of t on U, that is, the restriction of t to the set U. If U consists of one variable we denote the projection $t[\{z\}]$ just by $t[z]$.

Definition 1. *A set of variables $V = \{x_{i_0}, \ldots, x_{i_{p-1}}\}$ is a* determining set *for the partial function f if for every two tuples t and s from T_f, $t[V] = s[V]$ implies $t[y] = s[y]$.*

In other words, V is a determining set for the partial function f if $t = (a_0, \ldots, a_{n-1}, b)$ and $s = (c_0, \ldots, c_{n-1}, d)$ in T_f such that $a_{i_k} = c_{i_k}$ for $1 \le k \le p$ implies $b = d$. The collection of determining sets for f is denoted by $\mathsf{DS}(f)$.

V is a *minimal determining set for f* if V is a determining set for f and there is no strict subset of V that is a determining set for f. The set of minimal determining sets of f is denoted by $\mathsf{MDS}(f)$. Our main purpose is to present an algorithm that extracts the minimal determining sets for a partially specified function.

We introduce a partial order relation "$\sqsubseteq$" on the set of partial $\mathsf{PF}(\boldsymbol{r}^n, \boldsymbol{p})$ by defining $f \sqsubseteq g$ if $\mathrm{Dom}(f) \subseteq \mathrm{Dom}(g)$ and $f(a_1, \ldots, a_n) = g(a_1, \ldots, a_n)$ for every $(a_1, \ldots, a_n)$. In other words, we have $f \sqsubseteq g$ if g is an extension of f.

The following simple statement is crucial to the proposed algorithm.

Theorem 1. *Let f and g be two partial functions in $\mathsf{PF}(\boldsymbol{r}^n, \boldsymbol{p})$. If $V \in \mathsf{DS}(f)$ and $V \subseteq W$, then $W \in \mathsf{DS}(f)$. Furthermore, if $f \sqsubseteq g$, then $\mathsf{DS}(g) \subseteq \mathsf{DS}(f)$.*

Proof. If V and W are two sets of variables such that $V \subseteq W$ and t, s are two tuples in T_f, then $t[W] = s[W]$ implies $t[V] = s[V]$. Therefore, if V is a determining set for f and $t[W] = s[W]$, it follows that $t[V] = s[V]$, which implies $t[y] = s[y]$. Thus, W is a determining set for f.

For the second part of the theorem, observe that if $f \sqsubseteq g$ and $V \in \mathsf{DS}(g)$, then $t[V] = s[V]$ implies $t[y] = s[y]$, for every $t, s \in \mathrm{Dom}(g)$. Since $\mathrm{Dom}(f) \subseteq \mathrm{Dom}(g)$, the same implication holds for any two tuples in $\mathrm{Dom}(f)$, so $V \in \mathsf{DS}(f)$. $\square$

Note that if $f \sqsubseteq g$ and $V \in \mathsf{MDS}(g)$, then there exists $Z \in \mathsf{MDS}(f)$ such that $Z \subseteq V$.

3   An Apriori-Like Algorithm for Mining MDSs

Our algorithm uses Rymon trees (see [5,6]) also known as set enumeration trees, a device that is useful for the systematic enumeration of the subsets of a set S. The subsets of a set S (which constitute the power set of S, $\mathcal{P}(S)$) are listed using a pre-imposed total order on the underlying set S. The total order on S is specified by an one-to-one function $\mathsf{ind} : E \longrightarrow N$.

For every subset $U \subseteq S$, define its view as

$$\mathsf{view}(\mathsf{ind}, U) = \{s \in S \mid \mathsf{ind}(s) > \max_{u \in U} \mathsf{ind}(u)\}$$

Definition 2. *Let $\mathcal{F}$ be a collection of sets closed under inclusion (i.e., if $U \in \mathcal{F}$ and $U \subseteq V$, then $V \in \mathcal{F}$).*

The labelled tree $\mathcal{T}$ is a Rymon tree for $\mathcal{F}$ if

(i) *the root of $\mathcal{T}$ is labelled by the empty set $\emptyset$, and*
(ii) *the children of a node labelled U in $\mathcal{T}$ are $\{U \cup \{e\} \in \mathcal{F} \mid e \in \mathsf{view}(\mathsf{ind}, U)\}$.*

In Figure 1 we show the Rymon tree for the complete power set of a set $S = \{1, 2, 3, 4\}$. The proposed algorithm takes as input a partially defined function f and outputs a collection of sets with minimum number of variables that f depends on. The algorithm performs a breadth-first search on the Rymon tree for the power-set of the set of variables $E = \{x_1, x_2, ..., x_n\}$ of f. The search stops when all the sets with the minimum number of variables that functions f depends on are found; these sets are referred to as determining sets. The minimum number corresponds to the lowest level in the Rymon

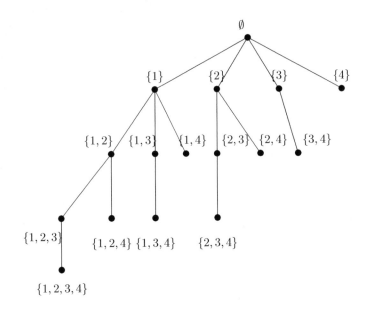

Fig. 1. The Rymon tree for the $\mathcal{P}(\{1, 2, 3, 4\})$

tree where the first solution set is found since all the levels below have nodes containing sets with a higher number of variable than any level above it.

In Algorithm 2 we denote the breadth first search queue by Q, the array of children of a node X by $\mathsf{Child}[X]$, and the tree level of the determining sets by dLevel.

The algorithm is using the following methods:

ENQUEUE(Q, V) inserts node V in queue Q;
DEQUEUE(Q)   removes the first element of queue Q;
LEVEL(V)   returns the level of node V in the tree;
IS_DSET(v)   informs if the set of variables corresponding
            to node V is a determining set for the
            partially defined function f;
ADD($\mathcal{D}, V$)   add the set of variables corresponding to node V to $\mathcal{D}$.

The core of the algorithm is the procedure IS_DSET(V) that has as an input argument a set of variables V and returns **true** if V is a determining set and false, otherwise. In principle, if T_f is a database table, the implementation of IS_DSET could be done using embedded SQL by running the query

select count(**distinct** y) **from** T_f **group by** V.

It is clear that if all values returned for count(**distinct** y) equal 1, then V is a determining set for f. In practice, the overhead entailed by using the database facility impacts negatively on the performance of the algorithm, which lead us to another solution that is based on storing T_f as a file and searching that file.

```
Input: A partially defined function f
Result: A collection 𝒟 of minimal determining variables sets
1 begin
2     dLevel ⟵ ∞
3     ENQUEUE(Q,∅)
4     while Q ≠ ∅ do
5         X ⟵ DEQUEUE(Q)
6         foreach V ∈ Child[X] do
7             ENQUEUE(Q,V)
8             if 𝒟 = ∅ or LEVEL(v) ≤ dLevel then
9                 if IS_DSET[V] then
10                    ADD(𝒟, V)
11                    if dLevel = ∞ then
12                        dLevel = LEVEL(V)
13            else
14                break
15 end
```

Fig. 2. Computing MDS(f)

The procedure *IS_DSET* shown in Figure 3 makes use of a hash table *MAP*, where the key is determined by the chosen set of variables. The following methods are used:

GET_VARIABLES(V): retrieves the set of variables corresponding to node V
GET_VALUES($tuple, S$): retrieves the values of the variables in S
ELEMENT(*MAP*, *key*): returns the object instance stored in *MAP*
that contains a certain *key*
GET_FVALUE(y): returns the function value of the object instance y
$F(tuple)$: returns the function value of a certain tuple
ADD(*MAP*, *key*, $F(tuple)$): adds an object instance containing
a key and function value to the *MAP*.
The following variables are used in the IS_DSET procedure:

S set of variables
v a node in the tree
$File$ input file containing the tuples of a partially defined function
$tuple$ a row of the input file
key a set with the values of the variables in S
MAP a hash structure that stores objects containing a key and a function value
y an object instance stored in the MAP

4   Experimental Results

We carried out experiments on a Windows Vista 64-bit machine with 8Gb RAM and 2 × Quad Core Xeon Proc E5420, running at 2.50 GHz with a 2×6Mb L2 cache. The algorithm was written in Java 6.

```
Input: A node containing a subset of the variables set
Output: true if the set is a determining one, false, otherwise
1 begin
2        S ⟵ GET_VARIABLES(V)
3        foreach tuple ∈ File do
4            key ⟵ GET_VALUES(tuple, S)
5            if key ∈ MAP then
6                y ⟵ ELEMENT(MAP, key)
7                if F(tuple) ≠ GET_FVALUE(y) then
8                    return false
9                    break
10           else
11               ADD(MAP,key,F(tuple))
12       return true
13 end
```

Fig. 3. Procedure IS_DSET(V)

We analyze the results in terms of running time, minimum number of variables of a determining set, and the number of determining sets as a function of the number of tuples in T_f.

A program that randomly generates comma separated text files representing partially defined functions with 8, 16 or 24 variables was developed. These values were chosen based on the experiments made in the related work of T. Sasao [4].

One hundred files were randomly generated for each type of partially defined function (with 8, 16, and 24 variables) using an input radix $r = 3$ and an output radix $p = 5$.

Note that a totally defined function with 8 variables and $r = 3$ has $3^8 = 6561$ tuples. In our experiments, we randomly generated 1000 tuples for partially defined functions with 8 variables. For functions that depend on 16 and 24 arguments we generated 5000 tuples because the number of tuples for completely defined functions with 16 or 24 variables is much higher.

In the experiments, we evaluate the performance of the algorithm with a varying number of tuples. By Theorem 1, if $(f_1, f_2, \ldots, f_k)$ is a sequence of functions such that

$$f_1 \sqsubseteq f_2 \sqsubseteq \cdots \sqsubseteq f_k,$$

we have

$$\mathsf{DS}(f_k) \subseteq \cdots \subseteq \mathsf{DS}(f_2) \subseteq \mathsf{DS}(f_1).$$

In other words, when we start with a partial function f_1 with a small specification table T_{f_k} and we expend sequentially the specification of the functions, the number of determining sets will decrease. The experiments compare the results for files with 8, 16 and 24 variables and they contain averages of the values corresponding to time, minimum number of variables the function depends on, and number of sets with minimum number of elements the function depends on as a function of the number of tuples. In our case, $k \in \{10, 15, 20, 30, 40, 50, 75, 90, 100, 200\}$. The averages are evaluated over 100 functions within each group of generated functions (8, 16 and 24 variables).

As shown in Fig. 4, the running time of the algorithm to find a solution increases with the number of tuples because in most cases, the algorithm needs to search deeper in the Rymon tree. Also, the time increases exponentially with the number of variables. The algorithm performs a breadth-first search and functions with more variables will have trees with a larger branching factor.

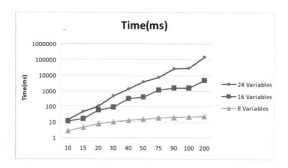

Fig. 4. Dependency of average time on number of tuples

Fig 5 shows that the minimum number of variables the function depends on is related to the number of tuples k. As k increases, the constraints imposed on the problem become more extensive, and the minimum number of variables that determine the value of the function increases.

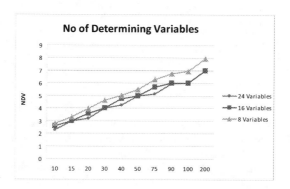

Fig. 5. Average size of minimal determining set for 8, 16 and 24 variables, as a function of the number of tuples

Finally, the experiments also show that the average number of minimal determining sets decreases as we extend the partial functions by introducing more tuples. Fig 6 illustrates this decrease for functions with 8 and 16 variables. The decrease is not as noticeable for functions with 24 variables because these functions have a large number of possible tuples and this behavior can only be observed for a much higher value of k than the maximum used in experiments presented here.

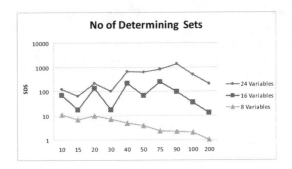

Fig. 6. Average size of $\mathsf{MDS}(f)$ for 8, 16 and 24 variables, as a function of the number of tuples

5   Concluding Remarks

This paper introduces an algorithm to determine the minimum number of variables on which a partially defined function depends on, as well as determine all sets of variables with minimum number of elements that define the function. The algorithm is based on traversing Rymon trees using a breadth-first search technique. We believe that the algorithm will be helpful for digital circuit design since it allows to determine the possible sets of variables on which a partial function depends starting from a tabular specification of the function

We intend to approach the same problem using a clustering technique for discrete functions starting from a semi-metric that measures the discrepancy between the kernel partitions of these functions.

References

1. Agrawal, R., Imielinski, T., Swami, A.N.: Mining association rules between sets of items in large databases. In: Buneman, P., Jajodia, S. (eds.) Proceedings of the 1993 International Conference on Management of Data, pp. 207–216. ACM, New York (1993)
2. Mannila, H., Toivonen, H.: Levelwise search and borders of theories in knowledge discovery. Technical Report C-1997-8, University of Helsinki (1997)
3. Zaki, M.J., Hsiao, C.: Efficient algorithms for mining closed itemsets and their lattice structure. IEEE Transactions on Knowledge and Data Engineering 17, 462–478 (2005)
4. Sasao, T.: On the number of variables to represent sparse logic functions. In: 17th International Workshop on Logic and Synthesis (IWLS 2008), Lake Tahoe, California, USA, pp. 233–239. IEEE-CS, Los Alamitos (2008)
5. Rymon, R.: Search through systematic set enumeration. In: Nebel, B., Rich, C., Swartout, W.R. (eds.) Proceedings of the 3rd International Conference on Principles of Knowledge Representation and Reasoning, Cambridge, MA, pp. 539–550. Morgan Kaufmann, San Mateo (1992)
6. Simovici, D.A., Djeraba, C.: Mathematical Tools for Data Mining – Set Theory, Partial Orders, Combinatorics. Springer, London (2008)

On the Integration of Neural Classifiers through Similarity Analysis of Higher Order Features

D.A. Karras[1] and B.G. Mertzios[2]

[1] Chalkis Institute of Technology, Automation Dept. and Hellenic Open University,
Rodu 2, Ano Iliupolis, Athens 16342, Greece
dakarras@teihal.gr, dakarras@ieee.org, dakarras@usa.net
[2] Technological Institute of Thessaloniki, Greece
mertzios@uom.gr

Abstract. A novel methodology is herein outlined for combining the classification decisions of different neural network classifiers. Instead of the usual approach for applying voting schemes on the decisions of their output layer neurons, the proposed methodology integrates higher order features extracted by their upper hidden layer units. More specifically, different instances (cases) of each such classifier, derived from the same training process but with different training parameters, are investigated in terms of their higher order features, through similarity analysis, in order to find out repeated and stable higher order features. Then, all such higher order features are integrated through a second stage neural network classifier having as inputs suitable similarity features of them. The herein suggested hierarchical neural system for pattern recognition shows improved classification performance in a computer vision task. The validity of this novel combination approach has been investigated when the first stage neural classifiers involved correspond to different Feature Extraction Methodologies (FEM) for shape classification. The experimental study illustrates that such an approach, integrating higher order features through similarity analysis of a committee of the same classifier instances (cases) and a second stage neural classifier, outperforms other combination methods, like voting combination schemes as well as single neural network classifiers having as inputs all FEMs derived features. In addition, it outperforms hierarchical combination methods non performing integration of cases through similarity analysis.

Keywords: Combination of Neural classifiers, similarity, committee machines, higher order features, feature extraction methods, hybrid and modular structures, shape recognition, statistical moments.

1 Introduction

In pattern recognition applications the features are extracted using a Feature Extraction Method (FEM), which produces a suitable set of features of the desired pattern, according to the requirements of each particular application. The selection of the appropriate FEM for a considered application depends on the specific conditions and requirements, in order to achieve the higher classification efficiency. To this end, it is

P. Perner (Ed.): ICDM 2009, LNAI 5633, pp. 361–371, 2009.
© Springer-Verlag Berlin Heidelberg 2009

essential in demanding applications to use a combination of different FEMs. The underlying idea is that multiple FEMs contribute different features of the same pattern that correspond to different levels of importance, carrying different information. We herein employ contour-based FEMs. A significant number of contour-based FEMs that are suitable for shape representation and analysis are reported in the literature, which are based on geometric characteristics [1], statistical moments [2]-[5], Fourier descriptors [6],[7], syntactic analysis [8] and polygon approximation techniques [9], [10]. Two different contour-based FEMs, whose features in general carry different and complementary information, are proposed for the development of the proposed hierarchical image classification scheme, which are appropriate for shape classification tasks. In the first FEM, the extracted feature vector is based on three Geometric Features (GF) of the contour of the shape [1]. The second FEM is based on the Scaled Normalized Central (SNC) moments [4], [5]. These methods are invariant of the position, the size and the orientation of the shape as well as they are invariant of the starting point of the contour of the shape.

However, the main question is how to efficiently combine these FEMs in a pattern classification system. An answer to this problem is the main goal of this paper. Actually, the main trend in the literature concerning such neural based classification systems is to use independent neural networks, one for each FEM, and then, attempting to combine their decisions through voting schemes mainly or through least squares etc. [11,12,13]. Such schemes try to combine output layer neurons decisions (outputs) involving a statistical procedure based on parametric or non-parametric statistics, or a procedure based on evidence theory, information theory etc.. However, we herein adopt a different approach since, although output layer neurons in single feedforward neural classifiers convey information, this information is their final class assignment. Such an assignment is based on higher order features extracted from their upper hidden layer units. However, this class assignment doesn't convey explicitly the rich information content of such higher order features. When attempting to combine only class assignments of different single feedforward neural classifiers we don't take explicitly into account this rich information content. Therefore, we adopt a methodology involving a direct combination of higher order features of different such neural classifiers in a second stage of an hierarchical neural system, instead of class assignment combination approaches mainly utilized in the literature. The experimental section of this paper shows that such a strategy outperforms other traditional rival neural decision combination approaches.

2   The Contour-Based Feature Extraction Methods (FEM)

The selection of FEMs for each recognition task depends on the conditions and requirements of the specific application. The FEMs, which are used to characterize a digitized shape, can be distinguished in two basic classes. The first class includes methods, which are applied to the 2-D representation of the shapes, while the second class includes methods, which are applied to the 1-D functional representation of the contour of the shapes.

In the sequel we consider only 1-D FEMs. Specifically, an 1-D representation of a contour, which is determined as an ordered sequence representing the Euclidean distance between the centroid of the contour and all the contour pixels of the digitized shape, is adopted. This ordered sequence provides a *signature* of the shape, which constitutes a unique representation of the shape [1]. Thus, if the set (x_i, y_i), $i = 1,2,...,N$, represents the coordinates of N ordered contour pixels, then the signature of the shape is constituted by the Euclidean distances z_i, $i=1,2,...,N$, of the vectors connecting the centroid $(\overline{x}, \overline{y})$ and the set of contour pixels (x_i, y_i).

The *Starting Point* is defined as the pixel of the contour (x_0, y_0) having the maximum Euclidean distance D_M from the centroid $(\overline{x}, \overline{y})$.

The area A, which is enclosed by the contour of the shape, is calculated by the formula:

$$A = \sum_{i=1}^{N} (x_i + x_{i-1})(y_i - y_{i-1})$$

Note that in computing the perimeter of a contour, whose points are given on a discretized grid, care should be taken to include the diagonal distances. The perimeter of the contour P is calculated by

$$P = N_p + \sqrt{2}N_d$$

where N_p is the number of all the elementary linear segments found on the horizontal or vertical direction and N_d is the number of elementary linear segments, when consecutive pixels are diagonally connected.

In the sequel, two different 1-D typical FEMs, which are suitable for generic shape discrimination applications, are described. These methods are implemented on the region boundaries and they are invariant of position, size and orientation of the shape. They also provide different types of features corresponding to various levels of importance and they are sensitive on different level of noise or distortion of the shape.

A. The geometric features (GF)

The geometric features of a shape are commonly used for pattern classification applications [1]. These features are useful in applications, where the patterns are not complicated and the desired discrimination efficiency is not very high. Specifically, these features are used (a) for non detailed recognition, (b) for an initial classification procedure and (c) in combination with other methods [1]. In the sequel three geometric features of different importance are described in this Section.

a) The *Compactness* C of the shape. Compactness provides a simple dimensionless measure of the shape complexity and is given by [1]

$$C = \frac{P^2}{A}, \quad 4\eth \leq C < \infty$$

where P is the length of the spape's perimeter and A is the area of the shape.

A more suitable form of the compactness measure may be given by the Normalized Inverse Compactness $\overline{C}$, which is given by:

$$\overline{C} = \frac{4\eth}{C} = \frac{4\eth A}{P^2}, \quad 0 < \overline{C} \leq 1$$

In the previous formula, the upper limit represents an ideal circle, while the lower limit represents an ideal line of infinite length. The above definition of $\overline{C}$ is more suitable as a classification feature, since $\overline{C}$ is bounded by finite limits. The normalized inverse compactness $\overline{C}$ (as well as C), is characterized by increased sensitivity to shape distortions and it presents better classification properties among shapes with complicated form, since it is related to the perimeter P of the region.

b) The *Normalized Area* $\overline{A}$ of the shape. This area is defined as the quotient of the area A of the shape divided by the area of the hypothetical circle with centre the centroid $(\overline{x},\overline{y})$ of the shape and radius the maximum distance D_M of the shape, and it is given by

$$\overline{A} = \frac{A}{\partial D_M^2}, \quad 0 < \overline{A} \leq 1$$

In this formula, the upper limit represents an ideal circle, while the lower limit represents an ideal line of infinite length. The normalized area $\overline{A}$ provides a simple measure of the shape's circularity, which constitutes a low discrimination efficiency feature among similar shapes. However, $\overline{A}$ constitutes a robust characteristic in shape distortions and therefore, it carries complementary information to compactness.

c) The *Normalized Length* $\overline{L}$ of the shape. This length constitutes a measure of the contour complexity and is defined as:

$$\overline{L} = \frac{\overline{z}}{D_M}, \quad 0 < \overline{L} \leq 1$$

where $\overline{z}$ is the mean of the set z_i, i=1,2,...,N. The upper limit represents an ideal circle, while an ideal line of infinite length is represented by a factor equal to 0.5. The normalized length $\overline{L}$ also constitutes a low discrimination efficiency feature among similar shapes and constitutes a robust characteristic in shape distortions, since it is related to the mean $\overline{z}$.

The above three features $\overline{C}$, $\overline{A}$ and $\overline{L}$ are dimensionless and are invariant of the position, the size, and the orientation of the shape. Also they carry different information, which results to a better description of a digitized shape.

B. The scaled normalized central set of moments (SNC)
Statistical moment-based methods are used in a large number of image analysis and pattern recognition applications. The features resulting from a moment based method usually provide good description of a shape and therefore have good classification properties. Many researcher efforts have been presented aiming to the improvement of the efficiency of the statistical moments regarding to the robustness, normalization and complexity reduction [2]-[5].

Given a $M \times N$ digital image, the *2-D geometric moment* of order (p,q) is defined as :

$$m_{pq} = \sum_{x=1}^{M}\sum_{y=1}^{N} x^p y^q f(x,y), \quad p,q = 0,1,2,...$$

where M and N are the horizontal and vertical dimensions of the image, and $f(x,y)$ is the intensity at a point (x,y) in the image. The *2-D central moment* of order (p,q) is invariant of translation and is defined by :

$$\grave{\imath}_{pq} = \sum_{x=1}^{M}\sum_{y=1}^{N}(x-\overline{x})^p(y-\overline{y})^q f(x,y), \quad p,q = 0,1,2,...$$

where $\overline{x}, \overline{y}$ are the coordinates of the centroid, which are given by :

$$\overline{x} = \frac{m_{10}}{m_{00}}, \quad \overline{y} = \frac{m_{01}}{m_{00}}$$

The above definition of 2-D moments provides robust characteristics and good shape description but the computation time is very high. An alternative definition of moments results by using the 1-D representation of the contour of the shape. The computation time of 1-D moments is severely reduced, at the expense of increased sensitivity in shape distortions. The 1-D geometric moment of order k is defined as:

$$m_k = \sum_{i=1}^{\acute{\imath}}(z_i)^k, \quad k = 0,1,2,...$$

and the 1-D central moment of order k is defined as :

$$\grave{\imath}_k = \sum_{i=1}^{\acute{\imath}}(z_i - \overline{z})^k$$

where

$$\overline{z} = \frac{m_1}{m_0} = \frac{1}{N}\sum_{i=1}^{N}z_i$$

Since high order moments increase exponentially and also describe the finer characteristics of a shape, the variations of small shape distortions are very high, thus resulting to low discrimination efficiency. In this application a special case of the general form of 1-D *scaled normalised central* (SNC) moments is used. The considered 1-D SNC moment of order k is defined as follows :

$$h_k = \left[\frac{\hat{a}}{m_0}\right]\acute{a}^k\sum_{i=1}^{N}(z_i - \overline{z})^k = \left[\frac{\hat{a}}{m_0}\right]\acute{a}^k\grave{\imath}_k, \quad k = 0,1,2,...$$

where α is the scaling factor corresponding to the h_k moment, which is used in order to avoid the exponential increase of the high order moments and β is the normalization factor. A good selection for the parameters α and β that produces robust characteristics and ensures that the high order moments are not exponentially increased, is given by $\acute{a} = (\overline{z}/2)^{-1}$ and $\beta = 1$. For this pair of parameters, the above formula results to:

$$h_k = \frac{1}{m_0} \left[\frac{2}{\bar{z}} \right]^k \sum_{i=1}^{N} (z_i - \bar{z})^k = \frac{1}{N} \left[\frac{2}{\bar{z}} \right]^k \hat{i}_k, \quad k = 0,1,2,\ldots$$

Also, the 1-D SNC moments are invariant of the position, size, orientation and the starting point of the shape. These properties render the 1-D SNC moments a good choice for shape recognition applications.

3   The Proposed Hierarchical System For Efficiently Combining Higher Order Features Extracted in Upper Hidden Layers via Case Based Similarity Analysis

A high performance Neural Network based hierarchical and hybrid Modular Image Classification Scheme is proposed in this paper, which is characterized by a great degree of modularity and flexibility, very efficient for demanding, large scale and generic pattern recognition applications. The classifier is composed of two stages. **The first stage is comprised of a number of different neural subnetworks/ classifiers that operate in parallel, while the second stage is a Coordinator Neural Network that performs the final image classification task combining the higher order features realized by all the neural subnetworks/ classifiers of the first stage**.

This is a major novelty of the present paper with regards to similar approaches found in the literature so far [11,12,13]. Namely, all previous similar research projects attempt to combine decisions of the modular subnetworks comprising the hierarchical classifier by applying different voting schemes on the decisions of such subnetworks realized by their output layers. However, in the herein presented approach we realize a completely different strategy.

The coordinator Neural Network, that is the second stage network in the hierarchy, combines the higher order features realized by the upper layer hidden units of every modular subnetwork of the proposed structure in order to enhance their single decisions. The rationale underlying such a novel scheme is that if all such higher order features are used to derive a unique decision be the Coordinator Neural Network then, the rich information conveyed by all of them is involved in the decision.

On the other hand, if the Coordinator Module combines only the decisions of such subnetworks acting independently one another, as it is usual in the literature, then, the different feature extraction underlying methods correlation characteristics are lost in the decision. Actually different feature extraction methods carry complementary, correlated information and depend on one another in a non analytical way. Our approach attempts to convey higher order correlation information between different feature extraction methods in the Coordinator Neural Network in order to enhance its decision capabilities.

Figure 1 below shows the schematic diagram of the proposed hierarchical architecture for image classification applications. In this figure Xi are the different feature extraction methods providing inputs to the different subnetworks, while Oj are the upper layer higher order features realized by such subnetworks. C-NET is the coordinator neural network

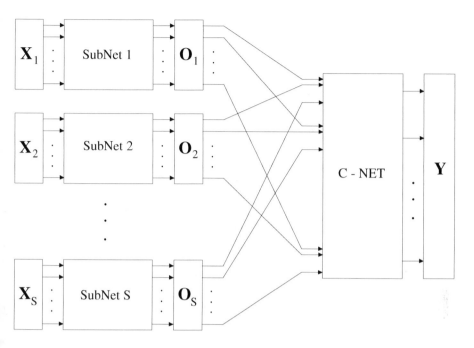

Fig. 1. The schematic diagram of the proposed hierarchical architecture for image classification

In general, each neural subnetwork operates as a classifier itself, with inputs the features that are extracted from different feature extraction methods and correspond to various levels of importance. The proposed hybrid hierarchical classifier is applied to a shape recognition task of 2-D digitized objects considering various levels of shape distortions. Two different kinds of features, which characterize a digitized object, are used, namely, geometric features and the 1-D scaled normalized central moments defined in the previous section. The two Subnets corresponding to these two FEMs constitute the first stage of the proposed system.

Each such subnet of figure 1 is trained independently of all other subnets based only on the corresponding FEM associated training vectors and their upper hidden layer neurons higher order features are extracted.

The main novelty, however, of the present paper lies in the fact that we don't consider a single instance of such a first stage subnetwork. We consider, instead, many cases (10 in our experimental study) of each such subnetwork which are derived from the same training process but with different training parameters. To this end, one could use different training subsets to train the classifier, or could change learning parameters, or even use different random starting points for the training process. Our strategy has been to select different initial weights for the training of the different cases of each neural subnetwork. The critical issue, however, is to answer why to increase complexity in the solution by considering such many cases for each subnetwork. The answer is that a single case of each subnetwork might convey not robust higher order features but features depending on training parameters! To alleviate this problem we attempt to derive training independent robust higher order features that will be integrated in the second stage with the coordinator network.

A first order integration is therefore, needed for such higher order features associated with the different cases. We herein propose that this can be achieved through similarity analysis. That is, we try to find the most similar, i.e the stable vectors of such higher order features. Such a task could be achieved through clustering and similar methodologies but our case is simpler, since we used only ten instances.

Suppose, $X1, X2, ... X10$ are such higher order feature vectors corresponding to the ten different cases of subnetwork of FEM X. It is easy, through calculating all such inner products $Xi\ Xj$ to estimate $\cos(Xi,Xj) = Xi\ Xj\ /\ (\|Xi\|\ \|Xj\|)$ as similarity metric of $Xi\ ,\ Xj$.

Then, by ordering in ascending order all these cosines we can have the maximum $\cos(Xi,Xj)$ which illustrates that the pair (Xi,Xj) is a pair of the most similar higher order feature vectors. This means they are stable vectors and indepen- dent of the subnetwork's training parameters.

In the sequel, we perform Singular Value Decomposition (SVD) for the matrix $Xi * Xj^T$ in order to extract the set of singular values that characterizes the specific pair in a unique way. This unique vector Si,j comprises the higher order features integration for all the cases of the specific subnetwork corresponding to FEM X.

The second stage of the presented methodology is a Coordinator Neural Network (C-NET) that performs the final classification task using a combination of all the neural subnetworks, so that the whole discrimination efficiency is optimized. The proposed architecture is hybrid in the sense that various neural network architectures, training algorithms and FEMs may be used for each individual subnetwork and modular since the same algorithm and architecture may be used for the implementation of more than one subnetworks. Its first stage may be composed from a number of building blocks (the subnetworks) that are separately designed.

The proposed novel hierarchical image classification architecture that uses in its first stage two feedforward neural subnetworks that operate with the above two FEMs, is applied on a shape recognition application. In this application it was assumed that the shape variations are those resulting from three quantized artificial types of bounded distortions, which produce a finite number of shape variations. The performance of the proposed methodology is illustrated and compared very favorably with those of single neural classifiers, which use one or more FEMs, as well as with the performance of other traditional classifier combination methods like voting techniques.

From the above discussion the proposed algorithm is as follows:

STEP 1: Prepare training and testing data sets.

STEP 2: Train all individual feedforward FEM neural networks based on their corresponding FEM derived training pattern vectors. Derive M (=10) cases of trained networks for each FEM subnetwork, starting from different initial weights

STEP 3: After the training phase is finished, run each subnet case through all training set again in its feedforward pass mode only, to extract its upper hidden layer outputs as a vector. Repeat this procedure for all cases for all subnetworks.

STEP 4: Perform similarity analysis for all pairs belonging to the instances (cases) of each subnetwork. Identify the most similar (stable) such pair and then, perform

SVD for the matrix corresponding to the specific pair in order to define the vector of singular values that comprise the stable higher order features for the specific subnetwork to be next integrated.

STEP 5: For each such vector form an augmented training vector having as components all such singular value vectors, for the different FEMs based subnetworks. A new training set is then, defined having as inputs such augmented vectors.

STEP 6: A multilayer feedforward (MLP) C-NET is trained using this set of higher order FEM feature training vectors. Of course, as implied by figure 1, C-NET as well as the FEM subnetworks have the same number of outputs, corresponding to all possible class assignments with regards to the given problem.

STEP 7: Run FEM subnetworks as well as the C-NET in the feedforward pass mode only, with regards to the test set of pattern vectors correspondingly.

4 Experimental Study

In order to justify the usefulness of the proposed method, a set of experiments have been carried out and their results are presented in this section. To perform the experiments some test objects (patterns) are initially selected. Figure 2 shows the 9 such shapes/ patterns used in our experiments. The (256x256) images of these parts are the initial nine patterns of our experiments.

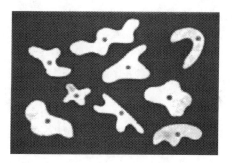

Fig. 2. The nine original patterns whose transformations produce the training and test sets

In order to produce the training and test sets out of these images 6750 variations of these patterns have been created (750 variations for each original image), in order to sufficiently represent the whole space of possible variations. The variations have been produced by adding different noise distributions to the original image (Gaussian etc.) as well as by applying rotations and distortions to them. The training set was consisted of 4050 randomly selected training patterns out of the total 6750 (450 for each prototype), while the rest were divided into a validation set of 675 patterns (75 for each prototype) and into a test set of 2025 patterns (225 for each prototype).

All feedforward neural networks have been trained using the Conjugate Gradient Algorithm.

Three combination methods are compared. The first one is derived by applying simple voting to the decisions of the two trained subnets, namely, NGF (Net of Geometric Features) and NSNC (Net of Scaled Normalized Central moments) as described in sections II, III.

NGF's performance has been optimized with a GF-60-9 architecture, while NSNC's performance has been optimized with a SNC-30-30-9 architecture.

The second method is the proposed one involving a 90 (60 hidden of NGF and upper 30 of NSNC subnets) -30-30-9 architecture.

The second combination method involves a single feedfoward neural network (MLP-Multilayer Perceptron) of GF+SNC input features -30-30-9 architecture

The third and the fourth method are basically the proposed one involving a 90 (60 hidden of NGF and upper 30 of NSNC subnets) -30-30-9 architecture. However, the third one does not involve but a single instance of each subnetwork, without similarity analysis, while the fourth one is the one described previously (section 3).

The results are as follows:

NGF alone TEST set accuracy: **87,75%**
NSNC alone TEST set accuracy: **98,32%**

First method's (voting) TEST set accuracy: **99,11%**

Second method's (single MLP) TEST set accuracy: **98.85%**

Third, the proposed methodology
but with no similarity analysis using
a single instance of each subnetwork
Training set success :   99.6296%
Validation set success: 98.6667%
TEST set success :   **99.7037%**

Fourth, the proposed methodology
with similarity analysis as described by
the steps 1-7 in section 3
Training set success :   99.90%
Validation set success: 98.97%
TEST set success :   **99.9037%**

It is clear that the proposed combination method outperforms the other rival approaches

5   Conclusions

A novel methodology has been presented for combining the decisions of different feedforward neural network classifiers by designing a hierarchical neural classification system based on similarity analysis of classifier cases regarding their higher order features, and a second stage neural system. Instead of the usual approach of applying voting schemes on the decisions of their output layer neurons, the proposed methodology integrates the higher order features extracted by their upper hidden layer units through applying similarity and SVD analysis, and through a second stage suitable coordinator neural model. The classification results yielded in a computer vision based experimental study show that the proposed approach clearly outperforms other rival combination approaches. More experimental work is needed, however, for more difficult real world tasks.

References

[1] Gonzalez, R.C., Wintz, P.: Digital Image Processing, 2nd edn. Addison-Wesley, Reading (1987)

[2] Dudani, S.A., Breeding, K.J., McGhee, R.B.: Aircraft identification by moment invariants. IEEE Trans. on Computers C-26, 39–46 (1977)

[3] Chen, K.: Efficient parallel algorithms for the computation of two-dimensional moments. Pattern Recognition 23, 109–119 (1990)

[4] Mertzios, B.G.: Shape discrimination in robotic vision using scaled normalized central moments. In: Proceedings of the IFAC Workshop of Computing Power and Control Theory, Prague, Chechoslavakia, September 1-2 (1992)

[5] Mertzios, B.G., Mitzias, D.A.: Fast shape discrimination with a system of neural networks based on scaled normalized central moments. In: Proceedings of the International Conference on Image Processing: Theory and Applications, San Remo, Italy, June 10-12, pp. 219–223 (1993)

[6] Lin, C.C., Chellapa, R.: Classification of partial 2-D shapes using Fourier descriptors. IEEE Trans. on Pattern Analysis and Machine Intelligence PAMI-8(5), 686–690 (1987)

[7] Zahn, C.T., Roskies, R.Z.: Fourier descriptors for plane closed curves. IEEE Trans. on Computers C-21(3), 269–281 (1972)

[8] Fu, K.S.: Syntactic Pattern Recognition and Application. Prentice-Hall, Englewood Cliffs (1982)

[9] Mitzias, D.A., Mertzios, B.G.: Shape recognition in robotic vision using a fixed size neural network. In: Proceedings of the Canadian Conference on Electrical and Computer Engineering, pp. WM4.28.1-4, Toronto, Canada (September 1992)

[10] Mitzias, D.A., Mertzios, B.G.: Shape recognition with a neural classifier based on a fast polygon approximation technique. Pattern Recognition 27(5), 627–636 (1994)

[11] Kittler, J., Alkoot, F.M.: Relationship of sum and vote fusion strategies. In: Kittler, J., Roli, F. (eds.) MCS 2001. LNCS, vol. 2096, pp. 339–348. Springer, Heidelberg (2001)

[12] Ghaderi, R., Windeatt, T.: Least squares and estimation measures via error correcting output code. In: Kittler, J., Roli, F. (eds.) MCS 2001. LNCS, vol. 2096, pp. 148–157. Springer, Heidelberg (2001)

[13] Sharkey, A.: Types of multinet system. In: Roli, F., Kittler, J. (eds.) MCS 2002. LNCS, vol. 2364, pp. 108–117. Springer, Heidelberg (2002)

On Cellular Network Channels Data Mining and Decision Making through Ant Colony Optimization and Multi Agent Systems Strategies

P.M. Papazoglou[1], D.A. Karras[2], and R.C. Papademetriou[3]

[1] Lamia Institute of Technology Greece, University of Portsmouth, UK, ECE Dept., Anglesea Road, Portsmouth, United Kingdom, PO1 3DJ
papaz@teilam.gr
[2] Chalkis Institute of Technology, Greece, Automation Dept., Psachna, Evoia, Hellas (Greece) P.C. 34400
dakarras@ieee.org
dakarras@teihal.gr
[3] University of Portsmouth, UK, ECE Department, Anglesea Road, Portsmouth, United Kingdom, PO1 3DJ

Abstract. Finding suitable channels to allocate in order to serve increasing user demands in a cellular network, which is a dynamical system, constitute the most important issue in terms of network performance since they define the bandwidth management methodology. In modern cellular networks these strategies become challenging issues especially when advanced services are applied. The effectiveness of decision making for channel allocation in a cellular network is strongly connected to current traffic and wireless environment conditions. Moreover, in large scale environments, network states change dynamically and the network performance prediction is a hard task. In the recent literature, the network adaptation to current real user needs seems it could be achieved through computational intelligence based channel allocation schemes mainly involving genetic algorithms. In this paper, a quite new approach for communication channels decision making, based on ant colony optimization, which is a special form of swarm intelligence, modelled through multi agent methodology is presented. The main novelty of this research lies on modelling this optimization scheme through multi agent systems. The simulation model architecture which includes network and ant agents are also presented as well as the performance results based on the above techniques. Finally, the current study, also, shows that there is a great field of research concerning intelligent techniques modelled through multi-agent methodologies focused on channels decision making and bandwidth management in wireless communication systems.

1 Introduction

1.1 The Need for More Efficient Bandwidth Management in Wireless Communications

The cellular network technology offers the opportunity for wireless communication (voice, data and video) between Mobile Users (MUs) within the network coverage

P. Perner (Ed.): ICDM 2009, LNAI 5633, pp. 372–387, 2009.
© Springer-Verlag Berlin Heidelberg 2009

area. According to the cellular principle, the whole coverage network area is divided in hexagonal areas called cells [1-4]. Each cell can serve a number of MUs through a Base Station (BS) which is placed in the center of the cell. A BS can allocate also a free channel for a MU that exists in a neighbour cell based on the corresponding channel allocation strategy. The available bandwidth is limited and at the same time the cellular network must serve numerous MUs which request different types of services. On the other hand, in large scale environments the bandwidth management is a difficult task due to dynamically changing network conditions [5,6]. Several channel allocation schemes have been proposed so far in the literature for supporting both voice and multimedia services. The most known channel allocation schemes are categorized in three types of strategies [7-16]. These strategies are based on constant and variable channel sets which are distributed in the available BSs. In Fixed Channel Allocation (FCA) [7-11] a fixed number of channels is allocated among the BSs of the cellular network. This strategy is based on predefined traffic calculation because the channel sets are permanently allocated to BSs during network operation. The channel availability can be adapted to current traffic conditions with the Dynamic Channel Allocation (DCA) [7,12-15] strategy. The approach offers higher flexibility as compared to the FCA solution. Due to changing traffic conditions within a wireless network, both strategies (FCA and DCA) can be applied. This combination forms the Hybrid Channel Allocation (HCA) [7,16].

1.2 Channel Allocation for Advanced Services in a Large Scale Wireless Network Environment Based on Conventional and Intelligence Approaches

Due to the limited bandwidth, the frequency reusability among the cellular network is a key issue. Thus, the network is divided in sub-areas called clusters which contain seven cells. Within each cell, unique frequencies are used. Based on the reusability technique, different MUs in different clusters can use the same frequency (same channel) and thus the signal interference is emerged (co-channel users) [1-4]. The signal quality is also affected by the distance between MU and BS and the propagation conditions. Thus, for a successful connection of a MU to a BS the signal quality in terms of carrier strength, noise and interference must be examined [1-4]. This quality affects the resulted channel capacity which supports advanced multimedia services such as real time video. Especially, when Quality of Service constraints exist the initial requested capacity must be guaranteed until the normal service termination [17-24,1-4]. Voice services require one single channel and so the corresponding allocation is based on classical signal measurements such as the signal to noise ratio plus interference factor [25,26]. On the other hand, the bandwidth management for supporting advanced services such as real time video is complicated due to the fact that multiple channels are needed for offering the initial requested capacity. Channel reservation and preferential treatment for high priority calls constitute the most known strategies among literature for supporting multimedia services [17-20]. In a large scale wireless network, traffic conditions are changing very rapidly and dynamically and so even small differences in the channel allocation procedure may largely affect network performance [5,6]. Thus, more efficient channel allocation schemes based on intelligent techniques have been proposed in the literature as compared to conventional approaches. These schemes are based on evolutionary algorithms such as Genetics. Genetic Algorithms (GAs) [27-29] have been thoroughly investigated in the literature for solving the channel allocation problem [30-34]. On the other hand, another intelligent technique such as swarm intelligence [35-37] has not been applied in the literature for solving the channel allocation problem except [5,6], which introduced for the very first time a special form of swarm intelligence called Ant Colony Optimization (ACO) [36] applied to channel allocation, but in a preliminary stage of development. A more elaborate ACO scheme is herein presented for the first time. The major issue, however, herein investigated are the benefits of modelling this approach through multi agent methodology.

The core idea of the swarm intelligence comes from the behaviour of swarms of birds, fish, etc. The concept that stems from this idea is that the group complexity can be based on individual agents without the need of any central control. A definition of swarm intelligence according to [42] is as follows:

"Swarm intelligence (SI) is the property of a system whereby the collective behaviours of (unsophisticated) agents interacting locally with their environment cause coherent functional global patterns to emerge". Swarm intelligence has been used in a wide range of optimization problems [35-37]. The ant colony optimization (ACO) is a specific form of swarm intelligence based on ant behaviour. Despite the simplicity of ant's behaviour, an ant colony is highly structured [36]. ACO has been used for routing problems [43] in networks, due to the fact that some ants search for food while others follow the shortest path to the food. The authors of the present paper were the first to introduce this methodology to the problem of resource allocation in wireless systems [5,6], but in a preliminary stage. This research is herein elaborated with multi-agent system strategies modelling of ACO, which improves the results of [5,6].

2   The Proposed Modelling Approach Based on Ant Colony Optimization and Multi-agent Technology

2.1   Experimental Simulation Model Components and Supported Network Services

A simulation model for wireless communication systems consists of some basic components which represent the wireless network functionality. In [44] the presented simulation model supports handoff procedures, simulating resource management, etc.

The whole simulation model consists of various individual components that correspond to different functions. These components can be also found in similar studies [44-47]. Figure 1 shows the layered structure of the simulation model.

Additionally, the simulation model supports the following network services:

- New call admission (NC)
- Call termination (FC)
- Reallocation call (RC)
- MU movement call (MC)

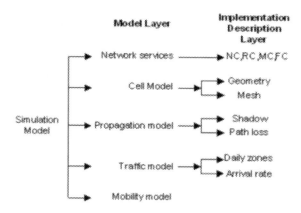

Fig. 1. Simulation model components

2.2 Multi-agent Technology in Wireless Communication Systems

An Agent is known to be a computational system that interacts autonomously with its environment and operates for the goal for which it has been designed [48] and are smaller than a typical application [49]. An Agent also perceives the environment conditions, acts according to these conditions, interprets perceptions and solves problems [50]. Among these, the most important attributes are (a) adaptability, (b) autonomy, (c) Collaboration and (d) interactivity. According to [51], a Multi-agent system consists of a number of agents which interact through communication. These agents act in an environment within which they have different areas of influence.

In several studies, the MAS technology has been used for the solution of the resource allocation problem. Within the developed models, various network components such as BSs, cells, etc have been modelled as agents [52-55]. A comprehensive simulation model for wireless cellular networks has been built by [55] who proposes a distributed channel allocation scheme using intelligent software agents. In the above study, intelligent collaborative software agents give autonomy to BSs, increase the network robustness, allow negotiation of network resources and improve resource allocation. For this purpose, several aspects of the cellular network infrastructure and operation have been modelled.

In the proposed approach, an agent represents an individual network service instead of a network node such as cell or BS as found in the literature.

2.3 Channel Allocation Based on a Novel ACO Algorithm

The ant colony optimization applied in the channel allocation problem was first introduced in [5,6]. This study, assumes that two MUs exist (U1,U2) in the center cell of the network. There are seven virtual paths between each MU (U1 or U2) and the neighbour BSs (including current cell). Thus, fourteen paths exist in the example cellular network of figure 2. Each path has a specific weight based on the signal and channel conditions of the BS that exists in its end. An ant starts from the MU position and decides which path to follow based on seven different probabilities that derive

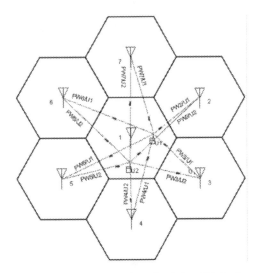

Fig. 2. Possible ant paths between MUs and existing BSs in the neighbour

from the corresponding weight of each path. After the channel allocation for a set of MUs, the signal and channel conditions are changed in the current cell and neighbour, therefore the path weights are changed also. Thus, the ant decisions are based on different probability distributions at every simulation step and every channel allocation procedure. As user activities are increasing the network conditions also changing. The resulting environment is a dynamic system where the network is self optimized

Figure 3 shows the abstract structure of the ACO algorithm and figure 4 shows the ACO algorithm operation.

Loop ($\forall$ *MU SubSet* $\in U$)

   Loop ($\forall$ *MU* $\left(\mathrm{Re}\,quest \in R \right) \subseteq MU$ *Subset* $\in U$)

      //Check and Store Channel Availability, 0=free, 1=occupied
      //Check and Store only the number of available channels in each BS
      //Check and Store Average CNIR in each Base Station
      //Calculate and Store Path Weights
      //Assign probability for each path
      //Channel allocation
      End Loop
  End Loop

Fig. 3. Abstract structure of the ACO Algorithm

2.4 Multi-agent Technology Adapted to Supported Network Services and Resource Management

The whole modelling methodology is based on the multi agent concepts and is divided in two hierarchical levels: Network services level and Ant colony level

Loop ($\forall$ *MU SubSet* $\in$ *U*)

    Loop ($\forall$ *MU* $\left(\text{Re}\,quest \in R\right) \subseteq$ *MU Subset* $\in$ *U*)

        //Check and Store Channel Availability, 0=free, 1=occupied
        AvMat[BS_i][Ch_j]=Ch_Availability $\forall$ $BS_i \in BS$ *and* $\forall$ $Ch_j \in Ch$

        //Check and Store only the number of available channels in each BS
        BSAvMat[BS_i]=$\sum_{j=1}^{z} AvMat[i][j]$

        //Check and Store Average CNIR in each Base Station
        AvCNIR[BS_i]=$\frac{1}{z}\sum_{j=1}^{z} CNIR\ Ch_j \in BSCh_i$

        //Calculate and Store Path Weights
        PathWMat[BS_i]=WeightFunction(BSAvMat[BS_i], AvCNIR[BS_i])

        //Assign probability for each path
        AntPathProb[$Path_i$]=ProbFunction(PathWMat[BS_i])

        //Channel allocation
        Try to allocate channels $\forall$ *MU* $\left(\text{Re}\,quest \in R\right) \subseteq$ *MU Subset* $\in$ *U*

    End Loop
End Loop

Fig. 4. General structure of the ACO algorithm for channel allocation

Figure 5 shows the hierarchical structure of the multiple agents.

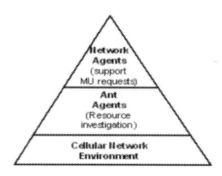

Fig. 5. Multiple agent hierarchical organization

Network and Ant Agent Definition

According to [56], "*An agent is a computer system that is situated in some environment, and that is capable of autonomous action in this environment in order to meet its design objectives*".

An agent interacts with its environment, gets input information from it and performs actions that affect this environment.

The NC and RC network agents (NCA, RCA) interact with the cellular network (environment), get input information (blocking probability-network performance,

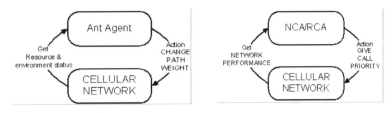

Fig. 6. Network Agent definition   **Fig. 7.** Ant Agent definition

dropping probability for RCA) and performs actions (gives priority to new calls or in handoffs) that affect the network performance [57,58] (Fig. 6). On the other hand, each ant agent (fig. 7) is associated with each path from a MU to the available BSs in the neighbour. An ant agent investigates the channel availability and the wireless environment in terms of CNIR conditions for the associated path to the specific BS. After the investigation completion, the ant agent informs the corresponding network agent (by changing the path weight) which is responsible for the service completion.

An ant agent measures the wireless network conditions and the channel availability. If current status of the path under investigation is in critical condition (e.g. a congested area) then the ant agent increases its priority in order to inform with priority the corresponding network agents to avoid the associated BS. In the case of more that one path in critical conditions, a priority based negotiation takes place between the active ant agents.

Multi agents distributed among the wireless network environment
The whole wireless network is divided in 7-cell areas called clusters. A set of network agents (NCA, RCA, MCA, FCA) and a set of ant agents (AA_1 through AA_7) are distributed in each cluster (figure 8).

Assuming that the cellular network has N cells distributed in cell clusters where each cluster contains i cells, the total number of clusters is N/i. Each set of the four agents is duplicated in every cluster.

Thus, the total required agents are $(4+7)*(N/i)=11*(N/i)$. In order to achieve acceptable adaptation of the multiple agents to the current traffic conditions and to improve the network performance the network agent negotiation takes place between NCA and RCA of a set of clusters that belong in the same influence area.

The Multi-agent / Multi-layered Architectural Model
The initial layered architecture was introduced in [59,57,58] Figure 9 shows a new layered architecture which include four operational layers focused on the multi-agent activities. The architectural layers are organized as follows:

- **Layer 1.** Represents the cellular network environment which is organized as cluster. Each cluster has its own network and ant agent set.
- **Layer 2.** Includes the ant agents which investigate the wireless environment (signal conditions and channel availability)
- **Layer 3.** Basic network agents for supporting MU requests inside current cluster
- **Layer 4.** Represents the simulation system environment which controls the agent activation.

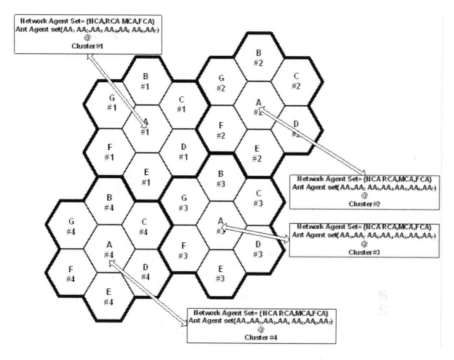

Fig. 8. Agents sets distributed among network clusters

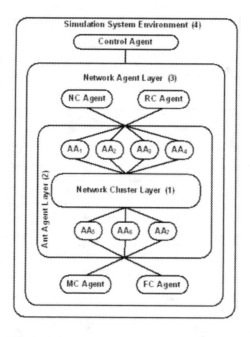

Fig. 9. Multi-Agent / Multi-Layered Architecture

Multi-agent Decision Making
Each agent has a specific goal and responsibility area. An agent takes decisions for satisfying its own goals and contributing to the network common goals such as the performance control and efficient resource management. Table 1 illustrates critical information about agent goals, responsibility areas, etc.

Table 1. Agent responsibilities, goals and actions

Agent	Responsibility	Goal	Actions
NCA	New call admission	Minimize blocking probability	Set self priority, negotiate with RCA
RCA	Call (channel) reallocation	Minimize dropping probability	Set self priority, negotiate with NCA
MCA	Movement call	Support MU movement	Collaborate with NCA/RCA
FCA	Call termination	Support normal service termination	Collaborate with NCA/RCA
Ant	Wireless environment investigation	Inform for resource availability	Set self priority, negotiate with other ant agents (in case of critical conditions: e.g. local area congestion)

Figure 10 shows how a network agent (e.g. NCA/RCA) takes decisions for self attributes (e.g. priority) and negotiation with other network agents.

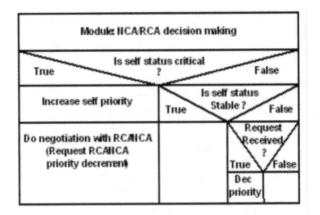

Fig. 10. Network agent decision making

The negotiation and the corresponding decision for the ongoing and new call priorities is based on current and previous network performance in terms of statistical behaviour of the wireless network. A comparison is made between current t and previous $t-1$ simulation steps. Thus the current network performance is measured. Based on this performance, each network agent negotiates in order to achieve the best terms in priority settings for supporting more efficient the corresponding calls in the influence area.

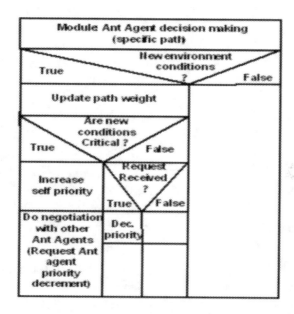

Fig. 11. Ant agent decision making

Figure 11 shows the ant agent behaviour based on current network environment conditions such as signal environment (CNIR conditions) and channel availability. The ant agent updates the corresponding path weight and increases its self priority in the case of critical wireless conditions (e.g. congested area, not acceptable CNIR conditions) for helping with priority the high priority network agent to complete successfully the MU service request. If critical conditions occurred within an ant agent activity area (path, BS), a negotiation procedure takes place in order to request priority decrement from the other ant agents to provide immediately useful information to the high priority network agent.

3   Simulation Model Evaluation and Indicative Simulation Results

An experimental study has been conducted to evaluate the ideas herein investigated. To this end, blocking and dropping probabilities have been considered as representative measures of network performance. Blocking and dropping probability represent the most widely used statistical metrics for the network performance evaluation [26, 60-63, 44,54, 25]. Blocking probability represents the MU unsuccessful attempts for call initiation as compared to the total new call events. On the other hand, dropping represents the number of unsuccessful attempts for handoffs of ongoing calls. Based on these two metrics the application of any proposed approach concerning the channel allocation can be clearly evaluated.

Figure 12 shows the wireless network performance in terms of blocking probability where the classical DCA and the ACO approach have been applied without any ant agents.

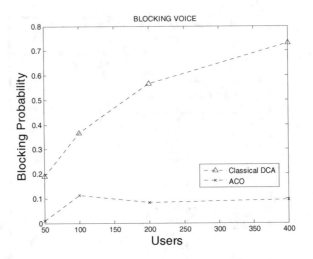

Fig. 12. Blocking probability of the proposed Ant Colony Optimization (ACO) algorithm against classical Dynamic Channel Allocation (DCA)

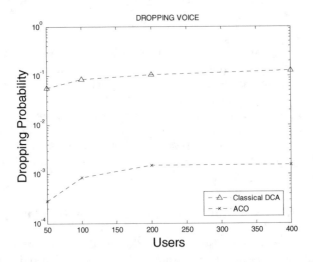

Fig. 13. Dropping probability of the proposed Ant Colony Optimization (ACO) algorithm against classical Dynamic Channel Allocation (DCA)

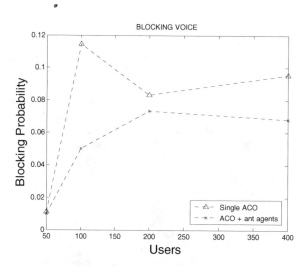

Fig. 14. Blocking probability of the traditional Ant Colony Optimization (ACO) algorithm against the proposed Ant Colony Optimization (ACO) based on ant agents

Figure 13 shows also the wireless network performance in terms of dropping probability where the classical DCA and the ACO approach have been applied without any ant agents.

Figure 14 shows that network performance is considerably improved when the ACO algorithm is modelled within the framework of multi agent systems.

4 Conclusions and Future Trends

This paper has investigated a multi agent based architecture for modelling the Ant Colony Optimization (ACO) algorithm towards its efficient application for an improved bandwidth management in cellular communication systems. Some preliminary results presented in the experimental study show quite promising and illustrate the effectiveness of the approach. However, more extensive evaluations are needed to clearly support these results towards incorporation of the proposed schemes to real world simulators of cellular communications. Moreover, the proposed approach should be compared with the GA approach extensively in order to find out its pros and cons with respect to this computational intelligence method.

References

[1] Rappaport, T.S.: Wireless Communications Principles and Practice. Prentice-Hall, Englewood Cliffs (2002)
[2] Goldsmith, A.: Wireless Communications. Cambridge University Press, Cambridge (2005)
[3] Milisch, A.: Wireless Communications. Wiley, IEEE Press (2005)
[4] Lee, W.C.Y.: Wireless and Cellular Telecommunications. McGraw-Hill, New York (2006)

[5] Papazoglou, P.M., Karras, D.A., Papademetriou, R.C.: On the Implementation of Ant Colony Optimization Scheme for Improved Channel Allocation in Wireless Communications. In: IEEE International Conference on Intelligent Systems (2008)

[6] Papazoglou, P.M., Karras, D.A., Papademetriou, R.C.: On Integrated Ant Colony Optimization Strategies for Improved Channel Allocation in Large Scale Wireless Communications. In: 10th WSEAS Int. Conf. on Mathematical Methods, Computational Techniques And Intelligent Systems (MAMECTIS 2008), Corfu, Greece (2008)

[7] Zhang, M., Yum, T.S.: Comparisons of Channel Assignment Strategies in Cellular Mobile Telephone Systems. IEEE Transactions on Vehicular Technology 38(4), 211–215 (1989)

[8] Lai, W.K., Coghill, G.C.: Channel Assignment through Evolutionary Optimization. IEEE Transactions on Vehicular Technology 45(1), 91–96 (1996)

[9] MacDonald, V.H.: The cellular Concepts. The Bell System Technical, Journal 58, 15–42 (1979)

[10] Elnoubi, S.M., Singh, R., Gupta, S.C.: A New Frequency Channel Assignment Algorithm in High Capacity Mobile Communication Systems. IEEE Transactions on Vehicular Technology VT-21(3), 125–131 (1982)

[11] Xu, Z., Mirchandani, P.B.: Virtually Fixed Channel Assignment for Cellular Radio-Telephone Systems: A Model and Evaluation. In: IEEE International Conference on Communications, ICC 1992, Chicago, vol. 2, pp. 1037–1041 (1982)

[12] Cimini, L.J., Foschini, G.J.: Distributed Algorithms for Dynamic Channel Allocation in Microcellular Systems. In: IEEE Vehicular Technology Conference, pp. 641–644 (1992)

[13] Cox, D.C., Reudink, D.O.: Increasing Channel Occupancy in Large Scale Mobile Radio Systems: Dynamic Channel Reassignment. IEEE Transanctions on Vehicular Technology VT-22, 218–222 (1973)

[14] Del Re, E., Fantacci, R., Giambene, G.: A Dynamic Channel Alloca-tion Technique based on Hopfield Neural Networks. IEEE Transanctions on Vehicular Technology VT-45(1), 26–32 (1996)

[15] Sivarajan, K.N., McEliece, R.J., Ketchum, J.W.: Dynamic Channel Assignment in Cellular Radio. In: IEEE 40th Vehicular Technology Conference, pp. 631–637 (1990)

[16] Kahwa, T.J., Georgans, N.D.: A Hybrid Channel Assignment Schemes in Large-Scale, Cellular Structured Mobile Communication Systems. IEEE Transactions on Communications 26, 432–438 (1978)

[17] Kim, S., Varshney, P.K.: An Integrated Adaptive Bandwidth-Management Framework for QoS-Sensitive Multimedia Cellular Networks. IEEE Transactions on Vehicular technology 53(3) (2004)

[18] Kim, H.B.: An Adaptive Bandwidth Reservation Scheme for Multimedia Mobile Cellular Networks. In: Proceedings of the IEEE International Conference on Communications (ICC 2005), Seoul, Korea (2005)

[19] Nasser, N., Hassanein, H.S.: Bandwidth Reservation Policy for Multimedia Wireless Networks and its Analysis. In: IEEE Internernational Conference on Communications (2004)

[20] Chen, H., Huang, L., Kumar, S., Kuo, C.C.J.: Radio Resource Management for Multimedia QoS Support in Wireless Cellular Networks. Kluwer Academic Publishers, Dordrecht (2004)

[21] Oliveria, C., Kim, J.B., Suda, T.: An adaptive bandwidth reservation scheme for high-speed multimedia wireless networks. IEEE J. Select Areas Commun. 16(6), 858–874 (1998)

[22] Das, S.K., Jayaram, R., Kakani, N.K., Sen, S.K.: A Call admission and control for quality-of-service (QoS) provisioning in next generation wireless networks. Wireless Networks 6, 17–30 (2000)

[23] Papazoglou, P.M., Karras, D.A., Papademetriou, R.C.: Modeling DCA Strategies for Supporting Multimedia Services with QoS over Cellular Communication Systems. In: 12th WSEAS Int. Conf. on Communications (2008)

[24] Papazoglou, P.M., Karras, D.A., Papademetriou, R.C.: Efficient Simulation Methodologies for Wireless Multimedia Communication Systems. In: HERCMA 2007, Athens, Greece, Athens University of Economics & Business (September 2007)

[25] Cherriman, P., Romiti, F., Hanzo, L.: Channel Allocation for Third-generation Mobile Radio Systems. In: ACTS 1998, vol. 1, pp. 255–261 (1998)

[26] Katzela, I., Naghshineh, M.: Channel assignment schemes for cellular mobile telecommunication systems: A comprehensive survey. IEEE Personal Communications, 10–31 (1996)

[27] Beasley, D., Bull, D.R., Martin, R.R.: An overview of Genetic Algorithms: Part I, Fundamentals. University Computing 15(2), 58–69 (1993)

[28] Holland, J.H.: Adaptation in Natural and Artificial Systems. MIT Press, Cambridge (1975)

[29] Goldberg, D.E.: Genetic Algorithms in Search, Optimization and Machine Learning. Addison Wesley, New York (1989)

[30] Lima, M.A.C., Araujo, A.F.R., Cesar, A.C.: Dynamic channel assignment in mobile communications based on genetic algorithms. Personal, Indoor and Mobile Radio Communications (2002)

[31] Yener, A., Rose, C.: Genetic Algorithms Applied to Cellular Call Admission Problem: Local Policies. Vehicular Technology IEEE 46(1), 72–79 (1997)

[32] Kendall, G., Mohamad, M.: Solving the Fixed Channel Assignment Problem in Cellular Communications Using An Adaptive Local Search. In: Burke, E.K., Trick, M.A. (eds.) PATAT 2004. LNCS, vol. 3616, pp. 219–231. Springer, Heidelberg (2005)

[33] Kim, J.S., Park, S., Dowd, P., Nasrabadi, N.: Channel Assignment in Cellullar Radio using Genetic Algorithms. Wireless Personal Communications 3(3), 273–286 (1996)

[34] Wang, L., Li, S., Lay, S.C., Yu, W.H., Wan, C.: Genetic Algorithms for Optimal Channel Assignments in Mobile Communications. Neural Network World 12(6), 599–619 (2002)

[35] Eberhart, R.C., Kennedy, J.: A New Optimizer Using Particle Swarm Theory. Presented at the Sixth International Symposium on Micro Machine and Human Science, Nagoya, Japan. pp. 39–43 (1995)

[36] Colorni, A., Dorigo, M., Maniezzo, V.: Distributed optimization by ant colonies. Presented at the First European Conference on Artificial Life (ECAL 1991), Paris, France. pp. 134–142 (1991)

[37] Bonabeau, E., Henaux, F., Guerin, S., Snyers, D., Kuntz, P., Theraulaz, G.: Routing in Telecommunications Networks with 'Smart' Ant-like Agents. In: Intelligent Agents for Telecommunications Applications. Frontiers in Artificial Intelligence & Applications, vol. 36 (1998)

[38] Papazoglou, P.M., Karras, D.A., Papademetriou, R.C.: Simulating and Evaluating Dynamic Channel Assignment Schemes in Wireless Communication Networks through an Improved Multi-Agent System. In: 3rd Indian International Conference on Artificial Intelligence (IICAI 2007) (2007)

[39] Papazoglou, P.M., Karras, D.A., Papademetriou, R.C.: High Performance Novel Hybrid DCA algorithms for efficient Channel Allocation in Cellular Communications modeled and evaluated through a Java Simulation System. WSEAS Transactions on Communications 11(5), 2078–2085 (2006)

[40] Papazoglou, P.M., Karras, D.A., Papademetriou, R.C.: A Critical overview on the recent advances in channel allocation strategies for voice and multimedia services in wireless communication systems and the Applicability of Computational Intelligence Techniques. In: 10th WSEAS Int. Conf. on Mathematical Methods, Computational Techniques And Intelligent Systems (MAMECTIS 2008), Corfu, Greece (2008)

[41] Papazoglou, P.M., Karras, D.A., Papademetriou, R.C.: Novel DCA algorithms for efficient Channel Assignment in Cellular Communications and their evaluation through a generic Java Simulation System. In: 6th WSEAS Int. Conf. on Simulation, Modelling And Optimization (SMO 2006) (2006)

[42] Bonabeau, E., Dorigo, M., Theraulaz, G.: Swarm Intelligence: From Natural to Artificial Systems. Artificial Life 7, 315–319 (2001)

[43] Bundgaard, M., Damgaard, T.C., Jacob, F.D., Winther, W.: Ant Routing System, IT University of Copenhagen (2002)

[44] Tripathi, N.D., Jeffrey, N., Reed, H., VanLandingham, H.F.: Handoff in Cellular Systems. IEEE Personal Communications (1998)

[45] Sobeih, A., Chen, W.-P., Hou, J.C., Kung, L.-C., Li, N., Lim, H., Tyan, H.-Y., Zhang, H.: J-Sim: A Simulation and Emulation Environment for Wireless Sensor Networks. IEEE Wireless Communications 13(4), 104–119 (2005)

[46] Zeng, X., Bagrodia, R., Gerla, M.: GloMoSim: A Library for Parallel Simulation of Large-scale Wireless Networks. In: Proceedings of the 12th Workshop on Parallel and Distributed Simulations (1998)

[47] Short, J., Bagrodia, R., Kleinrock, L.: Mobile wireless network system simulation. In: ACM Mobile Computing and Networking Conference (Mobicom 1995) (1995)

[48] Maes, P.: Artificial Life Meets Entertainment: Life like Autonomous Agents. Communications of the ACM 38(11), 108–114 (1995)

[49] Smith, D.C., Cypher, A., Spohrer, J.: KidSim: Programming Agents Without a Programming Language. Communications of the ACM 37(7), 55–67 (1994)

[50] Hayes-Roth, B.: An Architecture for Adaptive Intelligent Systems. Artificial Intelligence: Special Issue on Agents and Interactivity 72, 329–365 (1995)

[51] Jennings, N.R.: On agent-base software engineering. Artificial Intelligence 117, 277–296 (2000)

[52] Hayzelden, A., Bigham, J.: Software Agents for Future Communications Systems. Springer, Berlin (1999)

[53] Iraqi, Y., Boutaba, R.: A Multi-agent System for Resource Management in Wireless Mobile Multimedia Networks. In: Ambler, A.P., Calo, S.B., Kar, G. (eds.) DSOM 2000. LNCS, vol. 1960, pp. 218–229. Springer, Heidelberg (2000)

[54] Bigham, J., Du, L.: Cooperative Negotiation in a MultiAgent System for RealTime Load Balancing of a Mobile Cellular Network. In: AAMAS 2003, July 14–18 (2003)

[55] Bodanese, E.L.: A Distributed Channel Allocation Scheme for Cellular Networks using Intelligent Software Agents. PhD thesis, University of London (2000)

[56] Wooldridge, M., Jennings, N.R.: Intelligent agents: theory and practice. The Knowledge Engineering Review 10(2), 115–152 (1995)

[57] Papazoglou, P.M., Karras, D.A., Papademetriou, R.C.: On the Multi-Threading Approach of Efficient Multi-Agent Methodology for Modelling Cellular Communications Bandwidth Management. In: Nguyen, N.T., Jo, G.-S., Howlett, R.J., Jain, L.C. (eds.) KES-AMSTA 2008. LNCS, vol. 4953, pp. 431–443. Springer, Heidelberg (2008)

[58] Papazoglou, P.M., Karras, D.A., Papademetriou, R.C.: An Improved Multi-Agent Simulation Methodology for Modelling and Evaluating Wireless Communication Systems Resource Allocation Algorithms. Journal of Universal Computer Science 14(7), 1061–1079 (2008)

[59] Papazoglou, P.M., Karras, D.A., Papademetriou, R.C.: A Multi-Agent Architecture for Designing and Simulating Large Scale Wireless Systems Resource Allocation. In: Nguyen, N.T., Grzech, A., Howlett, R.J., Jain, L.C. (eds.) KES-AMSTA 2007. LNCS (LNAI), vol. 4496, pp. 405–415. Springer, Heidelberg (2007)
[60] Wong, S.H.: Channel Allocation for Broadband Fixed Wireless Access Networks. Unpublished doctorate dissertation, University of Cambridge (2003)
[61] Grace, D.: Distributed Dynamic Channel Assignment for the Wireless Environment. Unpublished Doctoral dissertation, University of York (1998)
[62] Haas, H.: Interference analysis of and dynamic channel assignment algorithms in TD–CDMA/TDD systems. Unpublished Doctoral dissertation, University of Edinburg (2000)
[63] Salgado, H., Sirbu, M., Peha, J.: Spectrum Sharing Through Dynamic Channel Assignment For Open Access To Personal Communications Services. In: Proc. of IEEE Intl. Communications Conference (ICC), pp. 417–422 (1995)

Responsible Data Releases

Sanguthevar Rajasekaran[1], Ofer Harel[2], Michael Zuba[1],
Greg Matthews[2], and Robert Aseltine[3]

[1] Department of CSE, University of Connecticut, Storrs, CT
[2] Department of Statistics, University of Connecticut, Storrs, CT
[3] University of Connecticut Health Center, Farmington, CT

Abstract. Data releases to the public should ensure the privacy of individuals involved in the data. Several privacy mechanisms have been proposed in the literature. One such technique is that of data anonymization. For example, synthetic data sets are generated and released. In this paper we analyze the privacy aspects of synthetic data sets. In particular, we introduce a natural notion of privacy and employ it for synthetic data sets.

1   Introduction

When releasing data about a population to the general public it is essential to ensure privacy of individuals in the population. There are two ways of releasing data: 1) **Interactive Mode:** In this case a user will pose a series of queries about the database; and 2) **Noninteractive Mode:** In this case a "sanitized" version of the database is released. We are interested in the noninteractive mode. It would be desirable (Dalenius [2]) if the sanitized database released is such that everything that can be learnt about an individual from this database can also be learnt without accessing the database. Dwork [3] shows that this form of security cannot be achieved.

The proof is based on the fact that for any version of the sanitized database one could always find relevant *auxiliary information* such that the database and the auxiliary information together will yield privacy information about an individual. For example, consider a sanitized database that has information about average annual salaries of different nationalities. This information alone is not enough to infer the salary of an individual. On the other hand if the following auxiliary information is available: "*Joe Smith's annual salary is $40K more than the average salary of an American*", then we can infer the salary of Joe Smith (if we also have access to the sanitized database).

Given that the idea of privacy alluded by Dalenius [2] is not possible, several other less stringent privacy mechanisms have been proposed in the literature. Examples include *k-anonymity* [12], *l-diversity* (see e.g., [6] and [1]), *differential privacy* [3], *probabilistic differential privacy* [5], etc.

All of these privacy mechanisms seem to have an underlying (unstated but implicit) model for the available auxiliary information. For example, they typically state scenarios like: "Consider the case when information X is available"

P. Perner (Ed.): ICDM 2009, LNAI 5633, pp. 388–400, 2009.
© Springer-Verlag Berlin Heidelberg 2009

(for some specific X). The proposed schemes will be good when the information they enumerate is available. Unfortunately it is impossible to enumerate all possible auxiliary information that may be available. Thus a privacy scheme that is applicable independent of the available auxiliary information is desirable. In this paper we propose one such scheme. The basic property that our scheme supports is the following. Let D be a given data set and let S be the corresponding synthetic data released. Let D' be the same as D except that it differs from D in only one record and S' be the corresponding synthetic data. The statistical difference between S and S' will be "small". In other words, a user having access to only S and S' will not be able to infer (with a "reasonable probability") that the underlying data for S and S' are different.

The rest of the paper is organized as follows. In Section 2 we survey some of the existing privacy approaches. In Section 3 we introduce our privacy scheme and illustrate it with two examples. In Section 4 we summarize our algorithms while in section 5 we present some simulation results. Section 6 concludes the paper.

2   Literature Survey

2.1   Synthetic Data Generation

One of the techniques that is being used to release data is with the generation of synthetic data (see e.g., [10], [9], and [7]). If $\mathcal{D}$ is the database under concern, we think of this as a sample from a population $\mathcal{P}$. It is assumed that $n = |\mathcal{D}|$ is much less than $|\mathcal{P}|$. Since $\mathcal{P}$ is usually unknown, one generates M synthetic populations $\mathcal{P}^l, 1 \leq l \leq M$ where each $\mathcal{P}^l$ is drawn from the posterior predictive distribution for $\mathcal{P}$.

The size of the population is too large to be released. Thus one typically releases $\mathcal{D}^l, 1 \leq l \leq M$, where $\mathcal{D}^l$ is a simple random sample from $\mathcal{P}^l$ of size k (for some appropriate integer k).

For a given $\mathcal{D}$, one way of generating the posterior predictive distribution for the corresponding $\mathcal{P}$ is to assume that the data follow a multivariate normal distribution with an unknown mean vector μ and an unknown covariance matrix Σ. Call this data releasing scheme *Normal Synthetic Data Generation (NSDG)*.

2.2   Differential Privacy

Differential privacy [3] captures the added risk of a disclosure when an individual participates in a database. If the risk of a disclosure does not change much whether or not an individual participates in the database, then we can consider this database as having good privacy. A formal definition of differential privacy follows [3]:

Definition 1. *A randomized function $\mathcal{F}$ has ϵ-privacy if $Pr[\mathcal{F}(D_1) \in S] \leq e^\epsilon \ Pr[\mathcal{F}(D_2) \in S]$ for any two databases D_1 and D_2 that differ by at most one element and for any subset S of the range of $\mathcal{F}$.*

The above concept can be extended to group privacy as well. For example, if there are g persons, then the multiplicative factor in the above inequality will become $e^{g\epsilon}$ (instead of e^ϵ).

The differential privacy is defined with respect to a query f on the database. In general, $f : \mathcal{D} \to \mathcal{R}^d$ (for some integer d). For any such f, its L1 sensitivity is defined as $\Delta f = \max_{\mathcal{D}_1,\mathcal{D}_2} ||f(\mathcal{D}_1) - f(\mathcal{D}_2)||_1$. Δf corresponds to the maximum difference in the query response between $\mathcal{D}_1$ and $\mathcal{D}_2$. For a query such as: "*How many people have a salary of* > $100K?", $\Delta f \leq 1$. The scheme of [3] works best when Δf is small.

The privacy mechanism $\mathcal{K}_f$ of [3] works as follws: It computes $f(\mathcal{D})$ on any database $\mathcal{D}$ and adds a noise that is exponentially distributed with a variance of σ^2 (for some appropriate σ value). In particular,

$$Pr[\mathcal{K}_f(\mathcal{D}) = r] \propto \exp(-||f(\mathcal{D}) - r||_1/\sigma). \qquad (1)$$

This distribution has independent exponentially distributed coordinates. In other words, an exponential noise is added to each coordinate of $f(\mathcal{D})$.

The following Theorem and Proof are from [3]:

Theorem 2. *The above mechanism $\mathcal{K}_f$ gives $(\Delta f/\sigma)$-differential privacy.*

Proof. Applying Equation 1 (replacing $\propto$ with equality) for $\mathcal{D}_1$ we get

$$Pr[\mathcal{K}_f(\mathcal{D}_1) = r] = \exp(-||f(\mathcal{D}_1) - r||_1/\sigma). \qquad (2)$$

Applying Equation 1 for $\mathcal{D}_2$ we get

$$Pr[\mathcal{K}_f(\mathcal{D}_2) = r] = \exp(-||f(\mathcal{D}_2) - r||_1/\sigma). \qquad (3)$$

From equations 2 and 3 and the triangular inequality we get

$$Pr[\mathcal{K}_f(\mathcal{D}_1) = r] \leq Pr[\mathcal{K}_f(\mathcal{D}_2) = r] \times exp(||f(\mathcal{D}_1) - f(\mathcal{D}_2)||_1/\sigma). \qquad (4)$$

Since $||f(\mathcal{D}_1) - f(\mathcal{D}_2)||_1$ is bounded by Δf, the theorem holds when S is a singleton. Using the union bound, the theorem can also be proven for any subset S of the range of $\mathcal{K}_f$. $\qquad \square$

If Q is a quantity to be estimated in $\mathcal{D}$, one could estimate Q in each synthetic data set $\mathcal{D}^l, 1 \leq l \leq M$ and using these estimates approximate Q with a normal distribution.

The data privacy scheme has several shortcomings: 1) The privacy measure is specific to each query. Given the large number of possible queries, it may be impractical to analyze the privacy for each such possible query; 2) If the same query (possibly in disguise) is asked again and again, it is possible to estimate $f(D)$ for any database D and any query f within a tight interval; and so on.

It is desirable to come up with a privacy measure that will be query-independent. In particular, in this paper we are interested in defining the privacy associated with a given data set. We present such a scheme. The scheme we offer assumes a specific way of releasing data. However the technique is general and could be extended to other data releasing schemes as well. In Section 3.3 we present details on the data releasing technique. The other sections introduce our ideas on data privacy.

2.3   Probabilistic Differential Privacy

In [5] a variant of the differential privacy scheme called *probabilistic differential privacy* has been proposed. This technique is explained with the example of commuters data. The origins and destinations are captured as k blocks and there is a histogram for each destination block. The histogram for any block d can be thought of as a vector $(n_1, n_2, \ldots, n_k)$ where n_i is the number of persons that commute from bock i to block d (for $1 \leq i \leq k$). The synthetic data corresponding to block d is obtained as a vector $(m_1, m_2, \ldots, m_k)$ such that $\sum_{i=1}^{k} m_i = O(\sum_{i=1}^{k} n_i)$.

The synthetic data is obtained as follows: Some noise (for example Laplacian) α_i is added to n_i, for $1 \leq i \leq k$. The synthetic data is nothing but a sample (of size $m = O(\sum_{i=1}^{k} n_i)$) from a multinomial distribution with a Dirichlet prior whose parameters are $(n_1 + \alpha_1, n_2 + \alpha_2, \ldots, n_k + \alpha_k)$. Let $\boldsymbol{\alpha} = (\alpha_1, \alpha_2, \ldots, \alpha_k)$. To calculate the differential privacy one can obtain using this scheme, we have to compute the maximum value of

$$\frac{Pr((m_1, m_2, \ldots, m_k) \mid (n_1, n_2, \ldots, n_k), \boldsymbol{\alpha})}{Pr((m_1, m_2, \ldots, m_k) \mid (n'_1, n'_2, \ldots, n'_k), \boldsymbol{\alpha})}.$$

Here $(n'_1, n'_2, \ldots, n'_k)$ is the histogram corresponding to D' such that D and D' differ in one point.

If D and D' differ in block d, the authors argue that, the worst case value of the above ratio happens when all the m points in the synthetic data fall into block d. The probability of this happening is very low. In this case the synthetic data is highly unrepresentative of the original data. The idea of *probabilistic differential privacy* is to compute the value of ϵ using a probable value of the above ratio (rather than the worst case maximum value).

As one could see, the probabilistic differential privacy is closely related to differential privacy. In particular, it is dependent on the query under concern.

3   A Novel Scheme for Data Privacy

3.1   Some Observations

Before presenting details on our approach, we enumerate some motivating observations.

1. We can think of the database $\mathcal{D}$ as a sample from the population $\mathcal{P}$. Equivalently, if $|\mathcal{D}| = n$, we can think of $\mathcal{D}$ as a collection of n draws from some appropriate probability distribution $\mathcal{U}$ (that models the population $\mathcal{P}$).
2. The released synthetic data is fully characterized by the predictive posterior distribution parameters (μ and Σ in the normal distribution case). In general, whatever synthetic data scheme one may employ, the released data is completely characterized by the parameters of an appropriate probability distribution $\mathcal{U}'$ (that models the data set $\mathcal{D}$).

3. In fact, the synthetic data sets $\mathcal{D}^l, 1 \leq l \leq M$ released in the synthetic data scheme of [10], [9], and [7] are completely characterized by the multivariate normal distribution (used to model $\mathcal{D}$). Note that the posterior distribution $\mathcal{U}'$ used to model $\mathcal{D}$ is indeed an estimate of $\mathcal{U}$.

4. What the user learns from $\mathcal{D}^l, 1 \leq l \leq M$ is at best the multivariate normal distribution that models $\mathcal{D}$. In general the user learns, at best, the parameters of the probability distribution $\mathcal{U}'$ that models $\mathcal{D}$.

5. In summary, the user knows $\mathcal{U}'$ and does not know $\mathcal{D}$. Can the user learn any point (i.e., row) of $\mathcal{D}$ using $\mathcal{U}'$ alone? For example, consider a binomial distribution with parameters m and p. Also assume that $\mathcal{D}$ is a collection of n draws from this distribution. Can one identify one of the points in $\mathcal{D}$ knowing m and p alone? This may not be possible.

   What can the user learn from $\mathcal{U}'$? Let $\mathcal{D}$ be a given data set. Let $\mathcal{D}'$ be another data set that is a subset of $\mathcal{D}$. One or more points of $\mathcal{D}$ are missing in $\mathcal{D}'$. Let S and S' be the synthetic data sets corresponding to $\mathcal{D}$ and $\mathcal{D}'$, respectively. If a user can infer (with some "reasonable" confidence) that S and S' correspond to two different data sets analyzing only S and S', then we have leaked some information. It is very much possible for the user to infer this using some auxiliary information (in addition to S and S'). However, we don't have any control over the auxiliary information that any user may have access to. In this case, we can call a data release scheme as a *responsible data release* if the user cannot infer, using only S and S', that S and S' correspond to two different data sets. In other words our data release scheme should be such that a user who has access only to S and S' will not be able to detect any significant statistical difference between S and S'. We refer to this property as *responsible data release property*.

6. The security of any data release scheme will depend on the type of data that we have as well as the data release technique used. The dependence on data type is very clear. For example, we may have a data set where only one point is (or a small number of data points are) far away from the rest of the points. Consider a community C where only one person is of height 7.5 feet and the others are of height less than or equal to 6 feet. Let X be a set of people from C that have a particular disease. Also assume that the person of height 7.5 feet is a member of X. If we employ the NSDG scheme of Section 2.1 (or any other scheme), X should be released only if it has the responsible data release property. In general, for any given dataset and any given data release scheme, the data should be released only if this property holds, as per our new ideas.

3.2   Our New Ideas

The above observations suggest the following technique for releasing data. Cluster the data points (for example using any hierarchical clustering algorithm). The number of clusters could be decided empirically to the one that yields the best security. If the number of clusters is chosen to be very large, then the size of each cluster will be very small. This is not desirable. If the number of clusters is too small, it will fail to identify the vulnerable points. Let $C_1, C_2, \ldots, C_k$ be the resultant clusters. We will not release data corresponding to those clusters that have a 'small' number of points in them since they run the risk of being breached. What is 'small' is again better decided empirically. Let $C'_1, C'_2, \ldots, C'_q, (q \leq k)$ be the surviving clusters. For each surviving cluster we will release synthetic data (using a scheme such as the NSDG protocol). For each data released we also compute a *risk* associated with the release. The risk is a quantification on how much the synthetic data deviates from the responsible data release property.

Let C be one of the surviving clusters and assume that we employ the NSDG scheme. Let $\bar{\mu}$ be the mean vector of the posterior distribution and let Σ be the covariance matrix. We are interested in quantifying the risk involved in releasing the data for C using this scheme. In particular, consider the example of quantifying the risk of revealing any one of the records of the database. Note that we will generate synthetic data sets $D_1, D_2, \ldots, D_n$ each one of these being a random sample from the posterior distribution.

The risk factor for C will depend on how 'homogeneous' the members of C are. If a member is very unique perhaps this member could be an outlier and hence runs the risk of being detected using the synthetic data sets. Let x be any member of C and let $C' = C - \{x\}$. Our question of interest is: If we were to release data for C' let $\bar{\mu}'$ and Σ' be the mean vector and covariance matrix, respectively. Let $D'_1, D'_2, \ldots, D'_n$ be the synthetic data for C'. Given the synthetic data sets $D_1, D_2, \ldots, D_n$; and $D'_1, D'_2, \ldots, D'_n$, what is the probability of detecting that these two data sets correspond to two different posterior distributions? We have to compute the maximum of this probability over all members x of C.

Clearly, this probability will depend on the 'distance' between the two distributions. There are many ways of defining the distance between these two sample data sets. For example, we can estimate the parameters (mean and covariance) from the synthetic data sets and compute the distance as one-norm between them.

3.3   Synthetic Data Sets Methodology

Using synthetic data sets the analyst would want to make inferences about some population quantity of interest, Q (e.g. regression coefficients). Q can be estimated by a point estimate q with an associated measure of uncertainty v. The intruder is not interested in the population quantity Q, but in the specific sample quantity q. For each synthetic data set $(1, 2, \ldots, M)$, q and v can be estimated, yielding M estimates $(q^{(l)}, v^{(l)})$, $l = (1, 2, \ldots, M)$. We have to use a slight variant of the scheme described in Raghunathan, *et al.* [9]. In particular, we approximate the population parameter, $Q|D_{Syn}$, by a normal distribution with mean $q_M = \sum_{l=1}^{M} q^{(l)}/M$ and variance $T_M = b_M + v_M$ where $v_M = \sum_{l=1}^{M} v^{(l)}/M$ and $b_M = \sum_{l=1}^{M} (q^{(l)} - q_M)^2/(M-1)$. The exact variance can be obtained using numerical procedures and evaluating the integrals described in Raghunathan, *et al.* [9].

To solidify the above ideas we provide two examples next.

3.4   An Example

Consider the example where each data record has only one (real-valued) attribute and where we use the NSDG protocol. Let μ and σ be the mean and variance of the posterior distribution for some cluster C. Let x be any member of C and let $C' = C - \{x\}$. Also, let μ' and σ' be the mean and variance of the posterior distribution for C'. If m synthetic data sets are released for C with n records in each, we can use the approximation in section 3.3 to approximate a normal distribution with mean $\mu = q_M$ and variance $\sigma = T_M$. Likewise, for C' also we have m synthetic data sets S' of size n each. Let the sample means and variances for S and S' be (μ_s, μ'_s) and (σ_s, σ'_s), respectively. To simplify the discussion assume that the pooled variance of C and C' is equal $\sigma_p = \frac{(n_1-1)T_M+(n_2-1)T'_M}{n_1+n_2-2}$.

Given S and S' what is the probability of inferring that μ and μ' are different? Clearly this will depend on how different μ and μ' are. If μ and μ' are nearly the same, then we may not be able to detect a statistical difference between the two samples. Any statistical statement is always made with respect to some confidence level. Let $\mathcal{P}$ be the confidence level of interest ($\mathcal{P}$ being close to 1). For a given confidence level $\mathcal{P}$ and a sample S, we can only estimate μ within an interval.

Appealing to the central limit theorem (see e.g., [11]), we can get confidence intervals for μ_s and μ'_s. For example, for large values of mn, we can assume that the distribution of μ_s is normal with mean μ and variance σ_p. As a result, it follows that $\frac{\mu_s-\mu}{\sigma_p}$ is the standard normal variate. Similarly, $\frac{\mu'_s-\mu'}{\sigma_p}$ can be thought of as the standard normal variate.

For the standard normal variate Z (with mean 0 and variance 1), it is well known that:

$$Pr[Z \geq z] \leq \frac{1}{\sqrt{2\pi}z} \exp\left(-\frac{z^2}{2}\right). \tag{5}$$

Applying Equation 5 for C, we realize that

$$\mu_s \in \mu \pm \sigma_p\sqrt{2\ln\left(\frac{1}{\sqrt{2\pi}(1-\mathcal{P})}\right)} \tag{6}$$

with probability $\geq \mathcal{P}$. On a similar token we can infer:

$$\mu'_s \in \mu' \pm \sigma_p\sqrt{2\ln\left(\frac{1}{\sqrt{2\pi}(1-\mathcal{P})}\right)} \tag{7}$$

with probability $\geq \mathcal{P}$.

If $A = \sigma_p\sqrt{2\ln\left(\frac{1}{\sqrt{2\pi}(1-\mathcal{P})}\right)}$, then μ_s is in the interval $[\mu - A, \mu + A]$ and μ'_s is in the interval $[\mu' - A, \mu' + A]$, both with probability $\geq \mathcal{P}$. Assume without loss of generality that $\mu' \geq \mu$ (the other case is similar). If μ'_s falls in the interval $[\mu + A, \mu' + A]$ then there is a high probability of inferring that $\mu' > \mu$ (under the confidence level $\mathcal{P}$). The probability of μ'_s falling in this interval can be taken as the risk of releasing C under the confidence level $\mathcal{P}$.

The probability of μ'_s falling in the interval $[\mu + A, \mu' + A]$ is nearly the same as $Pr[\mu'_s \geq \mu + A]$. $Pr[\mu'_s \geq \mu + A] = Pr[(\mu'_s - \mu') \geq (A - (\mu' - \mu))] = Pr\left[\frac{\mu'_s - \mu'}{\sigma_p} \geq \frac{A - (\mu' - \mu)}{\sigma_p}\right]$. Using Equation 5, this probability is

$\leq \frac{1}{\sqrt{2\pi}\left(\frac{A-(\mu'-\mu)}{\sigma_p}\right)} \exp\left(-\frac{1}{2}\left(\frac{A-(\mu'-\mu)}{\sigma_p}\right)^2\right)$. Upon simplification, this probability

is $\leq (1 - \mathcal{P})\frac{\sigma_p}{A-(\mu'-\mu)} \exp\left(\frac{2A(\mu'-\mu)-(\mu'-\mu)^2}{2\sigma_p^2}\right)$. RHS is nearly equal to

$(1 - \mathcal{P})\frac{\sigma_p}{A} \exp\left(\frac{A(\mu'-\mu)}{\sigma_p^2}\right)$.

As a result, the following definition of risk can be made.

Definition 3. *In the above example, the risk of releasing data C is* $(1 - \mathcal{P})\frac{\sigma_1}{A} \exp\left(\frac{A(\mu'-\mu)}{\sigma_p^2}\right)$ *under the confidence level $\mathcal{P}$.*

3.5   Example of Histograms

Machanavajjhala, *et al.* [5] have considered the example of commuting data. Space is divided into grids and each grid block serves both as an origin and a destination. Let the number of grid blocks be k. For every destination d, we have an associated array $A_d[1 : k]$, where $A_d[i]$ is the number of commuters whose origin is i and whose destination is d (for $1 \leq i \leq k$).

Synthetic data for this example of histograms can be released in a number of ways. In [5], noise is added to the data before releasing. They also introduce the idea of *probabilistic differential privacy*.

In this section we consider a different scheme for data release. Like in [5] we consider the individual destinations separately. From hereon we assume a specific destination without referring to it by name. Let $(n_1, n_2, \ldots, n_k)$ be the histogram corresponding to the destination under concern and let $n = \sum_{i=1}^{k} n_i$. The data released is a collection of m samples (for some appropriate value of m) drawn from the multinomial distribution with parameters $\left(\frac{n_1}{n}, \frac{n_2}{n}, \ldots, \frac{n_k}{n}\right)$.

Our idea of responsible data release suggests that the released data corresponding to $\mathcal{D}$ and $\mathcal{D}'$ be statistically indistinguishable (under some confidence level) where $\mathcal{D}$ is the original dataset and $\mathcal{D}'$ is a subset of $\mathcal{D}$. As in [5] consider the example where $\mathcal{D}'$ differs from $\mathcal{D}$ in exactly one point.

Consider a block g in $\mathcal{D}$ with i members. Let $\mathcal{D}'$ differ from $\mathcal{D}$ only in block g. $\mathcal{D}'$ has one less member in g than in $\mathcal{D}$. The multinomial distribution for $\mathcal{D}$ will have a probability of $\frac{i}{n}$ for block g and the multinomial distribution of D' will have the probability of $\frac{i-1}{n}$ for block g. Assume that we release m samples from the multinomial distributions corresponding to the data sets.

Let X be the total number of members that fall in block g in the m samples for $\mathcal{D}$. Let Y be the total number of members that fall into g in the m samples for $\mathcal{D}'$.

Using Chernoff bounds, $X \in (1 \pm \epsilon)m\frac{i}{n}$ with probability $\geq 1 - \exp\left(-\frac{\epsilon^2 mi}{3n}\right)$.

Likewise, $Y \in (1 \pm \epsilon)m\frac{(i-1)}{n}$ with probability $\geq 1 - \exp\left(-\frac{\epsilon^2 m(i-1)}{3n}\right)$.

Let $\mathcal{P}$ be a confidence level of interest. With probability $\geq \mathcal{P}$, $X \in (1 \pm \epsilon)m\frac{i}{n}$ where $\epsilon = \sqrt{\frac{3\ln\left(\frac{1}{1-\mathcal{P}}\right)}{mi/n}}$. Similarly, with probability $\geq \mathcal{P}$, $Y \in (1 \pm \epsilon_1)m\frac{(i-1)}{n}$ where $\epsilon_1 = \sqrt{\frac{3\ln\left(\frac{1}{1-\mathcal{P}}\right)}{m(i-1)/n}}$.

If the value of X is in the interval $\left[(1+\epsilon_1)m\frac{(i-1)}{n}, (1+\epsilon)m\frac{i}{n}\right]$, then we may be able to infer that $\mathcal{D}$ and $\mathcal{D}'$ differ in block g. The probability of this happening is no more than $Pr\left[X \geq (1+\epsilon_1)m\frac{(i-1)}{n}\right]$. This probability is no more than $\exp\left(-\left[(1+\epsilon_1)\frac{i-1}{i} - 1\right]^2 \frac{mi}{3n}\right)$. This probability can be thought of as the risk under confidence level $\mathcal{P}$. If i is large, the above probability is nearly equal to $(1-\mathcal{P})^{i/(i-1)}$.

Observation 3.1. *If one desires to keep the value of i large, one could introduce a constant bias to all the blocks. This is in contrast to stochastic noises introduced in the other schemes found in the literature.*

Observation 3.2. *In the two examples we have seen, $\mathcal{D}$ and $\mathcal{D}'$ differ in only one point. However, the paradigm and analyses can be extended to general cases where $\mathcal{D}'$ is any subset of $\mathcal{D}$.*

Observation 3.3. *We have seen only examples where we don't introduce any noise. However the paradigm is general and can be used also in cases where noises are introduced.*

Observation 3.4. *The paradigm we have introduced can be used to compute the risk associated with any data release (synthetic or otherwise). For example, a simple data release scheme is to eliminate some vital attributes (such as names, social security numbers, etc.) and release the data with the other attributes. In this case also we can associate a risk with this scheme as follows. Let k be the number of attributes remaining in the released data. One way of associating a risk with this data release is as follows.*

We can think of the database $\mathcal{D}$ released as a sample from a population $\mathcal{P}$. If $\mathcal{D}$ has n records, we can think of $\mathcal{D}$ as a collection of n draws from some appropriate probability distribution $\mathcal{U}$ (that models the population $\mathcal{P}$). One could estimate the parameters of $\mathcal{U}$ from $\mathcal{D}$. If $\mathcal{D}'$ is any subset of $\mathcal{D}$, one could also estimate the parameters of the distribution corresponding to $\mathcal{D}'$. Then one could compute the probability that these two parameters correspond to two different data sets. This probability can be used as the risk (similar to what we computed in Section 3.4).

4   Algorithms for Computing Risks of Data Sets

In this section we present psuedocodes for computing risk factors involved in releasing data sets. These psuedocodes basically encode the ideas that have been presented in the previous sections.

4.1   The Example of Section 3.3

Here we consider the example of each data record having only one attribute and the protocol used being NSDG. Consider a data set C that has N records. We assume that there is an algorithm called PD_Parameters that takes as input any data set and outputs the parameters of the posterior distribution corresponding to the data set. For the example of Section 3.3, there are two parameters of interest, namely, the mean μ and standard deviation σ.

Algorithm ComputeRisk1

> Let $C = \{x_1, x_2, \ldots, x_N\}$; Let $\mathcal{P}$ be the confidence level;
> The synthetic data is of size mn;
> Invoke PD_Parameters on C and get μ and σ;
> $Risk := 0$; $\sigma_p = \frac{(n_1-1)T_M + (n_2-1)T'_M}{n_1+n_2-2}$; $A = \sigma_p \sqrt{2\ln\left(\frac{1}{\sqrt{2\pi}(1-\mathcal{P})}\right)}$;
> **for** $i := 1$ **to** N **do**
>> Let $C' := C - \{x_i\}$;
>> Invoke PD_Parameters on C' to get μ' and σ';
>> $Risk := \max\{Risk, (1 - \mathcal{P})\,\frac{\sigma_p}{A}\,\exp\left(\frac{A(\mu'-\mu)}{\sigma_p^2}\right)\}$;
> Output $Risk$;

4.2   The General Case

In this section we consider a general scenario where for a given data set C, a parameter set $P = \{p_1, p_2, \ldots, p_q\}$ is generated, and P is then used to generate and release synthetic data sets. NSDG is an example of this paradigm. In this special case the parameters will be the mean vector and the covariance matrix. If there are m attributes, then the mean vector will be of size m and the covariance matrix will be of size $m \times m$. Here again we assume that there is an algorithm called Get_Parameters that takes as input any data set and returns the corresponding parameter set.

Given two parameter sets P_1 and P_2, let $d(P_1, P_2)$ be the *distance* between them. There are many possible ways of defining this distance. For example, this distance can be the Euclidean distance between these two vectors.

Under the above settings we can make use of the following algorithm.

Algorithm ComputeRisk2

> Let $C = \{x_1, x_2, \ldots, x_N\}$;
> Invoke Get_Parameters on C and get its parameter set P;
> $Risk := 0$;
> **for** $i := 1$ **to** N **do**
>> Let $C' := C - \{x_i\}$;
>> Invoke Get_Parameters on C' to get its parameter set P';
>> $Risk := \max\{Risk, d(P, P')\}$;
> $PercentageRisk := 100 * Risk/\|P\|$; Output $PercentageRisk$;

5   Experimental Results

We have tested our algorithm ComputeRisk2 with synthetic data sets. In particular, we have used several data sets using the following parameters as variables: number of observations (or points), the number of attributes, mean vector, and covaraince matrix. For each data set C we also generated another data set C' such that C and C' differ in only one point. In particular, we perturbed the value of one of the attributes in one of the points of C to get C'. If v is the value of C under concern, it is perturbed to different values (such as $v + 5\sigma$, $v + 10\sigma$, etc.). For each such perturbed value we get a new C'. We compute the risks of C and C' to see how different they are (as a function of the underlying parameters).

We have employed the R [8] function "mvrnorm", which produces one or more samples from a specified multivariate normal distribution. Please note that the length of μ is the number of attributes and Σ is of size $p \times p$ where p is the number of attributes. The diagonal elements of Σ will be the variances of each mean and the off diagonals are the covariances between a pair of variables.

For each data set we computed the risk using ComputeRisk2. Followed by this, we introduced an outlier in the same data set by perturbing one of the attributes by various amounts and calculated the risk again.

The results obtained are shown in the following tables.

Table 1. Risk factors (as percentages) for different data sets; Number of points $=100$

Mean	σ	Attributes	R_W/O_OL	R_W_OL $(\mu + 5\sigma)$	R_W_OL $(\mu + 10\sigma)$	R_W_OL $(\mu + 15\sigma)$
μ_1	0.5	2	8.81	14.48	18.09	35.12
μ_2	0.5	5	3.26	7.19	9.89	14.32
μ_3	0.5	10	3.30	4.22	17.23	16.87
μ_4	0.5	20	3.04	5.25	8.68	16.66
μ_1	0.4	2	16.23	17.10	17.36	33.77
μ_2	0.4	5	3.42	4.54	7.22	12.31
μ_3	0.4	10	4.88	4.95	10.85	8.59
μ_4	0.4	20	2.59	4.57	5.89	9.94
μ_1	0.25	2	22.31	22.27	27.55	30.79
μ_2	0.25	5	4.52	5.38	6.16	6.19
μ_3	0.25	10	4.93	6.91	7.86	7.17
μ_4	0.25	20	3.65	5.49	5.57	6.48
μ_1	0.1	2	27.12	40.89	49.48	66.87
μ_2	0.1	5	6.00	3.96	5.26	5.81
μ_3	0.1	10	5.09	7.05	8.30	5.48
μ_4	0.1	20	4.33	4.41	4.42	5.68

Table 2. Risk factors for different data sets; Number of points =1000

Mean	σ	Attributes	R_W/O_OL	R_W_OL $(\mu + 5\sigma)$	R_W_OL $(\mu + 10\sigma)$	R_W_OL $(\mu + 15\sigma)$
μ_1	0.5	2	1.05%	1.71%	1.94%	8.18%
μ_2	0.5	5	0.46%	0.50%	0.50%	1.22%
μ_3	0.5	10	0.53%	0.67%	1.24%	2.64%
μ_4	0.5	20	0.45%	0.65%	1.40%	1.88%
μ_1	0.4	2	1.34%	2.48%	3.03%	3.45%
μ_2	0.4	5	0.57%	0.60%	0.75%	1.76%
μ_3	0.4	10	0.61%	0.72%	0.68%	0.83%
μ_4	0.4	20	0.42%	0.81%	1.00%	1.72%
μ_1	0.25	2	2.65%	2.65%	2.70%	3.43%
μ_2	0.25	5	0.67%	0.64%	0.62%	0.78%
μ_3	0.25	10	0.70%	0.54%	0.73%	0.76%
μ_4	0.25	20	0.53%	0.54%	0.89%	0.91%
μ_1	0.1	2	4.35%	8.58%	8.65%	13.07%
μ_2	0.1	5	0.80%	0.93%	0.95%	1.02%
μ_3	0.1	10	1.16%	0.76%	0.75%	0.78%
μ_4	0.1	20	0.81%	0.78%	0.64%	0.81%

In Tables 1 and 2, $\mu_1 = (0,0)$; $\mu_2 = (2, -2.5, -5, 10)$; $\mu_3 = (1, -1, 6, -6, 8, 11, -14, 17, 20, -24)$; and $\mu_4 = (2, -2, 5, -6, 10, 14, -12, 18, -22, 25, 27, -30, 33, 36, -38, 40, 43, -45, 47, -47)$. The number of points in the data sets is 100. R_W/O_OL stands for "Risk without outliers" and R_W_OL stands for "Risk with an outlier".

As we can see from Table 1, in general, the risk factor increases significantly when an outlier is present. The exeption is when the value of σ is very small. Even when σ is small, the risk factor increases when the number of attributes is small. When σ is small and the number of attributes is large, even a large deviation in one attribute in one point is not sufficient to leak information. Also, in general, when the deviation is large the risk factor is also large.

This is what one would expect out of any risk computation.

In Table 2 also we see results similar to those in Table 1. The risk factors are in general smaller in Table 2. This is because the number of points is ten times more. As a result, the influence of one attribute in one point on the entire dataset tends to be smaller. However, as the deviation increases, we see a proportionate increase in the risk factor.

6   Conclusions

In this paper we have introduced a measure of data privacy that is only dependent on the data given (and is independent of the query). This measure overcomes the shortcomings present in the differential privacy measures of [3] and [5]. We have illustrated the new paradigm with two examples.

Acknowledgement

This research has been supported in part by an NSF Grant ITR-0326155.

References

1. Chen, B., Lefevre, K., Ramakrishnan, R.: Privacy skyline: Privacy with multidimensional adversarial knowledge. Journal of Very Large Databases (2007)
2. Dalenius, T.: Towards a methodology for statistical disclosure control. Statistisk Tidskrift 5, 429–444 (1977)
3. Dwork, C.: Differential Privacy (2008)
4. Horowitz, E., Sahni, S., Rajasekaran, S.: Computer Algorithms. Silicon Press (2008)
5. Machanavajjhala, A., Kifer, D., Abowd, J., Gehrke, J., Vilhuber, L.: Privacy: Theory meets Practice on the Map. In: Proc. 24th IEEE international Conference on Data Engineering, pp. 277–286 (2008)
6. Machanavajjhala, A., Kifer, D., Gehrke, J., Venkitasubramaniam, M.: l-diversity: Privacy beyond k-anonymity. ACM Transactions on Knowledge Discovery from Data 1(1) (2007)
7. Matthews, G.J., Harel, O., Aseltine, R.H.: Examining the Robustness of Fully Synthetic Data Techniques for Data with Binary Variables, Technical Report, UConn (June 2008)
8. R Development Core Team. R: A Language and Environment for Statistical Computing. R Foundation for Statistical Computing, Vienna, Austria (2007) ISBN 3-900051-07-0
9. Raghunathan, T.E., Reiter, J., Rubin, D.: Multiple imputation for statistical disclosure limitation. Journal of Official.Statistics 19, 1–16 (2003)
10. Rubin, D.B.: Discussion statistical disclosure limitation. Journal of Official Statistics 9(2) (1993)
11. Snedecor, G.W., Cochran, W.G.: Statistical Methods, 7th edn. The Iowa State University Press (1980)
12. Sweeney, L.: k-anonymity: a model for protecting privacy. International Journal on Uncertainty, Fuzziness and Knowledge-based Systems 10(5), 557–570 (2002)

Appendix A: Chernoff Bounds

If a random vraiable X is the sum of n iid Bernoulli trials with a success probability of p in each trial, the following equations give us concentration bounds of deviation of X from the expected value of np (see e.g., [4]). The first equation is more useful for large deviations whereas the other two are useful for small deviations from a large expected value.

$$Prob(X \geq m) \leq \left(\frac{np}{m}\right)^m e^{m-np} \tag{8}$$

$$Prob(X \leq (1 - \epsilon)pn) \leq exp(-\epsilon^2 np/2) \tag{9}$$

$$Prob(X \geq (1 + \epsilon)np) \leq exp(-\epsilon^2 np/3) \tag{10}$$

for all $0 < \epsilon < 1$.

Author Index